T0358348

EVEN ELECTRON MASS SPECTROMETRY WITH BIOMOLECULE APPLICATIONS

EVEN ELECTRON MASS SPECTROMETRY WITH BIOMOLECULE APPLICATIONS

BRYAN M. HAM

WILEY-INTERSCIENCE

A JOHN WILEY & SONS, INC., PUBLICATION

Library of Congress Cataloging-in-Publication Data:

Ham, Bryan M.
     Even electron mass spectrometry with biomolecule applications / Bryan M. Ham.
        p.   cm.
     ISBN 978-0-470-11802-3 (cloth)
     1. Mass spectrometry.   2. Biomolecules–Analysis.   I. Title
QP519.9.M3H36   2008
572'.36–dc22   2007035521

10   9   8   7   6   5   4   3   2   1

This book is dedicated to the three most important people in my life, my wife and my parents.

CONTENTS

PREFACE

Mass spectrometry is a rapidly changing field of science that has experienced substantial development of new instrumentation along with increased applications in the area of biomolecule analysis. This book is addressing these new aspects by presenting the latest developed mass spectrometers and ionization techniques. The greater majority of mass spectral analysis of biomolecules in the laboratory today use ionization sources that produce even electron ions as opposed to odd electron ion spectra. Traditionally, mass spectral interpretation books have focused solely on the treatment of odd electron spectra generated from electron ionization (EI) sources. While electron ionization is still used to some extent, the majority of ionization techniques used in today's mass spectrometers incorporate sources that generate even electron ions. Furthermore, techniques such as electrospray ionization and matrix-assisted laser desorption ionization are soft ionization techniques that produce primarily intact ions as opposed to the EI source where the molecular ion is often depleted and not observed spectrally. Even electron ions give a mass-to-charge ratio (m/z, molecular weight plus weight of adduct divided by the charge) of the biomolecule being analyzed, which is often subjected to collision-induced dissociation to produce an even electron product ion spectrum that is used for structural elucidation and unknown identification.

The book is also a combination of the detailed step-by-step methodology for mass spectral interpretation that is coupled to a complete coverage of mass spectrometry and the fundamentals that underlie this extremely useful but complex subject. For the interested researcher or student, mass spectrometers are complicated instrumentation and the field of the application of mass spectrometry to biomolecule analysis is broad, complex, and can be intimidating. The book fills a need for a comprehensive understanding of mass spectrometry in conjunction with an easily understandable methodology for spectra interpretation and structural elucidation. While the book does give a complete basic coverage of odd electron spectral interpretation, which is the focus of most mass spectrometry books used in classroom study (see Chapter 5), its main focus is upon even electron ion mass spectral interpretation. The book can be used in mass spectrometry classes giving the students a more up-to-date application of mass spectrometry. It can also be used by individuals who have had the traditional classroom introduction to mass spectrometry but who are interested in reviewing the subject and applying the technique of mass spectrometry to biomolecule analysis, which is now one of the major applications of mass spectrometry in research and industry. Upon reading and understanding this book, one can easily apply what is learned to the mass spectrometric analysis of daily routine work and to high-level research interpretation.

An introduction to mass spectrometry in conjunction with pertinent fundamentals to lay a foundation in which to build upon in later chapters is presented in Chapter 1. Chapter 2 discusses ionization techniques and sources including electron ionization and the more current soft ionization techniques of electrospray ionization and matrix-assisted laser desorption ionization at a fundamental level for an in-depth and comprehensive understanding of the ionization methods used in mass spectrometry. Chapter 3 covers the basics of the most commonly encountered mass analyzers geared toward a fundamental course in mass analyzer theory at either the undergraduate or graduate level. Gas-phase reaction rate theory and collision theory are covered in Chapter 4 to give a basic understanding behind gas-phase unimolecular reactions such as collision-induced dissociation (CID) that produce product ion spectra. A main focus of the book is presented in Chapters 5 and 6 to teach how to analyze mass spectra for unknown structure elucidation and identification. The approach presented uses techniques such as initial discussion of sample origin and spectra origination, the application of reasonable fragmentation pathways, and product ion structures. The teaching approach is to take structures of known compounds and break them apart into their basic building blocks in a kind of "exploded" representation and then relate this to a product ion spectrum by matching reasonable structures to the masses of the product ions produced in the collision-induced dissociation experiments. The following chapters are devoted to areas of biomolecule analysis by mass spectrometry including the analysis of proteins (Chapter 7), small-molecule analysis (Chapter 8), biopolymers (Chapter 9), and finally a detailed look at the postsource decay MALDI TOF/MS analysis of phosphorylated lipids including spectral interpretation (Chapter 10).

ACKNOWLEDGMENTS

I would like to acknowledge all those whose input, review, and criticism helped enormously in the early structuring and final content of this book. I would like to include in the acknowledgment my graduate advisor Dr. Richard B. Cole where in his mass spectrometry group at the University of New Orleans the idea of this book first came to me. Also included is Dr. Jean T. Jacobs at the Louisiana State University Health Sciences Center whose biological research models, which I worked on, helped to lay a strong foundation in biomolecule mass spectrometry. Next is The Ohio State University where I was a visiting scholar during the early stages of the writing and construction of this book while doing research in Dr. Kari B. Green-Churchs mass spectrometry and proteomics laboratory in conjunction with Dr. Jason J. Nichols and Dr. Kelly K. Nichols in the College of Optometry. I want to also include Pacific Northwest National Laboratory in the acknowledgment, where for the latter part of the book's construction and writing, I was conducting research in Dr. Richard D. Smith's proteomics group. Finally, and most important of all, is the acknowledgment of my wife Dr. Aihui Ma-Ham whose consultations, support, reviewing, and invaluable encouragement saw me through the entire process of this book with an unending presence of which the project would most certainly not have been completed to this level.

BRYAN M. HAM

CHAPTER 1

INTRODUCTION AND BASIC DEFINITIONS

1.1 DEFINITION AND DESCRIPTION OF MASS SPECTROMETRY

During the past decade mass spectrometry has experienced a tremendously large growth in its uses for extensive applications involved with complex biological sample analysis. Mass spectrometry is basically the science of the measurement of the mass-to-charge ratio (m/z) of ions in the gas phase. Mass spectrometers are generally comprised of three components: (1) an ionization source that ionizes the analyte of interest and effectively transfers it into the gas phase, (2) a mass analyzer that separates positively or negatively charged ionic species according to their mass-to-charge ratio (m/z), and (3) a detector used to measure the subsequently separated gas-phase ions. Mass spectrometers are computer controlled, which allows the collection of large amounts of data and the ability to perform various and complex experiments with the mass spectral instruments. Applications of mass spectrometry include unknown compound identification, known compound quantitation, structural determination of molecules, gas-phase thermochemistry studies, ion–ion and ion–molecule studies, and molecule–chemical property studies. Mass spectrometry is routinely used to determine elements such as Li^+, Na^+, Cl^-, Mg^{2+}, inorganic compounds such as $Li^+(H_2O)_x$ or $(TiO_2)_x^+$, and organic compounds including lipids, proteins, peptides, carbohydrates, polymers, and oligonucleotides [deoxyribonucleic acid (DNA) and ribonucleic acid (RNA)].

Even Electron Mass Spectrometry with Biomolecule Applications   By Bryan M. Ham
Copyright © 2008 John Wiley & Sons, Inc.

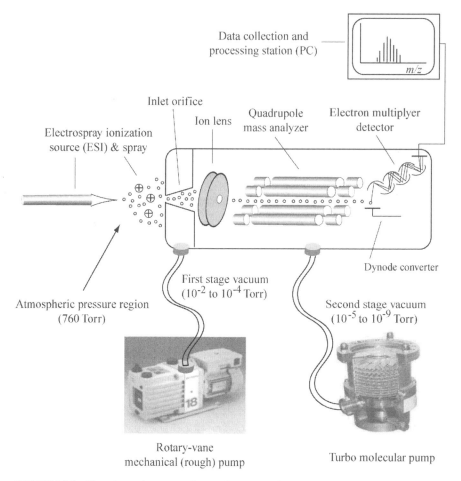

FIGURE 1.1   The six components that make up the fundamental configuration of mass spectrometric instrumentation consisting of: (1) inlet and ionization system, (2) inlet orifice (source), (3) mass analyzer, (4) detector, (5) vacuum system, and (6) data collection and processing station (PC). [See Wikipedia, Turbomolecular pump, http://en.wikipedia.org/w/index. php?title=Turbomolecular pump&oldid=71160479 (as of Aug. 24, 2006, 17:45 GMT).]

1.2   BASIC DESIGN OF MASS ANALYZER INSTRUMENTATION

Typical mass spectrometric instrumentation that are used in laboratories and research institutions are comprised of six components: (1) an inlet, (2) an ionization source, (3) a mass analyzer, (4) a detector, (5) a data processing system, and (6) a vacuum system. Figure 1.1 illustrates the interrelationship of the six components that make up the fundamental construction of a mass spectrometer. The *inlet* is used to introduce a sample into the mass spectrometer and can be a solid probe, a manual syringe or syringe pump system, a gas chromatograph, or a liquid chromatograph. The inlet system can be either at atmospheric pressure, as shown in Figure 1.1, or it can be at a reduced pressure under vacuum. The *ionization source* functions to convert neutral molecules into charged analyte ions, thus enabling their mass analysis. The ionization source can also be part of the inlet system. A typical inlet system and ionization source that is used with high-performance liquid chromatography (HPLC) is electrospray ionization (ESI). In an HPLC/ESI inlet system and ionization source, the effluent coming from the HPLC column is transferred into the ESI capillary, which has a high voltage applied to it, inducing the electrospray ionization process. In this configuration the inlet system and ionization source are located at atmospheric pressure outside of the mass spectrometric instrumentation, which is under vacuum. The spray that is produced passes through a tiny orifice that separates the internal portion of the mass spectrometer, which is under vacuum from its ambient surroundings, which are at atmospheric pressure. This orifice is also often called the *inlet* and/or the *source*. In the case of the coupling of a gas chromatograph to the mass spectrometer, the capillary column of the gas chromatograph is inserted through a heated transfer capillary directly into the internal portion of the mass spectrometer, which is under vacuum. This is possible because the species eluting from the capillary column are already in the gas phase, making their introduction into the mass spectrometer more straightforward as compared to the liquid eluant from an HPLC where analytes must be transferred from the solution phase to the gas phase. An example of an ionization process that takes place under vacuum in the front end of the mass spectrometer is a process called matrix-assisted laser desorption ionization, or MALDI. In this ionization technique a laser pulse is directed toward a MALDI target that contains a mixture of the neutral analytes and a strongly ultraviolet (UV) absorbing molecule, oftentimes a low-molecular weight organic acid such as dihydroxy benzoic (DHB) acid. The analytes are lifted off of the MALDI target plate directly into the gas phase in an ionized state. This is due to transference of the laser energy to the matrix and then to the analyte. The MALDI technique takes place within a compartment that is at the beginning of the mass spectrometer instrument and is under vacuum. The compartment in which this takes place is often called the ionization source, thus combining the *inlet system* and the *ionization source* together into one compartment. As illustrated in Figure 1.1 the analyte molecules (small circles), in an ionized state pass from atmospheric conditions to the first stage of vacuum in the mass spectrometer through an inlet orifice that separates the mass spectrometer, which is under vacuum from ambient conditions. The analytes are guided through a series of ion lenses into the mass analyzer. The *mass analyzer* is the heart of the system, which is a separation device that separates positively or negatively charged ionic species in the gas phase according to their respective mass-to-charge ratios. The mass analyzer gas-phase ionic species separation can be performed by an external field such as an electric field or a magnetic field, or by a field-free region such as within a drift tube. For the detection of the gas-phase separated ionic species, electron multipliers are often used as the *detector*. Electron multipliers are mass impact detectors that convert the impact of the gas-phase separated ionic species into a cascade of electrons, thereby multiplying the signal of the impacted ion many times.

The *vacuum system* ties into the inlet, the source, the mass analyzer, and the detector of the mass spectrometer at different stages of increasing vacuum as movement goes from the inlet to the detector (left to right in Fig. 1.1). It is very important for the mass analyzer and detector to be under high vacuum as this removes ambient gas, thereby reducing the amount of unwanted collisions between the mass-separated ionic species and gas molecules present. As illustrated in Figure 1.1, ambient, atmospheric conditions are generally at a pressure of 760 Torr. The first-stage vacuum is typically at or near 10^{-3} Torr, immediately following the inlet orifice and around the first ion transfer lenses. This stage of vacuum is obtained using two-stage rotary vane mechanical pumps that are able to handle high pressures such as atmospheric and large variation in pressures but are not able to obtain the lower pressures required further into the mass spectrometer instrument. The two-stage rotary vane mechanical pump has an internal configuration that utilizes a rotating cylinder that is off-axis within the pump's hollow body. The off-axis positioned rotor contains two vanes that are opposed and directed radially and are spring controlled to make pump body contact. As the cylinder rotates, the volume between the pump's body and the vanes changes, the volume increases behind each vane, which passes a specially placed gas inlet port. This will cause the gas to expand behind the passing vane while the trapped volume between the exhaust port and the forward portion of the vane will decrease. The exhaust gas is forced into a second stage and then is released by passing through the oil that is contained within the pump's rear oil reservoir. This configuration is conducive for starting up at atmospheric pressure and working toward pressures usually in the range of 10^{-3}–10^{-4} Torr.

The lower stages of vacuum are obtained most often using turbomolecular pumps as illustrated in Figure 1.1. Turbomolecular pumps are not as rugged as the mechanical pumps described previously and need to be started in a reduced-pressure environment. Typically, a mechanical pump will perform the initial evacuation of an area. When a certain level of vacuum is obtained, the turbomolecular pumps will then turn on and bring the pressure to higher vacuum. Using a mechanical vane pump to provide a suitable forepump pressure for the turbomolecular pump is known as roughing or "rough out" the chamber. Therefore, two-stage rotary vane mechanical pumps are often referred to as rough pumps. As illustrated in Figure 1.1 the turbomolecular pump contains a series of rotor–stator pairs that are mounted in multiple stages. The principle of turbomolecular pumps is to transfer energy from the fast rotating rotor (turbomolecular pumps operate at very high speeds) to the molecules that make up the gas. After colliding with the blades of the rotor, the gas molecules gain momentum and move to the next lower stage of the pump and repeat the process with the next rotor. Eventually, the gas molecules enter the bottom of the pump and exit through an exhaust port. As gas molecules are removed from the head or beginning of the pump, the pressure before the pump is continually reduced as the gas is removed through the pump, thus achieving higher and higher levels of vacuum. Turbomolecular pumps can obtain much higher levels of vacuum (up to 10^{-9} Torr) as compared to the rotary vane mechanical pumps (up to 10^{-4} Torr).

The final component of the mass spectrometer is a data processing system. This is typically a personal computer (PC) allowing the mass spectrometric instrumentation to be software controlled enabling precise measurements of carefully designed experiments and the collection of large amounts of data. Commercially bought mass spectrometers will come with their own software that is used to set the operating parameters of the mass spectrometer and to collect and interpret the data, which is in the form of mass spectra.

1.3 MASS SPECTROMETRY OF PROTEIN, METABOLITE, AND LIPID BIOMOLECULES

Proteins and lipids are two important classes of biological compounds that are found in all living species. Proteomics (the identification and characterization of proteins in cells, organisms, etc.) has now become a very important field of research in mass spectrometry, which is often used to describe a cell's or organism's proteome or to investigate a response to a stress upon a system through a change in protein expression. Global metabolic profiles (metabolomics) have also become a quickly advancing area of study utilizing mass spectrometric techniques. Lipidomics is a subclass of metabolomics that is also experiencing an increase in mass spectral applications. Lipids are often associated with proteins and act as physiological activators. Lipids constitute the bilayer components of biological membranes and directly participate in membrane protein regulation and function. In the past decade mass spectrometry has experienced increasing applications to the characterization and identification of these two important biological compound classes. Using proteomics, metabolomics, and lipidomics as examples of the application of mass spectrometry to biomolecular analysis we will briefly look at these areas in the following three sections.

1.3.1 Proteomics

The area of proteomics has been applied to a wide spectrum of physiological samples often based upon comparative studies where a specific biological system's protein expression is compared to either another system or the same system under stress. Often in the past the comparison is made using two-dimensional (2D) electrophoresis where the gel maps for the two systems are compared looking for changes such as the presence or absence of proteins and the up or down regulation of proteins. Proteins of interest are cut from the gel and identified by mass spectrometry. Figure 1.2 illustrates a one-dimensional (1D) sodium dodecylsulfate polyacrylamide gel electrophoresis (SDS-PAGE) gel (12% acrylamide) of rabbit tear from a normal versus a dry eye diseased state model and a protein molecular weight marker.[1] The technique of SDS-PAGE accomplishes the linear separation of proteins according to their molecular weights. In this process the proteins are first denatured and sulfide bonds are cleaved, effectively unraveling the tertiary and secondary structure of the protein. Sodium dodecylsulfate, which is negatively charged, is then used to

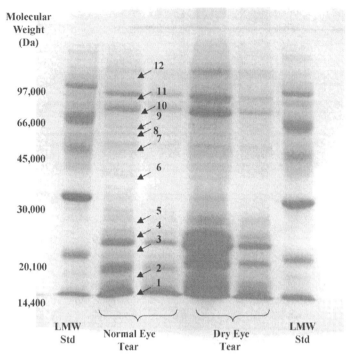

FIGURE 1.2 One-dimensional SDS-PAGE gel of rabbit tear proteins from a study of normal eye versus a dry eye model. Arrows and numbers represent bands of separated proteins that are used for protein identifications. (Reprinted with kind permission of Springer Science and Business Media from Ham, B. M., Jacob, J. T., and Cole, R. B. Single eye analysis and contralateral eye comparison of tear proteins in normal and dry eye model rabbits by MALDI-TOF mass spectrometry using wax-coated target plates. *Anal. Bioanal. Chem.* 2007, *3*, Figure 1, p. 891.)

coat the protein in a fashion that is proportional to the proteins' molecular weight. The proteins are then separated within a polyacrylamide gel by placing a potential difference across the gel. Due to the potential difference across the gel, the proteins will experience an electrophoretic movement through the gel, thus separating them according to their molecular weight with the lower molecular weight proteins having a greater mobility through the gel and the higher molecular weight proteins having a lower mobility through the gel. There are 12 clear bands observed and designated in Figure 1.2, as indicated by the numbering and associated arrows. These bands represent proteins of different molecular weights that have been separated electrophoretically. Often bands such as these are cut from the gel and the proteins are digested with an enzyme such as trypsin, which is an endopeptidase that cleaves within the polypeptide chain of the protein at the carboxyl side of the basic amino acids arginine and lysine (the trypsin enzyme has optimal activity at a pH range of 7–10 and requires the presence of Ca^{2+}). The enzyme-cleaved

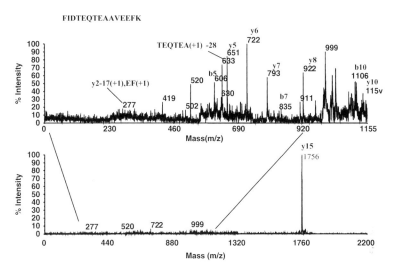

FIGURE 1.3   Matrix-assisted laser desorption ionization (MALDI) time-of-flight (TOF) postsource decay (PSD) mass spectrum illustrating the identification of the protein lipophilin CL2 recovered from the band 1 in Figure 1.2. (Reprinted with kind permission of Springer Science and Business Media from Ham, B. M., Jacob, J. T., and Cole, R. B. Single eye analysis and contralateral eye comparison of tear proteins in normal and dry eye model rabbits by MALDI-TOF mass spectrometry using wax-coated target plates. *Anal. Bioanal. Chem.* 2007, *3*, Figure 5, p. 895.)

peptides are then extracted from the gel and analyzed by mass spectrometry for their identification.

Figure 1.3 shows the identification of one of the proteins as lipophilin CL2[1] recovered from band 1 in Figure 1.2 using the mass spectrometric technique of MALDI time-of-flight (TOF) postsource decay (PSD) mass spectrometry. A recent example of a 2D SDS-PAGE approach was reported by Nabetani et al.[2] in a study of patients with primary hepatolithiasis, an intractable liver disease, where the authors reported the up-regulation of 12 proteins and the down-regulation of 21 proteins. Another recent approach for proteomic studies by mass spectrometry uses a technique known as multidimensional protein identification technology (MudPIT).[3] This is a gel-free approach that utilizes multiple HPLC–mass spectrometry analysis of in-solution digestions of protein fractions. Kislinger et al.[4] demonstrated the effectiveness of this approach in a study of biochemical (mal)-adaptations involving heart tissue that was both healthy and diseased. Other examples include the spatial profiling of proteins and peptides on brain tissue sections,[5] the analysis of viruses' capsid proteins for viral identification and posttranslational modifications,[6] and structural immunology studies including antibody and antigen structures, immune complexes, and epitope sequencing and identification.[7] The discipline of proteomics and the applications of mass spectrometry to protein analysis will be extensively covered in Chapter 7.

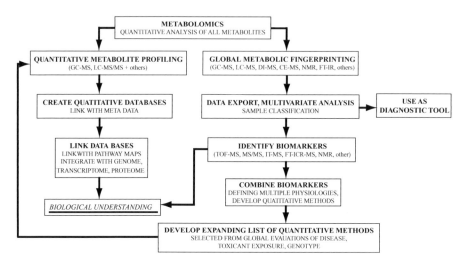

FIGURE 1.4 Strategies for metabolomic investigations. (Reprinted with permission of John Wiley & Sons, Inc. Dettmer, K., Aronov, P. A., and Hammock, B. D. Mass spectrometry-based metabolomics. *Mass Spectrom. Rev.* 2007, *26*, 51–78. Copyright 2007.)

1.3.2 Metabolomics

The study of global metabolite profiles that are contained within a system at any one time representing a set of conditions is metabolomics.[8] The system can be comprised of a cell, tissue, or an organism. Metabolomics has also been described as the measurement of all metabolite concentrations in cells, tissues, and organisms, while metabonomics involves the quantitative measurement that pathophysiological stimuli or genetic modification have upon the metabolic responses of multicellular systems.[9] The system's genomic interaction with the environmental system condition results in the production of metabolites. Metabolites represent an extremely diverse and broad set of biomolecule classes that include lipids, amino acids, nucleotides, and organic acids and cover widely different concentration ranges.[10] It has been estimated that there are from 2000 major metabolites[11] to 5000, with over 20,000 genes and 1 million proteins in humans.[12] Figure 1.4 illustrates the investigative strategies for two major approaches of mass-spectrometry-based metabolomics. The first strategy of the figure (left) is quantitative metabolic profiling, where generally predefined metabolites are targeted and measured to answer a biological question(s). The second strategy of the figure (right) is global metabolic fingerprinting, where the patterns or "fingerprints" of metabolites of a system are compared to measure changes due to a response of the system to disease, environmental changes, and the like. Table 1.1 lists some metabolomic-related definitions.

1.3.3 Lipidomics

Lipidomics has also experienced a dramatic increase in the amount of literature reported studies using mass spectrometry. Lipidomics involves the study of the lipid

TABLE 1.1   Metabolomic-Related Definitions

Metabolite   Small molecules that participate in general metabolic reactions and that are required for the maintenance, growth, and normal function of a cell.[a]

Metabolome   The complete set of metabolites in an organism.

Metabolomics   Identification and quantification of all metabolites in a biological system.

Metabolic Profiling   Quantitative analysis of set of metabolites in a selected biochemical pathway or a specific class of compounds. This includes target analysis, the analysis of a very limited number of metabolites, e.g., single analytes as precursors or products of biochemical reactions.

Metabolic Fingerprinting   Unbiased, global screening approach to classify samples based on metabolite patterns or "fingerprints" that change in response to disease, environmental, or genetic perturbations with the ultimate goal to identify discriminating metabolites.

Metabolic Footprinting   Fingerprinting analysis of extracellular metabolites in cell culture medium as a reflection of metabolite excretion or uptake by cells.

[a] This is in contrast to xenobiotic or foreign compound metabolites, which may overshadow natural metabolites in an analytical procedure. However, they can be very valuable in evaluating the physiological status of the organism.

Source: K. Dettmer, P. A. Aronov, and B. D. Hammock, Mass spectrometry-based metabolomics. Mass Spectrom. Rev., 26, 51–78, 2007. Copyright 2007. Reprinted with permission of John Wiley & Sons, Inc.

profile (or lipidome) of cellular systems and the processes involved in the organization of the protein and lipid species present,[13] including the signaling processes and metabolism of the lipids.[14] Lipids are a diverse class of physiologically important biomolecules that are often classified into three broad classes: (1) the simple lipids such as the fatty acids, the acylglycerols, and the sterols such as cholesterol, (2) the more complex polar phosphorylated lipids such as phosphatidylcholine and sphingomyelin, and (3) the isoprenoids, terpenoids, and vitamins. Recently, the Lipid Maps Organization has proposed an eight-category lipid classification (fatty acyls, glycerolipids, glycerophospholipids, sphingolipids, sterol lipids, prenol lipids, saccharolipids, and polyketides) that is based upon the hydrophobic and hydrophilic characteristics of the lipids.[15] Past methodology for the analysis of lipids often relied heavily on thin-layer chromatography, which generally requires a substantial amount of sample and the detection is nonspecific, done either visually or by spray phosphorus. For specific identifications the spot is usually scrapped from the plate, the lipids extracted, and then either analyzed directly or reacted with other reagents to increase volatility or detection ability with chromophore or fluorophore labeling. Often mass spectrometry can be used to separate biomolecules as illustrated in a single-stage spectrum that shows the molecular species present. Mass spectrometric experiments can then be performed where a single analyte species of interest observed in the single-stage mass spectrum is isolated from all of the other species present and collision-induced decomposition is performed. This effectively fragments the molecular species, which is recorded and illustrated in a product ion spectrum. The structure of the fragmented species can then be ascertained from the product ions produced. Figure 1.5 illustrates an electrospray ionization (ESI) single-stage mass spectrum of a complex biological extract where a number of molecular ion species

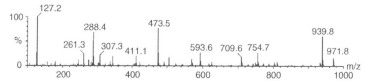

FIGURE 1.5   Electrospray ionization quadrupole–hexapole–quadrupole (ESI/QhQ) mass spectrum of the lithium adducts of a complex biological extract illustrating a number of biomolecules that are effectively separated in the gas phase.

are being observed as lithium adducts $[M + Li]^+$. The mass spectrometer has effectively separated these species according to their respective mass-to-charge ratios (*m/z*), producing a spectrum of peaks that represent individual biomolecular species. Using a type of mass spectrometer known as a quadrupole–hexapole–quadrupole mass spectrometer, the *m/z* 631 precursor ion species was isolated from all of the other species present and a tandem mass spectrometry experiment was performed as collision-induced dissociation (CID) of the precursor *m/z* 631 ion to produce the product ion spectrum illustrated in Figure 1.6. The quadrupole–hexapole–quadrupole mass spectrometer uses the inherent ability of a quadrupole mass analyzer to filter out all precursor ions except the one of interest, thus allowing it to pass through the system for the ensuing CID fragmentation study. Due to their filtering effect, quadrupoles are often referred to as mass filters. As illustrated in the product ion spectrum in Figure 1.6, there are two ranges of product ions, for the low range approximately *m/z* 43 to *m/z* 99 and for the high range approximately *m/z* 291 to *m/z* 365, that are used to deduce the structure and identity of the unknown biomolecule *m/z* 631. Using the mass spectrometric technique of CID and subsequent fragmentation, the unknown *m/z* 631 species was identified as 1,3-distearin. The process used in identifying unknowns such as the *m/z* 631 precursor biomolecule from its product ion spectrum will be extensively discussed in the ensuing chapters (see Chapters 5–10).

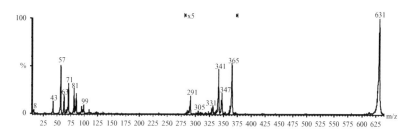

FIGURE 1.6   Electrospray ionization quadrupole–hexapole–quadrupole (ESI/QhQ) product ion mass spectrum of the lithium adduct precursor ion *m/z* 631 illustrated in Figure 1.5. Two product ion ranges, a high and a low, are observed that are used for structural identification of the unknown biomolecule. The *m/z* 631 species was identified as 1,3-distearin.

A recent review by Pulfer and Murphy[16] offers an informative discussion of the analysis of the phosphorylated lipids by mass spectrometry with a focus upon electrospray as the ionization technique, while Griffiths[17] covers the mass spectrometric analysis of the simple class of lipids (fatty acids, triacylglycerols, bile acids, and steroids). Recent examples of the application of mass spectrometry to biological sample lipid analysis includes a study by Lee et al.[18] where the relatively new technique of electron capture atmospheric pressure chemical ionization mass spectrometry was used in quantitative work of rat epithelial cell lipidomes, the analysis of human cerebellum gangliosides by nanoelectrospray tandem mass spectrometry,[19] the analysis of lyso-phosphorylated lipids in ascites from ovarian cancer patients,[20] the analysis of Amadori-glycated phosphatidylethanolamine in plasma samples from individuals with and without diabetes using a quadrupole ion trap mass spectrometer,[21] and the identification of nonpolar lipids and phosphorylated lipids in tear.[22,23]

A primary direction illustrated by these studies is for the understanding of biological systems in both normal and diseased states using the investigative technique of mass spectrometry. Proposed biological systems investigated can include studies of the expression of biomolecules from diseased states and stress models searching for relevant markers, the study of biological fluids for characterization, the study of biological processes intra- and extracellular, and the study of biomolecule interactions. Areas of mass spectrometric study of biomolecular systems include the characterization of the specific classes of proteins, lipids, carbohydrates, sugars, and so forth, the development of extraction, purification, and enrichment of biomolecules from complex biological samples, the development of quantitative analysis of biomolecules, the development of identification methods for biomolecules, and the identification and quantification of biomolecules in normal and diseased state expression studies and their function and importance. The discipline of lipidomics and the application of mass spectrometry to the analysis of lipids will be extensively covered in Chapter 8.

1.4 FUNDAMENTAL STUDIES OF BIOLOGICAL COMPOUND INTERACTIONS

The study of thermodynamic properties of complexes such as bond dissociation energies by CID tandem mass spectrometry for the measurement of the dissociation of gas-phase complexes has been a useful approach. For the study of the bond dissociation energy (BDE) of a cobalt carbene ion Armentrout and Beauchamp[24] introduced an ion beam apparatus and have since refined the measurement of noncovalent, collision-induced threshold bond energies of metal–ligand complexes by mass spectrometry. There are extensive examples of the use of CID tandem mass spectrometry[25] for the measurement of thermodynamic properties of gas-phase complexes including investigations of noncovalent interactions,[26–29] bond dissociation energies,[16] bond dissociation energies using guided ion beam tandem mass spectrometry,[30] gas-phase dissociative electron-transfer reactions,[31] critical energies for ion dissociation using a quadrupole ion trap,[26] or an electrospray triple quadrupole,[32,33] or a flowing afterglow triple quadrupole,[34,35] and gas-phase equilibria.[36–38]

During the ionization process, electrospray[39–42] is an ionization method that is well known to produce intact gas-phase ions with minimal fragmentation. Species originating from solution are quite accessible by electrospray mass spectrometry for studies of gas-phase thermochemical properties. Another "soft" ionization approach to gas-phase ionic complex analysis is matrix-assisted laser desorption ionization (MALDI); however, during crystallization with the matrix the complex's character in the solution phase can be lost. However, using a MALDI ion source coupled to a hybrid mass spectrometer, Gidden et al.[43] derived experimental cross sections for a set of alkali metal ion cationized poly(ethylene terephthalate) (PET) oligimers. With the advent of soft ionization techniques such as electrospray, it is now possible to study intact, wild-type associations such as lipid–metal/nonmetal, lipid–lipid, lipid–peptide, lipid–protein, protein–protein, and metabolite–protein adducts and complexes of both noncovalent and covalent-type bonding. Electrospray effectively allows the transfer of intact associations within solution phase (physiological/aqueous, or organic solvent modified) to the gas phase without interruption of the weak associations, thus preserving the original interaction (however, specific and nonspecific interactions need to be deciphered). Methods for determining the dissociation energies of these types of associations that can be applied to these studies have been developed.[44]

Figure 1.7 is the product ion spectrum of the diacylglycerol 1-stearin,2-palmitin illustrating the product ions produced upon collision-activated dissociation measured by an electrospray quadrupole–hexapole–quadrupole mass spectrometer (ESI/QhQ-MS). Using energy-resolved studies, it is possible to estimate the energies that were required to initiate bond breakage to form the product ions that are illustrated in Figure 1.7.

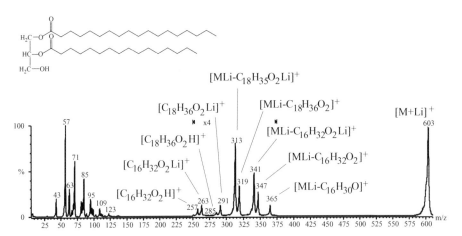

FIGURE 1.7 Product ion spectrum of the lithium adduct of the diacyglycerol 1-stearin,2-palmitin at m/z 603 illustrating the product ions generated by collision-induced dissociation.

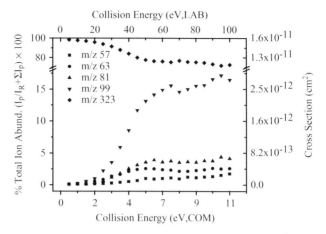

FIGURE 1.8   Energy resolved breakdown graph illustrating product ion formation with increasing internal energy for the collision-induced dissociation of monopentadecanoin.

Figure 1.8 is a energy-resolved breakdown graph of the production of lower molecular weight ions in CID experiments of monopentadecanoin showing increased production with increasing internal energy. The experiments used to construct the plot in Figure 1.8 are performed by measuring the products produced by CID such as those illustrated in Figure 1.7 at increasing internal energies. These results can then be used to relate the measured products of the CID experiments to the dissociation energies involved in the bond cleavages. Figure 1.9 is an energy diagram describing the production of a selected number of low-molecular-weight product ions from the CID of monopentadecanoin. Energy-resolved experiments and the application of mass spectrometry to gas-phase thermochemistry will be covered further in Chapter 4.

1.5   MASS-TO-CHARGE RATIO (*m/z*): HOW THE MASS SPECTROMETER SEPARATES IONS

For an analyte to be separated by a mass analyzer and subsequently measured by a detector, it must first be in an ionized state. Often the analyte is in a neutral state whether in a solid or liquid matrix (e.g., in a soil sample, solid tissue sample, an aqueous solution, or as an organic solvent extract). Using various ionization techniques, the analyte is ionized and transferred into the gas phase in preparation for introduction into the mass analyzer (see Chapters 2 and 3). The ionization of the analytes and transfer into the gas phase are typically done in an apparatus called a source, which acts as a front-end preparation stage for the mass analysis of the analyte as a positively or negatively charged ion. An analyte M can be in the form of a protonated molecule giving the analyte species a single positive charge

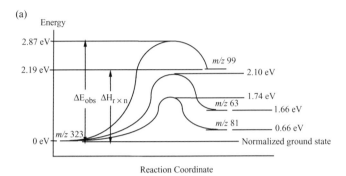

FIGURE 1.9 Energy diagram describing the formation of selected product ions from the collision-induced dissociation of monopentadecanoin.

$[M + H]^+$, a metal adduct such as sodium giving the molecular species a single positive charge $[M + Na]^+$, a chloride adduct giving the molecular species a single negative charge $[M + Cl]^-$, a deprotonated state giving the molecular species a single negative charge $[M - H]^-$, through electron loss giving the molecular species a single positive charge as a radical $M^{+\bullet}$ (also known as a molecular ion), and so on. All mass analyzers that are used in mass spectrometric analysis possess a characteristic means of relating an analyte species' mass to its respective charge (m/e), where m is the species mass and e is the fundamental charge constant of 1.602×10^{-19} C. The time-of-flight (TOF) mass analyzer affords a simple and direct example of the mass analyzer's physical relationship of analyte mass to analyte charge. The TOF mass analyzer separates analytes by time that have drifted through a preset described length drift tube. In the ionization source the analytes have all been given the same initial kinetic energy (KE); therefore, their physical description of $\frac{1}{2}mv^2$ are virtually constant for all of the ions for their drift through the drift tube. Using the relationship of $KE = zeV = \frac{1}{2}mv^2$, where V is the acceleration potential, and v is the velocity of the ion, an analyte with a larger mass *m* will possess a lower velocity *v*. The time (*t*) for the ions to arrive at the detector at the end of the drift tube is measured and is converted to a mass scale:

$$KE = zeV = \tfrac{1}{2}mv^2 \tag{1.1}$$

$$v = \left(\frac{2zeV}{m} \right)^{1/2} \tag{1.2}$$

$$t = \frac{L}{v} \qquad L = \text{length of drift tube} \tag{1.3}$$

$$t = L \left(\frac{m}{2zeV} \right)^{1/2} \qquad V \text{ and } L \text{ are fixed} \tag{1.4}$$

Solving for *m/z*:

$$t^2 = L^2 \left(\frac{m}{2zeV} \right) \tag{1.5}$$

$$\frac{m}{z} = \frac{2eVt^2}{L^2} \tag{1.6}$$

Thus, we have related the mass-to-charge ratio (*m/z*) of the analyte species as a positively or negatively charged ionic species to the drift time (*t*) of the ion through the drift tube while keeping the acceleration potential (*V*) and the length (*L*) of the drift tube fixed. This effectively allows the TOF mass analyzer to separate ionized species according to their mass-to-charge ratio. In general, an ion with a smaller mass and a larger charge will have the shorter drift time *t*. All mass analyzers must possess the physical characteristic of separating ionized species according to their mass-to-charge ratio. This characteristic aspect of general mass analyzers will be further discussed and illustrated in Chapter 3.

An example of a typical mass spectrum illustrating one usual design used is illustrated in Figure 1.10. The *x* axis of the mass spectrum represents the mass-to-charge (*m/z*) ratio measured by the mass analyzer, and the *y* axis is a percentage scale that has been normalized to the most abundant species in the spectrum. Often the

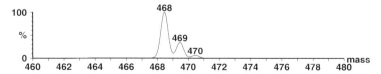

FIGURE 1.10   Typical mass spectrum illustrating three peaks at *m/z* 468, *m/z* 469, and *m/z* 470. The *y* axis is in percentage response and the *x* axis represents the mass-to-charge ratio (*m/z*).

y axis may also be presented as an intensity scale of arbitrary units. In this particular mass spectrum there are three peaks illustrated at *m/z* 468, *m/z* 469, and *m/z* 470.

1.6   EXACT MASS VERSUS NOMINAL MASS

Mass analyzers with high enough resolution have the ability to separate the naturally occurring isotopes of the elements. Therefore, in mass spectra there is often observed a series of peaks for the same species. This is illustrated in Figure 1.11 for an *m/z* 703.469 species (the farthest most left peak with lowest mass). The isotopic peaks are at *m/z* 704.465 and *m/z* 705.480. The *m/z* 703.469 peak is known as the monoisotopic peak where the particular species is made up of ^{12}C, and, if containing hydrogen, oxygen, nitrogen, phosphorus, and the like, it will be comprised of the most abundant isotope for that element: ^{1}H, ^{16}O, ^{14}N, ^{31}P, and so forth. Organic compounds are primarily comprised of carbon and hydrogen. Therefore, it is these two elements that contribute the most to the isotopic pattern observed in mass spectra (mostly carbon contribution) and to the mass defect of the molecular weight of the species (mostly hydrogen contribution). The mass defect of a species is the difference between the molecular formula's nominal value and its exact mass. A molecular formula nominal mass is calculated using the integer value of the most abundant isotope and

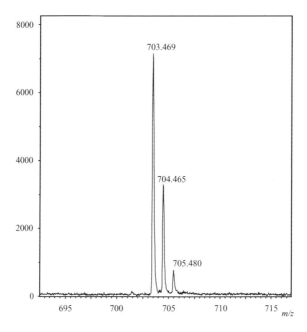

FIGURE 1.11   Mass spectrum of isotopically resolved peaks where *m/z* 703.469 species (the farthest most left peak of lowest mass) contains only ^{12}C and is known as the monoisotopic peak, the *m/z* 704.465 peak contains one ^{13}C, and the *m/z* 705.480 peak contains two ^{13}C.

a molecular formula exact mass (monoisotopic mass) is calculated using the most abundant isotopes' exact masses. For a formula of $C_{10}H_{20}O_3$, the nominal mass is 10×12 amu $+ 20 \times 1$ amu $+ 3 \times 16$ amu $= 188$ amu, and the exact monoisotopic mass is 10×12 amu $+ 20 \times 1.007825035(12)$ amu $+ 3 \times 15.99491463(5)$ amu $= 188.141245$ amu. Therefore, the mass defect is equal to 188.141245 amu $- 188$ amu $= 0.141245$ amu. For the mass spectrum shown in Figure 1.11 the second peak at m/z 704.465 is due to the inclusion of one ^{13}C into the structure of the measured species. The peak at m/z 705.480 is present due to the inclusion of two ^{13}C into the structure of the measured species.

The periodic table that the student is most familiar with lists the average weighted mass for the various elements listed in the table. For example, in the periodic table carbon is listed as 12.0107 amu, while in exact mass notation used in mass spectrometry carbon has an atomic weight of 12 amu. The abundances of the natural isotopes of carbon are ^{12}C (12.000000 amu by definition)[45] at 98.93(8)% and ^{13}C [13.003354826(17) amu] at 1.07(8)%. Therefore, the weighted average of the two carbon isotopes gives a mass of 12.0107 amu as is found in periodic tables. This is true of all the elements where the average atomic weight values in the periodic table represent the sum of the mass of the isotopes for that particular element according to their natural abundances (see Appendix 1 for a listing of the naturally occurring isotopes).

Some of the elements such as chlorine or bromine if present will also have a significant influence on the isotopic pattern observed in mass spectra. For example, chlorine has two isotopes: ^{35}Cl [34.968852721(69) amu at 75.78(4)% natural abundance] and ^{37}Cl [36.96590262(11) amu at 24.22(4)% natural abundance] that produces a more complex isotope pattern than that illustrated in Figure 1.10. The contribution of a chlorine atom to the isotopic pattern seen in mass spectra is illustrated in Figure 1.12 where the top spectrum is the isotope pattern for a molecular formula of $C_{30}H_{60}O_3$ (nominal mass of 468). In this isotope pattern there are three peaks observed for the ^{12}C monoisotopic peak at m/z 468, the inclusion of one ^{13}C at m/z 469, and the inclusion of two ^{13}C at m/z 470. If we take the same formula and add one chlorine

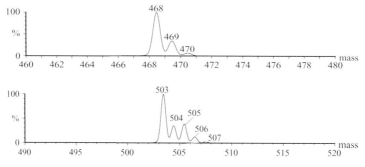

FIGURE 1.12   Isotope patterns for molecular formulas $C_{30}H_{60}O_3$ (top with m/z 468 nominal mass) and $C_{30}H_{60}O_3Cl$ (bottom with m/z 503 nominal mass).

atom, giving a molecular formula of $C_{30}H_{60}O_3Cl$ (nominal mass of 503), we will see a more complex isotope pattern. This is illustrated in the bottom mass spectrum of Figure 1.12 where there are now five clear isotopic peaks present. The m/z 503 peak represents the ^{12}C and ^{35}Cl containing species, the m/z 504 peak represents the ^{12}C and ^{35}Cl containing species with one ^{13}C, the m/z 505 peak represents a mixture of a ^{12}C and ^{35}Cl containing species with two ^{13}C and a ^{12}C and ^{37}Cl containing species, the m/z 506 peak represents a ^{12}C and ^{37}Cl containing species with one ^{13}C, and the m/z 507 peak represents a ^{12}C and ^{37}Cl containing species with two ^{13}C. This can, in fact, be a useful tool when a more complex isotopic series is observed that contains more than three peaks in the identification of elements present. Most software packages used with mass spectrometers have the ability to calculate and display isotopic series for a given atomic or molecular formula and to perform isotopic pattern matching.

1.7 MASS ACCURACY AND RESOLUTION

Mass accuracy in mass spectrometry is related to the calibration of the mass analyzer to properly assign the true mass-to-charge ratio to a detected ion and to the resolution of the detector response, which is in the form of an intensity spike or peak within a mass spectrum. In calibrating a mass spectrometer a series of standard compounds, usually in the form of a multicomponent standard, are measured and related to the fundamental properties of the mass analyzer such as the drift time within a TOF mass spectrometers drift tube. An example of an electrospray TOF mass analyzer calibration is illustrated in Figure 1.13 for sodium iodide (NaI) clusters. The top

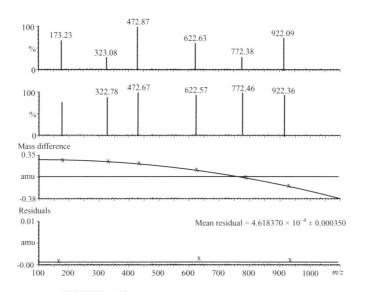

FIGURE 1.13   Calibration output for NaI clusters.

spectrum is the experimental spectrum obtained on the instrument in the laboratory. The second to top spectrum is a reference spectrum of a sodium iodide standard. A mass difference plot and residual plot are also displayed, showing the statistical results of the calibration fit. Calibrations of mass analyzers can be linear or polynomial; however, if more than two points are used for the calibration of the mass analyzer, a polynomial fit is generally used.

The mass accuracy in mass spectrometry is usually calculated as a parts-per-million (ppm) error where the theoretical mass (calculated as the monoisotopic mass) is subtracted from the observed mass, divided by the observed mass and multiplied by a 10^6 factor. Mass accuracy (ppm error) can thus be represented as

$$\text{Mass accuracy (ppm)} = \frac{m_{\text{observed}} - m_{\text{theoretical}}}{m_{\text{observed}}} \times 10^6 \qquad (1.7)$$

In mass spectrometry mass resolution is generally calculated as $m/\Delta m$ where m is the m/z value obtained from the spectrum and Δm is the full peak width at half maximum (FWHM). This is illustrated is Figure 1.14 for the m/z 703.469 ion, which has an FWHM of 0.1797 m/z. The resolution of the m/z 703.469 would be calculated as $m/\Delta m = 703.469/0.1797 = 3914$. The narrower and sharper a peak is will result in a higher resolution value and a better estimate of the apex of the Gaussian-shaped peak, which thus results in a better estimate of the true value of the mass-to-charge ratio and therefore higher accuracy. In contrast to this, as a peak becomes

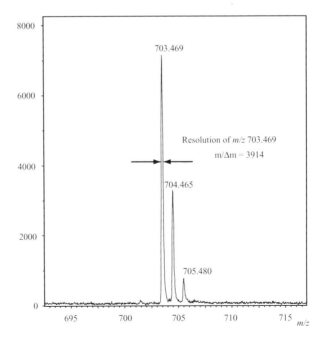

FIGURE 1.14 Full width at half maximum (FWHM) calculation of the resolution of m/z 703.469 as $m/\Delta m = 3914$.

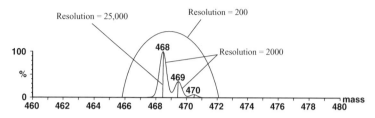

FIGURE 1.15   Effect of mass resolving power on the mass spectrum.

broader and wider, the apex of the peak becomes more distorted, resulting in a lower resolution value and thus reducing the accuracy of the estimate of the mass-to-charge ratio (m/z).

To get a better idea of the impact that the mass resolving power of a mass analyzer has upon mass spectra, the illustration in Figure 1.15 shows a comparison of a mass resolution of approximately 200 for the overall curve where none of the isotopic peaks have been resolved. A resolution of 200 will give an average mass value for the species being measured with a high degree of error for the exact mass (>500 ppm). The three isotope peaks are resolved at a resolution of 2000 and an intermediate accurate exact mass is obtainable. At a resolution of 25,000 the three isotopic peaks are completely baseline resolved and a mass accuracy typically less than 5 ppm can be obtained.

1.8   HIGH-RESOLUTION MASS MEASUREMENTS

There is a useful property that can be used when measuring the monoisotopic peak previously discussed in Section 1.6. For low-molecular-weight species (usually less than 2000 amu) measured as mass-to-charge ratio ions (m/z) in the gas phase using mass spectrometry, the monoisotopic peak will be the left most peak (lowest m/z value) of the isotopic peaks for that particular species, the most abundant peak in the isotopic series, and will contain only ^{12}C isotopes. Again, by definition and used as a standard the ^{12}C isotope has been assigned a mass of 12.00000000 amu. In bioorganic molecules the isotopic species most often encountered are carbon 12 as $^{12}C = 12.00000000$ amu, hydrogen 1 as $^{1}H = 1.007825035$ amu, nitrogen 14 as $^{14}N = 14.00307400$ amu, phosphorus 31 as $^{31}P = 30.973763$ amu, and oxygen 16 as $^{16}O = 15.99491463$ amu. When adding up the isotopes' mass contributions for a particular molecule (known as the theoretical molecular weight), such as a lithium adduct of cis-9-octadecenamide as $[C_{18}H_{35}NOLi]^+$, a cation in positive ion mode mass spectrometry, the molecular weight will include the mass error that is associated with the masses just listed for hydrogen (+0.007825035 amu), nitrogen (+0.00307400 amu), phosphorus (−0.026237 amu), and oxygen (−0.0.00508537 amu) where phosphorus and oxygen actually have a negative mass defect. The theoretical molecular weight for the lithium adduct of cis-9-octadecenamide as $[C_{18}H_{35}NOLi]^+$ is m/z 288.2879 (shown as a mass-to-charge m/z ion measured by mass spectrometry). To a certain

TABLE 1.2   Possible Formulas for *m/z* 288.2879 at a Mass Accuracy of 50 ppm

Theoretical Mass (Da)	Calculated Mass (Da)	Mass Difference (Da)	Mass Error (ppm)	Formula
288.2879	288.2879	0.0	0.0	$C_{18}H_{35}LiNO$
	288.2903	−2.4	−8.2	$C_{17}H_{38}NO_2$
	288.2855	2.4	8.4	$C_{19}H_{32}Li_2N$
	288.2815	6.4	22.3	$C_{14}H_{32}Li_2N_3O_2$
	288.2991	−11.2	−38.9	$C_{17}H_{35}LiN_3$
	288.3015	−13.6	−47.1	$C_{16}H_{38}N_3O$

accuracy we can derive the only molecular formula that will fit the theoretical mass of 288.2879 Da. At a mass accuracy of 50 ppm and using the most common atoms found in biomolecules (and also lithium 7 at $^7Li = 7.016005$ Da as we already have the luxury of knowing what type of cationized species it is), there exists six molecular formulas for a mass of 288.2879 Da. The six molecular formulas are listed in Table 1.2.

As can be seen in Table 1.2 a mass accuracy of 25 ppm will include the top four entries, at a mass accuracy of 10 ppm the top three will be included, and finally at a mass accuracy of 5 ppm only the top formula will be included, which is in fact the correct one. In this example we knew the true formula beforehand. However, oftentimes the mass spectrometrist will be working with unknown samples. In this case, the most accurate measurement of the monoisotopic peak is essential for reasonable molecular formula determination or confirmation. A benchmark or rule-of-thumb value of ≤5 ppm for the mass error is often used for consideration of an unknown's molecular formula derived from a high-resolution mass measurement. Figure 1.16 shows the results of a high-resolution mass measurement of the lithium adduct of cis-9-octadecenamide. The spectrum has been smoothed, centroided, and a lock mass technique has been used for fine adjustment of the calibration curve within the vicinity of the monoisotopic peak of interest. In this particular case the NaI cluster peak at *m/z* 322.7782 has been used as an internal standard for fine adjustment of the calibration used in collecting this spectrum.

The monoisotopic peak of cis-9-octadecenamide has a high-resolution mass measurement value of *m/z* 288.2876, as illustrated in Figure 1.16. Using the mass accuracy equation of Equation (1.7), we calculate an error of 1.0 ppm for the high-resolution mass measurement of cis-9-octadecenamide. Using the mass spectrometric approach

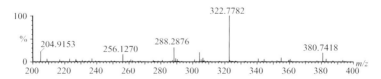

FIGURE 1.16   High-resolution mass measurement of cis-9-octadecenamide at *m/z* 288.2876 as the lithium adduct $[C_{18}H_{35}NOLi]^+$. The NaI cluster peak at *m/z* 322.7782 was used for the internal calibration standard.

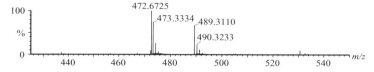

FIGURE 1.17 High-resolution mass measurement of unknown species at m/z 489.3110 as the sodium adduct $[M + Na]^+$. The NaI cluster peak at m/z 472.6725 was used for the internal calibration standard.

of high-resolution mass measurement and an error tolerance of $\leq$5 ppm, there is only one molecular formula that will match, demonstrating the utility of this methodology.

At higher molecular weights the possible molecular formulas that will match a high-resolution mass measurement will increase greatly for a mass error tolerance of $\leq$10 ppm. For example, for a high-resolution mass measurement of m/z 489.3110, illustrated in Figure 1.17 using the NaI cluster at m/z 472.6725 as the internal standard calibrant, gives 23 possible molecular formulas. The 23 molecular weight possibilities for the unknown species at m/z 489.3110 are listed in Table 1.3 for a mass error

TABLE 1.3 Possible Formulas for m/z 489.3110 at a Mass Accuracy of 10 ppm

Theoretical Mass (Da)	Calculated Mass (Da)	Mass Difference (Da)	Mass Error (ppm)	Formula
489.3110	489.3110	0.0	0.0	$C_{22}H_{51}O_7P_2$
	489.3110	0.0	0.1	$C_{27}H_{47}O_4PNa$
	489.3109	0.1	0.1	$C_{32}H_{43}ONa_2$
	489.3116	−0.6	−1.1	$C_{24}H_{49}N_2ONaP_2$
	489.3117	−0.7	−1.5	$C_{31}H_{41}N_2O_3$
	489.3099	1.1	2.2	$C_{21}H_{48}N_4O_3NaP_2$
	489.3099	1.1	2.2	$C_{26}H_{44}N_4Na_2P$
	489.3123	−1.3	−2.7	$C_{28}H_{43}N_4NaP$
	489.3123	−1.3	−2.7	$C_{23}H_{47}N_4O_3P_2$
	489.3094	1.6	3.4	$C_{24}H_{46}N_2O_6P_2$
	489.3093	1.7	3.4	$C_{29}H_{42}N_2O_3Na$
	489.3133	−2.3	−4.8	$C_{34}H_{42}ONa$
	489.3134	−2.4	−4.9	$C_{29}H_{46}O_4P$
	489.3086	2.4	5.0	$C_{25}H_{48}O_4Na_2P$
	489.3136	−2.6	−5.3	$C_{19}H_{45}N_4O_{10}$
	489.3140	−3.0	−6.1	$C_{26}H_{48}N_2ONaP_2$
	489.3077	3.3	6.8	$C_{26}H_{41}N_4O_5$
	489.3147	−3.7	−7.6	$C_{30}H_{42}N_4P$
	489.3069	4.1	8.3	$C_{22}H_{47}N_2O_6NaP$
	489.3069	4.1	8.4	$C_{27}H_{43}N_2O_3Na_2$
	489.3152	−4.2	−8.6	$C_{22}H_{46}N_2O_8Na$
	489.3064	4.6	9.5	$C_{25}H_{45}O_9$
	489.3157	−4.7	−9.7	$C_{36}H_{41}O$

tolerance of ≤10 ppm. The most abundant atoms that make up biomolecules were included in the molecular formula search and are C, H, N, O, P, and Na.

If the mass error tolerance is decreased to ≤5 ppm, it can be seen in Table 1.3 that there are still 14 possibilities for the identification of the unknown species molecular weight. If the mass error tolerance is decreased to ≤2 ppm, there are now still 5 possibilities, illustrating the utility of a high mass accurate measurement for decreasing the number of possible molecular formulas of an unknown species. Finally, the inclusion of the theoretical isotopic distribution can help to reduce the number of possible formulas associated with an unknown's molecular weight even further.

1.9 RINGS PLUS DOUBLE BONDS (r + db)

Due to the valences of the elements that make up most of the biomolecules that are analyzed by mass spectrometry, there is a very useful tool known as the "total rings plus double bonds" (r + db) that can be applied to molecular formulas such as $C_{12}H_{24}O_2$. The most general representation of the rings plus double bonds relationship is

$$r + db = \sum GroupIVA - \frac{1}{2}\sum (H + GroupVIIA) + \frac{1}{2}\sum GroupVA + 1$$

(1.8)

In mass spectrometry this expression is often simplified to only include the most common elements that typically make up the molecular ion species measured. This is generally presented as

$$r + db = x - \frac{1}{2}y + \frac{1}{2}z + 1$$

(1.9)

for the molecular formula $C_xH_yN_zO_n$. For example, normal hexane has no rings plus double bonds, cyclohexane contains one ring plus double bonds and benzene contains four rings plus double bonds, as illustrated in Figure 1.18.

The rings plus double bonds relationship presented in Equations (1.8) and (1.9) are applicable to organic compounds and to odd electron (OE) molecular ions ($M^{+\cdot}$) that are produced by electron ionization mass spectrometry (EI-MS). However, for even-electron (EE) ions that are produced by electrospray ionization mass spectrometry (ESI-MS) or matrix-assisted laser desorption ionization mass spectrometry (MALDI-MS) where the species has been ionized in the form of an acid adduct $[M + H]^+$, a metal adduct such as sodium $[M + Na]^+$, or a chloride adduct $[M + Cl]^-$, Equations (1.8) and (1.9) must be followed by the subtraction of 1/2. This is also a useful tool for interpreting whether a molecular ion is even electron or odd electron, as will be applied in the next section.

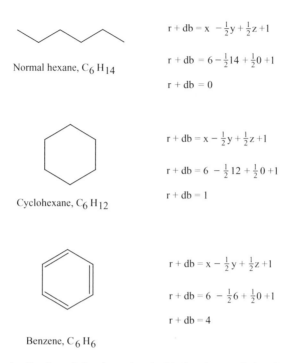

$$r + db = x - \tfrac{1}{2}y + \tfrac{1}{2}z + 1$$

$$r + db = 6 - \tfrac{1}{2}14 + \tfrac{1}{2}0 + 1$$

$$r + db = 0$$

Normal hexane, C_6H_{14}

$$r + db = x - \tfrac{1}{2}y + \tfrac{1}{2}z + 1$$

$$r + db = 6 - \tfrac{1}{2}12 + \tfrac{1}{2}0 + 1$$

$$r + db = 1$$

Cyclohexane, C_6H_{12}

$$r + db = x - \tfrac{1}{2}y + \tfrac{1}{2}z + 1$$

$$r + db = 6 - \tfrac{1}{2}6 + \tfrac{1}{2}0 + 1$$

$$r + db = 4$$

Benzene, C_6H_6

FIGURE 1.18 Application of the rings plus double bonds association for normal hexane, cyclohexane, and benzene using the simplified form of the equation for molecular formulas represented by $C_xH_yN_zO_n$.

1.10 NITROGEN RULE IN MASS SPECTROMETRY

Another useful tool used in mass spectrometry of biomolecule analysis is the so-called nitrogen rule. The nitrogen rule for organic compounds states that, if a molecular formula has a molecular weight that is an even number, then the compound contains an even number of nitrogen (e.g., 0N, 2N, 4N, ..., etc.). The same applies to an odd molecular weight species, which will thus contain an odd number of nitrogen (e.g., 1N, 3N, 5N, ..., etc.). The nitrogen rule is due to the number of valences that the different elements possess. The number of valences is the number of hydrogen atoms that can be bonded to the element. Most of the elements encountered in mass spectrometry either have an even mass and subsequent even valence or an odd mass and an odd valence. For example, the even mass/even valence elements are:

Carbon: ^{12}C has a valence of 4, $CH_4 = 16$ Da (even mass and even number of N = 0).

Oxygen: ^{16}O has a valence of 2, $H_2O = 18$ Da (even mass and even number of N = 0).

Silicon: ^{28}Si has a valence of 4, $H_4Si = 32$ Da (even mass and even number of $N = 0$).

Sulfur: ^{32}S has a valence of 2, $H_2S = 34$ Da (even mass and even number of $N = 0$).

The odd mass/odd valence elements are:

Hydrogen: ^{1}H has a valence of 1, $H_2 = 2$ Da (even mass and even number of $N = 0$).

Fluorine: ^{19}F has a valence of 1, $HF = 20$ Da (even mass and even number of $N = 0$).

Phosphorus: ^{31}P has a valence of 3 or 5, $H_3P = 34$ Da (even mass and even number of $N = 0$).

Chlorine: ^{35}Cl has a valence of 1, $HCl = 36$ Da (even mass and even number of $N = 0$).

However, nitrogen has an even mass but odd valence:

Nitrogen: ^{14}N has a valence of 3, $NH_3 = 17$ Da (odd mass and odd number of $N = 1$).

An odd number of nitrogen results in an odd mass such as ammonia NH_3, which has a formula weight of 17 Da. An example of a N_2-containing compound would be urea, CH_4N_2O with a nominal mass of 60 Da. The nitrogen rule can be directly applied to odd-electron (OE) molecular ion species $(M^{+\cdot})$ that are generated with electron ionization (EI; see Chapters 2 and 3) because the ionization process has not changed the species molecular formula or molecular weight. However, this is not the case with processes such as electrospray ionization mass spectrometry (ESI-MS) or matrix-assisted laser desorption ionization mass spectrometry (MALDI-MS) where the species has been ionized in the form of an acid adduct $[M + H]^+$, a metal adduct such as sodium $[M + Na]^+$, or a chloride adduct $[M + Cl]^-$. In these cases the precursor ion is an even electron (EE) ion and the weight of the original molecule has changed while the valence has not. Here the exact opposite to the above-stated nitrogen rule that applied to organic compounds and odd electron molecular ions is the case. In other words the nitrogen rule for even electron precursor ions states that an ion species with an odd mass will have an even number of nitrogen (e.g., 0N, 2N, 4N, ..., etc.), and an ion species with an even mass will have an odd number of nitrogen (e.g., 1N, 3N, 5N, ..., etc.).

1.11 PROBLEMS

1.1. What are the six components that comprise a mass spectrometer?

1.2. To what does mass analyzer separate charged ionic species in the gas phase?

1.3. Why is it important for the mass analyzer and detector to be under high vacuum?

1.4. What two types of pumps are used with mass spectrometric instrumentation, and what pressures do they obtain approximately?

1.5. To what three areas has mass spectrometry currently found an increased application?

1.6. Give a definition of proteomics, metabolomics, and lipidomics.

1.7. How is it that the soft ionization technique of electrospray allows gas-phase analysis of solution-phase associations? What are some examples of these associations?

1.8. What are a couple of examples of ionized analytes that are measured using mass spectrometry?

1.9. Given an acceleration voltage (KE) of 2000 eV, calculate the mass-to-charge ratio (m/z) of a biomolecule measured by a TOF mass spectrometer with a drift tube length of 2 m and a flight time of 68 μs.

1.10. A biomolecule has a mass of 129 amu and a charge of 2+. If it is given an acceleration voltage of 1800 eV, what would be its drift time through a drift tube of length 2.3 m?

1.11. Calculate the exact mass (using the "relative atomic mass" of the highest abundance isotope in Appendix 1) and the weighted average mass (using the "standard atomic weight" in Appendix 1) for each of the following molecular formulas. What is the associated mass defect for each of the molecular formulas?
  a. $C_{10}H_{32}O_2$
  b. $C_{15}H_{28}N_2OPNa$
  c. $C_{29}H_{42}N_2O_3K$
  d. $C_{40}H_{75}N_2OP_3$
  e. $C_5H_{10}O_3Cl_3$

1.12. Calculate the mass accuracy of the following observed masses for the following biomolecules according to their proposed formula:
  a. $C_{24}H_{45}O_3N$ observed at m/z 395.4511
  b. $C_{64}H_{120}O_{10}N_4P_2$ observed at m/z 1166.8398
  c. $C_{48}H_{72}O_6$ observed at m/z 744.5310
  d. $C_{78}H_{130}O_{12}N_8$ observed at m/z 1370.9932

1.13. Using the following data, calculate the resolution of the following observed mass spectral peaks:
  a. m/z 342.4766 with peak width of m/z 0.6624
  b. m/z 1245.6624 with peak width of m/z 0.0573
  c. m/z 245.1191 with peak width of m/z 1.2444

1.14. Using the rings plus double bonds formula decide whether the following ions are even-electron or odd-electron ions and draw their formulas as such (as an example the first one is done). Using the nitrogen rule list whether you would expect each ion to have an even or odd mass-to-charge (m/z) ratio value. What is that value?

a. CH_4

r + db = 0, odd electron (OE) as $CH_4^{+\bullet}$, even mass ion with $m/z = 16$ Th.

b. NH_3

c. NH_4

d. H_3O

e. C_2H_7O

f. $C_{14}H_{28}O_2$

g. $C_6H_{14}NO$

h. $C_4H_{10}SiO$

i. $C_{29}H_{60}N_2O_6P$

REFERENCES

1. Ham, B. M., Jacob, J. T., and Cole, R. B. *Anal. Bioanal. Chem.* 2007, *3*, 889–900.
2. Nabetani, T., Tabuse, Y., Tsugita, A., and Shoda, J. *Proteomics* 2005, *5*, 1043–1061.
3. Wolters, D. A., Washburn, M. P., and Yates, J. R. *Anal. Chem.* 2001, *73*, 5683–5690.
4. Kislinger, T., Gramolini, A. O., MacLennan, D. H., and Emili, A. *J. Am. Soc. Mass Spectrom.* 2005, *16*, 1207–1220.
5. Pierson, J., Norris, J. L., Aerni, H. R., Svenningsson, P., Caprioli, R. M., and Andren, P. E. *J. Proteome Res.* 2004, *3*, 289–295.
6. Siuzdak, G. *J. Mass Spectrom.* 1998, *33*, 203–211.
7. Downard, K. M. *J. Mass Spectrom.* 2000, *35*, 493–503.
8. Rochfort, S. *J. Nat. Prod.* 2005, *68*, 1813–1820.
9. Nicholson, J. K., and Wilson, I. D. *Nat. Rev. Drug Discov.* 2003, *2*, 668–676.
10. Dettmer, K., Aronov, P. A., and Hammock, B. D. *Mass Spectrom. Rev.* 2007, *26*, 51–78.
11. Beecher, C. W. W. In *Metabolic Profiling: Its Role in Biomarker Discovery and Gene Function Analysis.* New York: Springer, 2003.
12. Ginsburg, G. S., and Haga, S. B. *Expert Rev. Mol. Diagn.* 2006, *6*, 179–191.
13. Han, X., and Gross, R. W. *Mass Spectrom. Rev.* 2005, *24*, 367–412.
14. Han, X., and Gross, R. W. *J. Lipid Res.* 2003, *44*, 1071–1079.
15. Fahy, E., Subramaniam, S., Brown, H. A., Glass, C. K., Merrill, A. H., Murphy, R. C., Raetz, C. R. H., Russell, D. W., Seyama, Y., Shaw, W., Shimizu, T., Spencer, F., van Meer, G., Van Nieuwenhze, M. S., White, S. H., Witztum, J. L., Dennis, E. A. *J. Lipid Res.* 2005, *46*, 839–862.
16. Pulfer, M., and Murphy, R. C. *Mass Spectrom. Rev.* 2003, *22*, 332–364.
17. Griffiths, W. J. *Mass Spectrom. Rev.* 2003, *22*, 81–152.
18. Lee, S. H., Williams, M. V., DuBois, R. N., and Blair, I. A. *Rapid Commun. Mass Spectrom.* 2003, *17*, 2168–2176.

19. Zamfir, A., Vukelic, Z., Bindila, L., Peter-Katalinic, J., Almeida, R., Sterling, A., Allen, M. *J. Am. Soc. Mass Spectrom.* 2004, *15*, 1649–1657.

20. Xiao, Y., Schwartz, B., Washington, M., Kennedy, A., Webster, K., Belinson, J., and Xu, Y. *Anal. Biochem.* 2001, *290*, 302–313.

21. Nakagawa, K., Oak, J. H., Higuchi, O., Tsuzuki, T., Oikawa, S., Otani, H., Mune, M., Cai, H., and Miyazawa, T. *J. Lipid Res.* 2005, *46*, 2514–2524.

22. Ham, B. M., Jacob, J. T., Keese, M. M., and Cole, R. B. *J. Mass Spectrom.* 2004, *39*, 1321–1336.

23. Ham, B. M., Jacob, J. T., and Cole, R. B. *Anal. Chem.* 2005, *77*, 4439–4447.

24. Armentrout, P. B., and Beauchamp, J. L. *J. Chem. Phys.* 1981, *74*, 2819–2826.

25. Shukla, A. K., and Futrell, J. H. *J. Mass Spectrom.* 2000, *35*, 1069–1090.

26. Daniel, J. M., Friess, S. D., Rajagopala, S., Wendt, S., and Zenobi, R. *Int. J. Mass Spectrom.* 2002, *216*, 1–27.

27. Graul, S. T., and Squires, R. R. *J. Am. Chem. Soc.* 1990, *112*, 2517–2529.

28. Sunderlin, L. S., Wang, D., and Squires, R. R. *J. Am. Chem. Soc.* 1993, *115*, 12060–12070.

29. Pramanik, B. N., Bartner, P. L., Mirza, U. A., Liu, Y. H., and Ganguly, A. K. *J. Mass Spectrom.* 1998, *33*, 911–920.

30. Ervin, K. M., and Armentrout, P. B. *J. Chem. Phys.* 1985, *83*, 166–189.

31. Dougherty, R. C. *J. Am. Soc. Mass Spectrom.* 1997, *8*, 510–518.

32. Anderson, S. G., Blades, A. T., Klassen, J., and Kebarle, P. *Int. J. Mass Spectrom. Ion Processes* 1995, *141*, 217–228.

33. Klassen, J. S., Anderson, S. G., Blades, A. T., and Kebarle, P. *J. Phys. Chem.* 1996, *100*, 14218–14227.

34. Graul, S. T., and Squires, R. R. *J. Am. Chem. Soc.* 1990, *112*, 2517–2529.

35. Sunderlin, L. S., Wang, D., and Squires, R. R. *J. Am. Chem. Soc.* 1993, *115*, 12060–12070.

36. Nielsen, S. B., Masella, M., and Kebarle, P. *J. Phys. Chem. A* 1999, *103*, 9891–9898.

37. Kebarle, P. *Int. J. Mass Spectrom.* 2000, *200*, 313–330.

38. Peschke, M., Blades, A. T., and Kebarle, P. *J. Am. Chem. Soc.* 2000, *122*, 10440–10449.

39. Whitehouse, C. M., Dreyer, R. N., Yamashita, M., and Fenn, J. B. *Anal. Chem.* 1985, *57*, 675–679.

40. Fenn, J. B. *J. Am. Soc. Mass Spectrom.* 1993, *4*, 524–535.

41. Cole, R. B. *J. Mass Spectrom.* 2000, *35*, 763–772.

42. Cech, N. B., and Enke, C. G. *Mass Spec. Rev.* 2001, *20*, 362–387.

43. Gidden, J., Wyttenbach, T., Batka, J. J., Weis, P., Jackson, A. T., Scrivens, J. H., and Bowers, M. T. *J. Am. Soc. Mass Spectrom.* 1999, *10*, 883–895.

44. Ham, B. M., and Cole, R. B. *Anal. Chem.* 2005, *77*, 4148–4159.

45. Commission on Atomic Weights and Isotopic Abundances Report for the International Union of Pure and Applied Chemistry, in *Isotopic Compositions of the Elements 1989, Pure and Applied Chemistry*, 1998, *70*, 217.

CHAPTER 2

IONIZATION IN MASS SPECTROMETRY

2.1 IONIZATION TECHNIQUES AND SOURCES

There are numerous ionization techniques for the production of molecular cations (e.g., $M^{+\cdot}$, a radical cation) or adducted ionized species (e.g., $[M + H]^+$ or $[M + Cl]^-$) from neutral biomolecules in use today. Closely related are the instrumental sources used for the introduction of the ionized species into mass spectrometers that are available to the mass spectrometrist, and both will be covered in this chapter. Before an analyte can be measured by a mass analyzer, it must first be in an ionized state. Biomolecules for mass analysis are often in a neutral state and must be converted to an ionized state, which may exist as a cation or as an anion depending upon the molecule to be mass analyzed. This is a very important fundamental aspect of mass spectrometry, and much developmental work has been devoted to processes that can be used to convert a neutral molecule into an ionized species. The source, which is used to transfer the newly formed ionized analyte into the mass spectrometer, is the second important step in mass analysis of ionized analyte species. The source is often considered synonymous with the ionization technique. This is often due to the ionization technique taking place within what would be considered the source. Source/ionization systems include electron ionization (EI), electrospray ionization (ESI), chemical ionization (CI), atmospheric pressure chemical ionization (APCI), atmospheric pressure photoionization (APPI), and matrix-assisted laser desorption ionization (MALDI). These ionization techniques produce ions of analyte molecules (often designated as M for molecule), which includes molecular ions $M^{+\cdot}$ (from EI), protonated molecules

Even Electron Mass Spectrometry with Biomolecule Applications   By Bryan M. Ham
Copyright © 2008 John Wiley & Sons, Inc.

([M + H]$^+$), deprotonated molecules ([M − H]$^-$), and metal ([M + metal]$^+$, e.g., [M + Na]$^+$) or halide ([M + halide]$^-$, e.g., [M + Cl]$^-$) adduct (all possible from ESI, CI, APCI, APPI, and MALDI). All of these ionization techniques and sources are used to effectively ionize a neutral molecule and transfer the formed ionized molecule into the gas phase in preparation to introduction into the low-vacuum environment of the mass analyzer.

2.2   ELECTRON IONIZATION (EI)

Electron ionization is the oldest ionization source that has been used in mass spectrometry and is still widely in use today. This design basically stems from the work of J. J. Thompson (1856–1940) who is considered the first mass spectrometrist to win the Nobel Prize. In 1906 he was awarded the Nobel Prize in physics for his studies involving the discharge of electricity into gases producing ionized species. Indeed, the recently introduced unit for the mass-to-charge (*m/z*) value in mass spectrometry, the Thompson (Th), has been proposed in his honor. Typically, electron ionization is used as an ionization source after analytes have been separated within a gas chromatograph. The most common mass analyzer used in this configuration is the single quadrupole mass filter, which will be covered in Chapter 3, Section 3.5. Figure 2.1 illustrates the basic design of an electron ionization source. Incoming into the source from the left are the neutral analytes from some type of inlet system such as a gas chromatography column. The electron ionization source is not at atmospheric pressure (760 Torr or 101.325 kPa) but at a reduced pressure of 10^{-3} Pa (10^{-5} Torr), eliminating the ionization of ambient species by the electron beam and collisions with the gaseous analyte molecules. The neutral analytes in the gas phase pass through

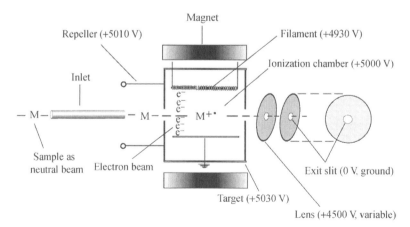

FIGURE 2.1   Basic design of an electron ionization (EI) source including potentials involved in moving the sample through the source. Movement of sample beam from neutral to ionized is from left to right. Sample progresses through the exit slit to the mass spectrometer.

an electron beam that is generated by heating a filament that subsequently boils off electrons. Opposite the filament is a target collector for the boiled off electron beam at a potential of $+5030$ V while the filament is kept at a potential of $+4930$ V. This potential difference between the filament and the target is important and is kept constant from instrumentation to instrumentation. The potential difference between the filament and the target imparts to the electron beam energy of approximately 70 eV. By keeping this imparted energy constant at 70 eV from instrument to instrument, standardized spectra are obtained that are basically universal. This allows the searching of unknown EI spectra against EI spectral libraries due to reproducible spectra obtained from the standardized 70 eV electron beam impact-induced EI spectra. As the analyte passes through the electron beam, energy is transferred from the fast moving electrons within the beam to the analyte. The excited analyte releases a portion of the absorbed energy through ejection of an electron, thus producing singly positive charged molecular ions, $M^{+\cdot}$, that are odd electron (OE) radical ions.

$$M + e^-(70\,eV) \rightarrow M^{+\cdot} + 2e^- \tag{2.1}$$

Typical organic molecules have ionization potentials ranging from approximately 6 to 10 eV, which is the energy required to free an electron from the molecule. The excess energy remaining in the molecular ion produced from the electron ionization process generally causes fragmentation to take place, often severe enough to deplete the precursor molecular ion. The process of electron ionization is a "hard" ionization process resulting in considerable fragmentation of the molecular ion due to the excess energy imparted to the analyte (** and * represent excited states) from the electron beam. This process is illustrated in Equation (2.2) where the first step is transfer of energy from the electron beam to the analyte, producing an excited state in the neutral analyte. The second step is the ejection of an electron from the excited analyte, producing an odd electron molecular ion, but still in an excited state due to excess energy imparted in the ionization process. The final step is fragmentation of the molecular ion, producing both even and odd electron molecular ions:

$$
\begin{aligned}
ABCDE + e^-(70\,eV) &\rightarrow [ABCDE]^{**} + e^-(<70\,eV) \\
[ABCDE]^{**} &\rightarrow [(ABCDE)^{+\cdot}]^* + e^- \\
[(ABCDE)^{+\cdot}]^* &\rightarrow AB^{+\cdot} + CDE \quad \text{(OE radical product plus neutral)} \\
&\searrow \\
&\quad\; ABC^+ + \cdot DE \quad \text{(EE product plus radical neutral)}
\end{aligned}
\tag{2.2}
$$

Located above and below the filament and the target trap used for the electron beam are the opposite poles of a magnet. The small magnetic field induces the electrons in the beam to follow a spiral trajectory from filament to target. This in effect causes the electrons to follow a longer path from filament to target, producing a greater chance of ionizing the neutral analyte molecules passing through the electron beam. Behind the electron beam is a repeller that is kept at $+5010$ V used to give the newly formed molecular ions a push out of the ionization chamber, which is kept at $+5000$ V. The

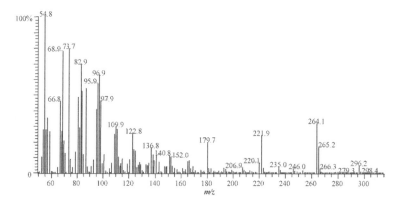

FIGURE 2.2 Odd electron product ion spectrum of metyl oleate ($C_{19}H_{36}O_2$) at m/z 296.2 as $M^{+\cdot}$ collected by electron impact ionization/single quadrupole mass spectrometry (EI/MS).

newly formed molecular ions then pass through a series of focusing lenses that are kept at +4500 V (but variation is allowed for ion beam focusing). Finally, the newly formed molecular ions pass through an exit slit that is held at 0 V (ground). The whole source has a potential gradient made up of a slope in voltages from the repeller to the exit slit that the molecular ion beam will follow leading out of the source and into the mass analyzer. Figure 2.2 illustrates a typical electron ionization mass spectrum where the molecular ion, $M^{+\cdot}$, is observed at m/z 296.2. The y axis of the EI mass spectrum is in relative abundance (%). Typical of mass spectra the y axis has been normalized to the most abundant peak in the spectrum such as the m/z 54.8 peak in Figure 2.2. The x axis is the mass-to-charge ratio (m/z) moving to increasing values from left to right. Notice in the EI mass spectrum the large degree of fragmentation taking place, depleting the molecular ion at m/z 296.2.

2.3 CHEMICAL IONIZATION (CI)

A complementary ionization technique to electron ionization (IE) is chemical ionization (CI), an ionization technique that uses gas-phase ion-molecule reactions to ionize the neutral analytes. Chemical ionization often can be chosen as a softer ionization technique rather than electron impact (electron ionization, EI), which is known as a hard ionization technique often depleting the molecular ion produced during the ionization. Therefore, a greater abundance of the analyte in an ionized state is observed in chemical ionization as compared to electron ionization. Chemical ionization in turn can add confidence in the determination of the molecular weight of the ionized analyte while electron ionization gives structural information through extensive fragmentation of the molecular ion. This is illustrated in Figure 2.3 for the CI mass spectrum of methyl oleate. Here we see the precursor ion at m/z 297.2 as a protonated species $[M+H]^+$, giving molecular weight (MW) information for the analyte. In chemical ionization the source acts to create a very reactive, high-pressure

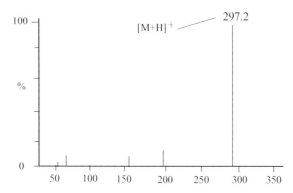

FIGURE 2.3   Chemical ionization (CI) mass spectrum of methyl oleate at m/z 297.2 as a protonated species $[M+H]^+$, giving MW information for the analyte without extensive fragmentation.

region containing the ionized reagent gas known as the plasma region. It is in this plasma region that the analyte passes through allowing ionization and subsequent mass analysis and detection. The source design for chemical ionization is similar to that used for electron ionization, however, with a few operational differences required for the technique. The slits and apertures are smaller in the chemical ionization source to help maintain the high pressure that is created in the source due to the addition of a reactive gas into the source used for the chemical ionization. In chemical ionization the emission current leaving the filament is measured instead of the target current as in electron ionization. The repeller voltage is also kept lower in CI as compared to EI in order to maximize the number of collisions between the analyte and the CI gas. In the CI source design a reagent gas is introduced into the "tight" source and maintained at a pressure that is higher than the mass analyzer but lower than atmospheric pressure. A maintained higher pressure in the source of the reagent gas serves to maximize the number of collisions between the reagent gas and the analyte. The higher pressure also serves in promoting dampening collisions between the analyte and the reagent gas. Dampening collisions are slow, low-energy collisions that help to relax the ionized analyte molecules by transferring excess energy from the ionized analyte to the gas molecule. These dampening collisions help to decrease the amount of fragmentation that can take place due to excess energy from the ionization process. This helps to maintain a softer ionization, producing a greater abundance of intact ionized analyte molecules. However, the amount of energy transferred in the ionization step can also be controlled by the analyst through the use of different chemical reagent gases as will be discussed next.

In the chemical ionization process the first step is the ionization of the reagent gas by the electron beam from the filament/target setup, which is similar to that found in EI sources. The ionization of the reagent gas by the electron beam often produces a protonated reagent gas that then transfers a proton to the analyte, producing the ionized form of the analyte, $[M + H]^+$. This gives even electron (EE) analyte ions that are generally more stable then the radical, odd electron (OE) molecular ions $(M^{+\cdot})$

that are produced with EI. In positive chemical ionization that results in protonation the analyte ion is 1 mass unit greater than the neutral analyte (M + H). In negative chemical ionization that results in proton abstraction the analyte ion is 1 mass unit less than the neutral analyte (M − H).

$$
\begin{aligned}
\text{Positive ion:} \quad & GH^+ + M \rightarrow MH^+ + G \\
\text{Negative ion:} \quad & (G - H)^- + M \rightarrow (M - H)^- + G
\end{aligned}
\tag{2.3}
$$

where G is the reagent gas and M is the analyte molecule.

2.3.1 Positive Chemical Ionization

Chemical Ionization processes are ion–molecule reactions between the analyte molecules (M) and the ionized reagent gas ions (G) that produce the analyte ions. These are gas-phase acid–base reactions according to the Bronsted–Lowrey theory. In general, these are exothermic reactions taking place in the gas phase. In positive chemical ionization the three most common reagent gasses used are methane (CH_4), isobutane (i-C_4H_{10}), and ammonia (NH_3). The first step in the process is electron ionization of the reagent gas by the filament-produced electron beam in the source. This is illustrated for the reactive gas formation of methane as follows (the species designated with * are the most reactive forms of the ionized gas molecules):

$$
\begin{aligned}
CH_4 + e^- &\rightarrow [CH_4]^{+\bullet} + 2e^- \\
&\quad\ \hookrightarrow \quad CH_3^+ + H^\bullet \\
CH_4^{+\bullet} + CH_4 &\rightarrow (CH_5^+)^* + CH_3^\bullet \\
CH_4^{+\bullet} + CH_4 &\rightarrow (C_2H_5^+)^* + H_2 + H^\bullet
\end{aligned}
\tag{2.4}
$$

In the ionization of the reagent gas a very reactive, high-pressure region in the source containing the reagent ionized gas has been created that is called the plasma region. The protonated form of methane, $(CH_5^+)^*$, is a very reactive reagent gas that transfers the proton to the analyte with a high degree of excess energy. This is a very exothermic gas-phase reaction that is a "hard" ionization of the analyte, producing an ionized analyte that often to some degree fragments. Methane is able to accommodate the extra proton in the gas phase through a solvation of the proton by the σ electrons of the four covalent hydrogen bonds (illustrated in Fig. 2.4). The protonated form of methane, however, is very unstable and highly reactive.

   A second example of a reactive chemical ionization gas used is isobutane. Again, the first step in the process of creating the reactive gas is the electron ionization of the reagent gas by an electron-emitting filament. The electron ionization produces the radical cation form of the isobutane reagent gas:

$$
\begin{aligned}
\text{i-}C_4H_{10} + e^- &\rightarrow \text{i-}C_4H_{10}^{+\bullet} + 2e^- \\
\text{i-}C_4H_{10}^{+\bullet} + \text{i-}C_4H_{10} &\rightarrow (\text{i-}C_4H_9^+)^* + C_4H_9^\bullet + H_2
\end{aligned}
\tag{2.5}
$$

The second step is the same as observed for methane in Equation (2.4), the reactive radical cation form of the reagent gas reacts with other reagent gas molecules present

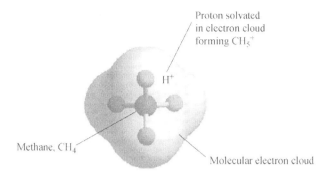

FIGURE 2.4 Protonated form of methane. The proton is solvated by the σ bonds of the covalently bound hydrogen, producing a highly reactive and unstable species. The protonated methane will transfer a proton to a neutral analyte (exothermic reaction), producing the ionized form of the analyte as a protonated species $[M + H]^+$.

to form the reactive chemical ionization species. However, unlike methane isobutane produces primarily only one reactive species as compared to two that are produced when using methane as the reagent gas. In the case of isobutane, however, the reactive species is not a protonated form of isobutane but the cation gaseous species i-$C_4H_9^+$, which is highly reactive and able to transfer a proton (H^+) for analyte ionization.

Finally, we have ammonia as a third example of a gaseous reagent used in chemical ionization. Like methane ammonia also produces two reactive species that may be used in the chemical ionization of neutral analyte molecules. In the second step of the ionization of ammonia to produce the reactive reagent gas species, a protonated form of ammonia is produced. This species is more stable than the protonated form of methane

$$\begin{align}
NH_3 + e^- &\rightarrow NH_3^{+\cdot} + 2e^- \\
NH_3^{+\cdot} + NH_3 &\rightarrow (NH_4^+)^* + NH_2^{\cdot} \\
NH_4^+ + NH_3 &\rightarrow (N_2H_7^+)^*
\end{align} \tag{2.6}$$

due to the lone pair of electrons available on the nitrogen atom in the ammonia molecule, which helps to delocalize the positive charge (see Fig. 2.5). The two reactive species produced from ammonia are both protonated species that have an extra proton that is available to transfer to a neutral analyte during the chemical ionization process.

The reactivity in chemical ionization is dependent upon the proton affinities (PA) of the ionized reagent gas (RG) and the analyte (M). The transfer of the proton from the reagent gas to the analyte must be an exothermic reaction for this to occur spontaneously in the chemical ionization source. The change in enthalpy (ΔH°) for the following reaction must be less than 0 for the transfer of the proton to take place spontaneously:

$$RGH^+ + M \rightarrow MH^+ + RG \quad \Delta H^\circ < 0 \tag{2.7}$$

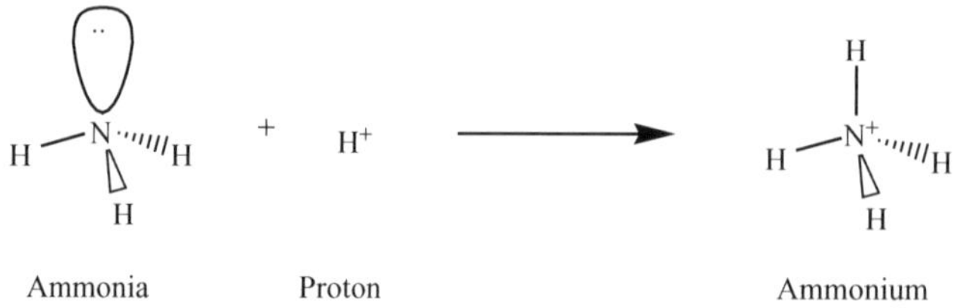

FIGURE 2.5 Reaction for the protonation of ammonia. Ammonia has available a lone pair of electrons in which to donate to the proton, forming the protonated reactive reagent gas species ammonium (NH_4^+). The proton shares the lone pair of electrons on the nitrogen, producing a reactive species that is more stable than protonated methane. The protonated ammonia will transfer a proton to a neutral analyte (exothermic reaction), producing the ionized form of the analyte as a protonated species.

The proton affinity of the analyte molecule must be greater than the proton affinity of the reactive reagent gas; $PA_{molecule} > PA_{reagent\ gas}$. This is almost always the general case as the reagent gas is a highly reactive species with a very low relative proton affinity (in other words, the reagent gas is looking to lose the extra proton it possesses). The proton affinity is equal to the change in enthalpy of the deprotonation reactions of the reagent gas and the analyte:

$$
\begin{array}{ll}
RGH^+ \rightarrow RG\ + H^+ & \Delta H = PA_{reagent\ gas} \\
\underline{M + H^+ \rightarrow MH^+} & \underline{\Delta H = -PA_{molecule}} \\
RGH^+ + M \rightarrow MH^+ + RG & \Delta H = PA_{RG} - PA_M
\end{array}
\qquad (2.8)
$$

If $PA_M > PA_{RG}$, then the change in enthalpy is negative, the reaction is exothermic and can proceed spontaneously to the right, thus protonating the analyte molecule.

The reagent gases used in chemical ionization possess different proton affinities. Therefore, the proper choice of reagent gas must be made according to the amount of fragmentation desired. The greater the exothermicity of the proton transfer reaction will dictate the amount of fragmentation that will take place in the chemical ionization process. The diagram in Figure 2.6 illustrates the relationship between the proton

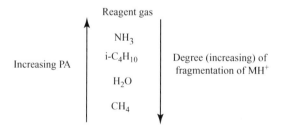

FIGURE 2.6 Illustration of the inverse relationship between the proton affinity of reagent gasses versus the degree of fragmentation observed upon protonation of the analyte molecule within the chemical ionization source. Ammonia (NH_3) has the greatest proton affinity, producing the lowest amount of fragmentation of the protonated analyte species.

affinity of the reagent gas and the degree of fragmentation observed upon ionization (proton transfer) of the analyte molecule.

As illustrated in Figure 2.6 there is an inverse relationship between proton affinity of the reagent gas and the degree of fragmentation of the analyte molecule. Ammonia (NH_3) has the greatest proton affinity owing to its lone pair of electrons on the nitrogen. This in turn equates to a lesser degree of fragmentation of the analyte molecule during the proton transfer reaction in the chemical ionization source. At the bottom of the figure we see that methane has the lowest proton affinity and therefore produces the greatest amount of fragmentation of the analyte molecule during the proton transfer reaction in the chemical ionization source. The effect of the choice of reagent gas is also illustrated in Figure 2.7 for the mass analysis of 4-hydroxy-pentanoic acid methyl ester ($C_6H_{12}O_3$). In Figure 2.7a ammonia (NH_3) was chosen

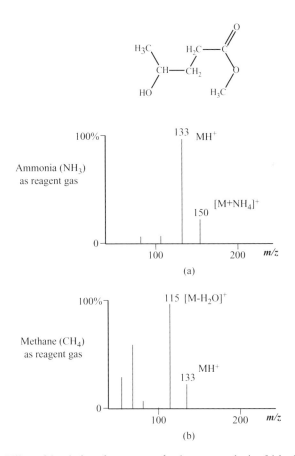

FIGURE 2.7   Effect of the choice of reagent gas for the mass analysis of 4-hydroxy-pentanoic acid methyl ester ($C_6H_{12}O_3$). (*a*) Ammonia (NH_3) chosen as the reagent gas where very little fragmentation is observed to be taking place. (*b*) Substantial fragmentation with the use of methane as the reagent gas (CH_4).

TABLE 2.1   Proton Affinities of Chemical Ionization Reagent Gases

Reagent Gas	Ionized Reagent Gas	Proton Affinity (kJ·mol^{-1})
H_2	H_3^+	418.4
CH_4	CH_5^+	531.4
H_2O	H_3O^+	690.4
i-C_4H_{10}	i-$C_4H_9^+$	815.9
NH_3	NH_4^+	857.7

as the reagent gas where very little fragmentation is observed to be taking place. Figure 2.7*b* illustrates substantial fragmentation with the use of methane as the reagent gas (CH_4).

The proton affinities for typical chemical ionization reagent gases are listed in Table 2.1. The proton affinities for a number of organic compounds are listed in Table 2.2. The protonated form of diatomic hydrogen (H_3^+) has a PA value of 418.4 kJ·mol^{-1} while the protonated form of ammonia has a PA value of 857.7 kJ·mol^{-1}. Again, the PA of a substance is the amount of energy required to pull off the positively charged proton (H^+). It takes much less energy to remove the proton from protonated diatomic hydrogen than it does for protonated ammonia. As discussed previously, the lone pair of electrons on the nitrogen of ammonia is available to delocalize the charge on the proton, thus stabilizing the ion. The charge on the proton adducted to diatomic hydrogen only has available to it the σ electrons of the hydrogen–hydrogen covalent bonds, making this species unstable and highly reactive. From Table 2.2 we see that the proton affinity of alcohols in general is greater than the proton affinity of the protonated form of methane (CH_5^+); therefore, CH_5^+ will always protonate alcohols. For example, if we select the middle of the PA range for alcohols as 774.0 kJ·mol^{-1}, the overall change in enthalpy for the chemical ionization reaction with protonated methane would be:

$$\begin{array}{ll} CH_5^+ \rightarrow CH_4 + H^+ & \Delta H = 531.4 \text{ kJ·mol}^{-1} \\ \underline{\text{Alcohols} + H^+ \rightarrow \text{Alcohols } H^+} & \underline{\Delta H = -774.0 \text{ kJ·mol}^{-1}} \\ CH_5^+ + \text{Alcohols} \rightarrow \text{Alcohols } H^+ + CH_4 & \Delta H = -242.6 \text{ kJ·mol}^{-1} \end{array} \qquad (2.9)$$

The change in enthalpy for the reaction is -242.6 kJ·mol^{-1}, indicating that the reaction is exothermic and will proceed to the right spontaneously.

TABLE 2.2   Approximate Proton Affinities for Some Typical Organic Compounds

Neutral Species	Proton Affinity (kJ·mol^{-1})
Alcohols	732.2–815.9
Acids	736.4–811.7
Aldehydes	740.6–807.5
Esters	795.0–845.2
Ethers	807.5–841.0
Ketones	836.8

2.3.2 Negative Chemical Ionization

Negative chemical ionization, while less common than positive chemical ionization presented in Section 2.3.1, can also be used as a method for analyte ionization in preparation for mass analysis. As we saw earlier, the most common form of positive chemical ionization is the protonation of a neutral analyte molecule by a highly reactive reagent gas. In contrast to this one form of negative chemical ionization is the production of a negatively charged analyte species through proton exchange or abstraction. In this process a reactive ion that has a large affinity for a proton is used to remove a proton from the analyte. Some examples of commonly used reactive ions are fluoride (F^-), chloride (Cl^-), oxide radical ($O^{-\bullet}$), hydroxide (OH^-), and methoxide (CH_3O^-). This form of chemical ionization is an acid–base reaction that can generally be represented as

$$X^- + M \rightarrow XH + [M - H]^- \tag{2.10}$$

where X^- represents the reactive reagent gas ion and M the neutral analyte. Due to the high degree of negative charge on the reactive reagent gas ion the proton exchange from the analyte to the reactive ion is exothermic and will proceed to the right spontaneously. An actual example is as follows where a proton is abstracted from the commonly used ketone solvent acetone by the reactive chloride anion:

$$Cl^- + H_3C-\overset{\overset{\textstyle O}{\|}}{C}-CH_3 \rightarrow HCl + H_2C^--\overset{\overset{\textstyle O}{\|}}{C}-CH_3 \tag{2.11}$$

Electron capture has also been used in negative chemical ionization. In this technique, electrons emitted from a filament are captured by neutral analyte molecules present in the source, producing a radical anion species. These bare electrons emitted from the filament are unstable by themselves and will readily attach to the analyte. The introduction of a nonreactive gas into the source will help to slow down the electrons through collisions and enhance the ionization of the neutral analyte molecules:

$$N_2 + e^- (70 \text{ eV}) \rightarrow N_2 + e^- (\ll 70 \text{ eV}) \tag{2.12}$$

Some examples of nonreactive gases used are argon (Ar), nitrogen (N_2), methane (CH_4), and isobutane (i-C_4H_{10}). Fast moving electrons that have not undergone any collisional cooling have cross sections (probabilities) that are small for capture to take place. This often meant a very low yield in negatively ionized analytes, resulting in poor sensitivity.

$$\text{Neutral compound (M)} + e^- (\text{slow}) \xrightarrow{\overset{\text{Electron}}{\text{capture}}} \text{Ionized compound (M}^{-\bullet}) \tag{2.13}$$

The slower (relaxed or cooled) electrons thus have an increased probability of being captured, resulting in higher yields and greater sensitivity for the negative chemical ionization technique. It is often observed that certain groups have a higher affinity for capturing electrons as compared to others. Compounds such as pesticides that contain chlorine atoms often will exhibit a very high affinity for electron capture.

Another process used for negative chemical ionization is known as dissociative electron capture. In this process enough energy has been absorbed by the analyte during the electron capturing process to initiate a small amount of fragmentation to take place. This process generally does not give a large amount of structural information and can be a problem if the initial analyte is an unknown compound.

In general CI gives a much higher abundance of the intact analyte ion versus the process of electron ionization (EI). The intact analyte ions produced by CI exist as, for example, in positive ion mode, as protonated molecules $[M + H]^+$ (sometimes referred to as acid adducts), and in the negative ion mode as either deprotonated molecules $[M - H]^-$ or radical anions $M^{-\cdot}$. The ionized analyte species formed using chemical ionization in the positive ion mode are almost exclusively EE ions. In contrast electron ionization almost exclusively produces OE ions ($M^{+\cdot}$) during the in-source analyte ionization process. Oftentimes the sensitivity of the detection of the ionized analyte species can be functional group dependent when using chemical ionization. Electron capture can be difficult without the presence of a moiety within the molecular structure of the analyte that possesses a high electron affinity. Often the CI spectrum will contain a single peak representing the m/z value of the intact precursor ion, allowing the determination of the molecular weight of the analyte species. The EI spectrum usually does not contain the precursor m/z value to any appreciable amount but contains a significant amount of product ions giving structural information. The combination of the two techniques can be complementary in determining a species molecular weight and structure.

2.4 ATMOSPHERIC PRESSURE CHEMICAL IONIZATION (APCI)

The previous sections introducing EI and CI as two techniques for transforming an analyte molecule from a neutral state to an ionized state both shared a common design where the ionization takes place within a source that is either at a much lower pressure (EI, 10^{-5} Torr) than ambient (760 Torr) or at a somewhat slightly lower pressure than ambient (CI, 0.1–2 Torr) due to input of reactive reagent gas. Figure 2.8 illustrates the basic design of an atmospheric pressure ionization (API) source.

As can be seen in Figure 2.8, the API source is directly suitable for a liquid stream such as that obtained from a separation science technique such as high-performance liquid chromatography (HPLC). In the atmospheric pressure chemical ionization (APCI) technique the ions are formed at atmospheric pressure using either a beta particle (β^-) emitting source (e.g., ^{63}Ni foil, old technique that is not in use much anymore) or a corona discharge needle. In corona discharge a needle is held near the source with a very high negative voltage applied to it. Right on the sharp tip of the needle there will be a very high electron (e^-) density buildup. The electrons will jump

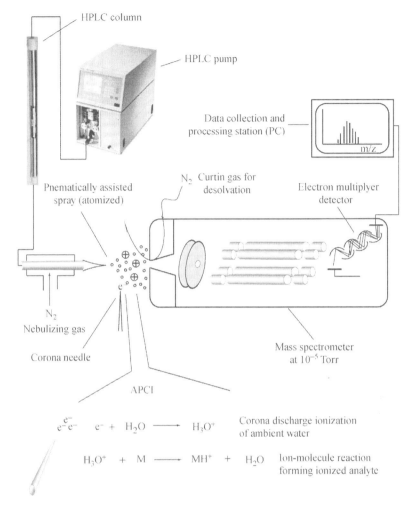

FIGURE 2.8 Design of a general atmospheric pressure chemical ionizaton (APCI) inlet system for mass spectrometric analysis.

off of the needle tip and be accelerated forward by the negative field associated with the discharge needle. The discharged electrons then ionize the surrounding ambient species present in the surrounding air such as oxygen or water. The ionized ambient gaseous species (H_2O^- and $O^{-\cdot}$) will then undergo true ion–molecule reactions with the analyte species present. This process is illustrated in the expanded view of the corona needle in Figure 2.8. In this way an introduced reactive reagent gas is not necessary for the ionization of the analyte species to take place; thus, a low-pressure source is not needed. During the ion–molecule reactions taking place, an adduct is formed between the neutral analyte species and the surrounding ionized ambient gases. The reactive intermediate ions produced will then interact further with other neutral analyte molecules present, producing ionized species. There are a number of ion–molecule reactions that are possible during the APCI process as follows:

$$
\begin{array}{lll}
M + e^- & \to M^{-\cdot} & \text{(associative electron capture)} \\
MX + e^- & \to M^{\cdot} + X^- & \text{(dissociative electron capture)} \\
R^{+/-} + M & \to R + M^{+/-} & \text{(charge transfer)} \\
MH + R^+ & \to M^+ + RH & \text{(hydride abstraction)} \\
MH + R^- & \to M^- + RH & \text{(proton transfer)} \\
RH^+ + M & \to R + MH^+ & \text{(protonation)} \\
M + R^{+/-} & \to MR^{+/-} & \text{(adduct formation)} \\
MX + R^+ & \to M^+ + X + R & \text{(charge transfer with dissociation)}
\end{array}
\tag{2.14}
$$

where M = neutral analyte
X = any species
R = reactive species formed during the APCI process

The ionization takes place at atmospheric pressure, making this technique well suited for a liquid stream such as that eluting from a liquid chromatography column. There is an inherent incompatibility between liquids and low pressure; if the liquid is directly introduced into the low-pressure inlet region, the liquid will expand and vaporize instantaneously, thus producing high pressure within the source. This is too difficult to maintain and control when the desired pressure in the source inlet region is to be kept at or near 10^{-5} Torr. It has been observed that near 100% ionization efficiency can be obtained with APCI for compounds that are present at very low concentrations, thus increasing the sensitivity of the method. However, oftentimes adduct and cluster ions of the ionized ambient air and water species are observed to be formed during the APCI process, causing a high chemical background noise reducing the analyte signal-to-noise (S/N) ratios (decreased sensitivity). To overcome this problem a dry nitrogen (N_2) curtain gas (see Fig. 2.8) is used to break up the clusters through collisions between the N_2 gas and the clusters. The clusters will then tend to fall apart, producing less interference at higher masses. The nitrogen curtain gas also aids in the evaporation of the solvent prior to introduction into the instrument source. This helps to reduce the problem of a large liquid stream evaporating and expanding within the source, producing higher pressures. Finally, as can be seen in Figure 2.8, nitrogen is also used as a nebulizing gas that helps to atomize the eluant spray.

2.5 ELECTROSPRAY IONIZATION (ESI)

Electrospray ionization is a process that enables the transfer of compounds in solution phase to the gas phase in an ionized state, thus allowing their measurement by mass spectrometry. The use of ESI coupled to mass spectrometry was pioneered by Whitehouse et al.[1] and Fenn[2] in 1985 and 1993, by extending the work of Dole

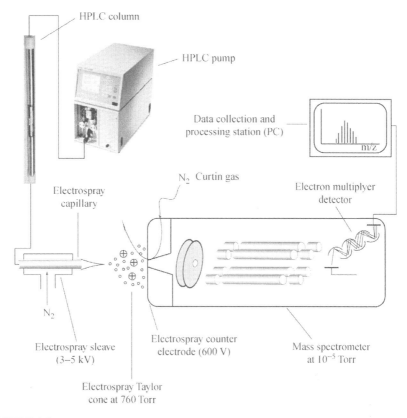

FIGURE 2.9 General setup for ESI when measuring biomolecules by electrospray mass spectrometry.

et al.[3] in 1968, who demonstrated the production of gas-phase ions by spraying macromolecules through a steel capillary that was electrically charged and subsequently monitoring the ions with an ion-drift spectrometer. The process by which ESI works has received much theorization, study, and debate[4-10] in the scientific community, especially the formation of the ions from the Taylor[11] cone droplets and offspring droplets. Figure 2.9 shows the general setup for ESI when measuring biomolecules by electrospray mass spectrometry. The ESI process is done at atmospheric pressure, and the system is quite similar to that shown for APCI in Figure 2.8. Both APCI and ESI are shown in-line with HPLC and a mass analyzer; however, the two differ extensively in the area of the spray right before introduction into the mass analyzer. The electrospray process is achieved by placing a potential difference between the capillary and a flat counterelectrode. This is illustrated in Figure 2.10 where the "spray needle" is the capillary and the "metal plate" is the flat counterelectrode. The generated electric field will penetrate into the liquid meniscus and create an excess abundance of charge at the surface. The meniscus becomes unstable and protrudes out, forming a Taylor cone. At the end of the Taylor cone a jet of emitting

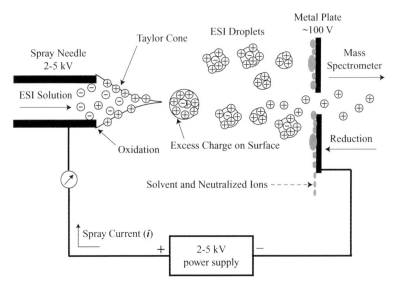

FIGURE 2.10   Electrospray ionization process illustrated in positive ion mode.[6] (Reprinted with permission of John Wiley & Sons, Inc. Cech, N. B., and Enke, C. G. Practical implications of some recent studies in electrospray ionization fundamentals. *Mass Spectrom. Rev.* 2001, *20*, 362–387. Copyright 2001.)

droplets (number of drops estimated at 51,250 with radius of 1.5 μm) will form that contain an excess of charge. Pictures of jets of offspring droplets are illustrated in Figure 2.11. As the droplets move toward the counterelectrode, a few processes take place. The drop shrinks due to evaporation, thus increasing the surface charge until columbic repulsion is great enough that offspring droplets are produced. This is known as the Rayleigh limit, producing a columbic explosion. The produced offspring droplets have 2% of the parent droplets mass and 15% of the parent droplets charge. This process will continue until the drop contains one molecule of analyte and charges that are associated with basic sites (positive ion mode). This is referred to as the *charged residue model*, which is most important for large molecules such as proteins. This process is illustrated in Figure 2.12. As the droplets move toward the counterelectrode a second process also takes place known as the *ion evaporation model*. In this process the offspring droplet will allow evaporation of an analyte molecule from its surface along with charge when the charge repulsion of the analyte with the solution is great enough to allow it to leave the surface of the drop. This usually takes place for droplets with a radius that is less than 10 nm. This type of ion formation is most important for small molecules.

In the ensuring years since its introduction, electrospray mass spectrometry has been used for structural elucidation and fragment information[13–15] and noncovalent complex studies,[16, 17] just to name a few recent examples of its overwhelmingly wide range of applications.

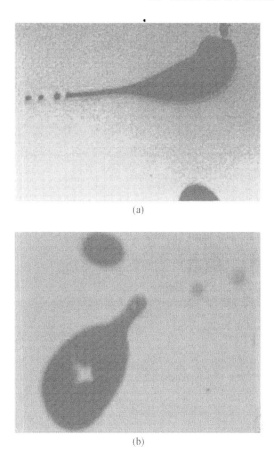

(a)

(b)

FIGURE 2.11   Pictures illustrating the jet production of offspring droplets.[12] (Reprinted with permission from Alessandro Gomez. *Phys. of Fluids* 1994, *6*, 404. Copyright 1994, American Institute of Physics.)

Electrospray[18–21] is an ionization method that is now well known to produce intact gas-phase ions with very minimal, if any, fragmentation being produced during the ionization process. In the transfer process of the ions from the condensed phase to the gas phase, several types of "cooling" processes of the ions are taking place in the source: (1) cooling during the desolvation process through vibrational energy transfer from the ion to the departing solvent molecules, (2) adiabatic expansion of the electrospray as it enters the first vacuum stage, (3) evaporative cooling, and (4) cooling due to low-energy dampening collisions with ambient gas molecules. The combination of these effects, and the fact that electrospray can effectively transfer a solution-phase complex to the gas phase with minimal interruption of the complex, makes the study of noncovalent complexes from solution by electrospray ionization mass spectrometry attractive.

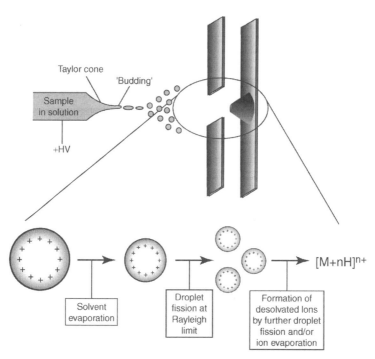

FIGURE 2.12 Gas-phase ion formation process from electrospray droplets.[7] (From Gaskell, S. J. Electrospray: Principles and practice. *J. Mass Spectrom.* 1997, *32*, 677–688. Copyright John Wiley & Sons, Inc., 1997. Reproduced with permission.)

2.6 NANOELECTROSPRAY IONIZATION (NANO-ESI)

A major application of biomolecule analysis using mass spectrometry has been the ability to allow liquid flows to be introduced into the source of the mass spectrometer. This has enabled the coupling of HPLC to mass spectrometry where HPLC is used for a wide variety of biomolecule analysis. Normal electrospray ionization, introduced in the preceding section, typically has flow rates on the order of microliters per minute (~1–500 µL/min). Traditional analytical HPLC systems designed with UV/Vis detectors generally employ flow rates in the range of milliliters per minute (~0.1–1 mL/min). A recent advancement in the electrospray ionization technique has been the development of nanoelectrospray where the flows employed are typically in the range of nanoliters per minute (~1–500 nL/min). Following the progression of the development of electrospray from Dole's original reporting in 1968 through Fenn's work reported in 1984 and 1988, a more efficient electrospray process was reported by Wilm and Mann,[22] employing flows in the range of 25 nL/min. This early reporting of low flow rate electrospray was initially termed as microelectrospray by Wilm and Mann but was later changed to nanoelectrospray.[23] At the same time that Wilm and Mann[22] had reported the microelectrospray, Caprioli et al.[24] had also reported a miniaturized ion source that they had named microelectrospray. The name *nanoelectrospray* for Wilm's source is actually more descriptive due to flow rates used in the nanoliter per minute range and the droplet sizes that are produced in the nanometer range. Conventional electrospray sources before the introduction of nanoelectrospray produced droplets on the order of 1–2 μm. The nanoelectrospray source produces droplets in the size range of 100–200 nm, which is 100–1000 times smaller in volume. When spraying standard solutions at concentrations of 1 pmol/μL, it is estimated that droplets of the nanometer size contain only one analyte molecule per droplet.

The original nanoelectrospray sources that were used were comprised of pulled fused silica capillary tips 3–5 cm long with orifices of 1–2 μm in diameter. The tips also have a thin gold plating that allows current flow. The tips are loaded with 1–5 μL of sample directly using a pipette[25] and coupled to the electrospray source completing the closed circuit required for the production of the applied voltage electrospray Taylor cone generation. This is illustrated in Figure 2.13 where in the top portion of the figure a sample is being loaded into the nanospray tip using a pipette. The tip is then placed into the closed-circuit system for the electrospray to

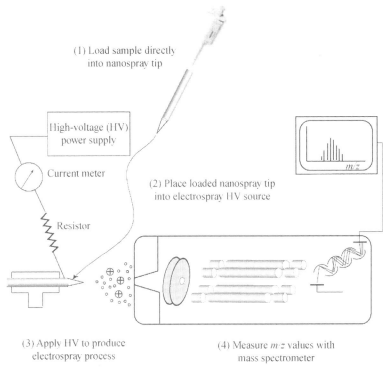

FIGURE 2.13 Top of figure illustrates the loading of a nanoelectrospray tip. Bottom of figure illustrates the coupling of the nanoelectrospray tip to the closed-circuit system.

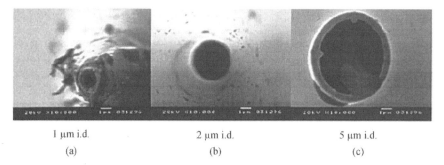

1 μm i.d.   2 μm i.d.   5 μm i.d.

(a)   (b)   (c)

FIGURE 2.14 Illustration of different nanoelectrospray tip orifice diameters.[30] Scanning electron microscopy images of employed nanospray emitters: (*a*) 1-, (*b*) 2-, and (*c*) 5-μm tip. Images were obtained after 2 h of use. (Reprinted with permission from Li, Y., and Cole. R. B. *Anal. Chem.* 2003, *75*, 5739–5746. Copyright 2003 American Chemical Society.)

take place. The sample flow rate is very low, using the nanospray tips allowing the measurement of a very small sample size over an extended period of time. It has also been observed that nanospray requires a lower applied voltage for the production of the electrospray that helps to reduce problems with corona electrical discharges that will interrupt the electrospray. In nanoelectrospray the flow rate is lower than in conventional electrospray and is felt to have a direct impact on the production of the droplets within the spray and the efficiency of ion production. The lower flow rate produces charged droplets that are reduced in size as compared to conventional electropsray. This has been described in detail by Wilm and Mann,[22] by Fernandez de la Mora and Loscertales,[26] and by Pfeifer and Hendricks.[27] There are fewer droplet fission events required with smaller initial droplets in conjunction with less solvent evaporation taking place before ion release into the gas phase.[28,29] A result of this is that a larger amount of the analyte molecule is transferred into the mass spectrometer for analysis. Though the efficiency of ionization is increased with nanoelectrospray, the process is also influenced by the size and shape of the orifice tip.[30,31] Pictures of nanoelectrospray orifice tips are illustrated in Figure 2.14. Figure 2.15 shows an example of the production and observance of an ESI Taylor cone.

While Figure 2.15*d* does show a Taylor cone formed, Figure 2.16 gives a good picture of an array of Taylor cones formed from a microelectrospray emitter. In the picture multiple cones can be seen along with their associated spray produced from the electrospray process.

As mentioned previously nano-HPLC is increasingly being coupled to nano-electrospray for biomolecule analysis. A nano-HPLC-ESI system is illustrated in Figure 2.17. The flow involved in nano-HPLC-ESI often ranges between 10 to 100 nL/min. The fused silica capillary columns that are used in nano-HPLC have very small diameters often around 50 μm. These small-diameter columns can often create high backpressures in the HPLC system. One way to achieve the very low flow rate through the fused silica nano-HPLC column is to use a flow splitter that is located in-stream between the column and the HPLC pump as illustrated in Figure 2.17. The

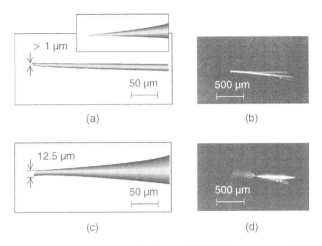

FIGURE 2.15 Examples of nanospray tip sizes and the influence upon the ESI Taylor cone. The cone is not observed in (*b*) at a diameter of >1 μm. The cone is observed in (*d*) for a diameter of 12.5 μm. (Reprinted with permission. Schmidt, A., and Karas, M. Effect of different solution flow rates on analyte ion signals in nano-ESI MS, or: when does ESI turn into nano-ESI? *J. Am. Soc. Mass Spectrom.* 2003, *14*, 492–500. Copyright Elsevier, 2003.)

tubing from the splitter to waste is called a restrictor and is used to regulate the flow through the nanocolumn. A smaller diameter restrictor will increase the backpressure, forcing more mobile phase through the nanocolumn. If a larger diameter restrictor is used, the backpressure will be lower, resulting is less flow being directed through the column. The nanocolumns have a nano-ESI tip coupled to them (diameters can range

FIGURE 2.16 Photograph of nine stable electrosprays generated from the nine-spray emitter array.[32] (Reprinted with permission from Tang et al. *Anal. Chem.* 2001, *73*, 1658–1663. Copyright 2001 American Chemical Society.)

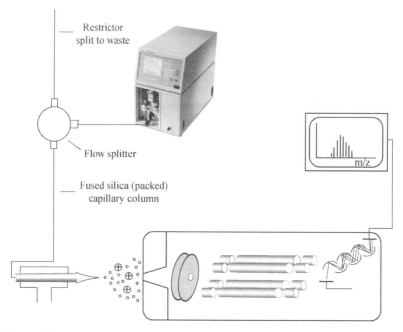

FIGURE 2.17   Design of a nano-HPLC nano-ESI system for mass spectrometric analysis of biomolecules.

from 1 μm up to 100 μm) to produce the electrospray. Another difference observed here as compared to the atmospheric pressure source is the absence of a nebulizing gas or a drying gas. These are not needed or used in nano-ESI.

2.7   ATMOSPHERIC PRESSURE PHOTOIONIZATION (APPI)

A recent addition to ionization sources that are useful in mass spectrometric measurement of biomolecules is the atmospheric pressure photoionization (APPI) source. As the name indicates the photo-induced ionization of analytes is taking place under atmospheric conditions. As we will see, the technology has been applied to analytes that are within a liquid effluent such as that coming out of an HPLC column making the ionization methodology directly applicable to biomolecule analysis by mass spectrometry. Photoionization of analytes is actually a technique that has been in use for a number of years. The photoionization technique was primarily used in conjunction with gas chromatography (GC) using a detection system measuring a change in current through a collection electrode induced by the photoionized analyte species eluting from the GC column.[33,34] The photoionization was initiated using discharge lamps that produce vacuum-ultraviolet (VUV) photons. The technique of photoionization for the detection of analyte species has also been explored with the use of liquid chromatography.[35–38] However, with both GC and LC analysis of postcolumn photoionized biomolecular species, there is no information obtained regarding the analyte's mass and no structural information either. This is where the mass spectrometer contributes a clear advantage with the addition of measured information regarding analyte mass and structural information as compared to the other two analytical techniques.

2.7.1 APPI Mechanism

The process of photoionization involves the absorption of radiant energy from a UV source where the incident energy is greater than the first ionization potential of electron loss from the analyte. In Figure 2.18 two single-photon processes are illustrated for a molecule of benzene. In Figure 2.18*a* a photon is absorbed with an energy that is less than the ionization potential (IP) of the benzene molecule, resulting

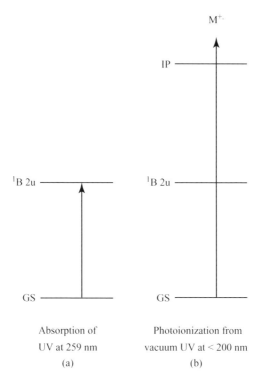

Absorption of
UV at 259 nm

(a)

Photoionization from
vacuum UV at < 200 nm

(b)

FIGURE 2.18 Single-photon processes for a molecule of benzene: (*a*) a photon is absorbed with an energy $E < $ IP, resulting in the elevation of an electron from the ground state (GS) to an excited state ($^1B_{2u}$) but with no ionization in the form of electron ejection. (*b*) A photon is absorbed with an energy $E > $ IP, resulting in electron ejection from the benzene molecule forming a radical cation ($M^{+\bullet}$) of the benzene molecule.

in the elevation of an electron from the ground state (GS) to an excited state ($^1B_{2u}$) but with no ionization in the form of electron ejection. In process Figure 2.18b a photon is absorbed with an energy that is greater than the first ionization potential of the benzene molecule resulting in electron ejection from the benzene molecule. Photoionization in this example is a single-photon process that results in the liberation of an outer valence shell electron forming a radical cation ($M^{+\cdot}$) of the benzene molecule. The process of photoionization at atmospheric pressure usually results in the production of molecular ions ($M^{+\cdot}$) often with minimal fragmentation taking place during the ionization. Most organic compounds have ionization potentials that range between 7 and 10 eV; therefore, photon sources are needed that will be able to supply photons with sufficient energy to induce photoionization. Two ionization processes are common in photoionization.[39] The first involves the absorption of radiant energy with subsequent liberation of an electron producing the radical cation:

$$M + hv \rightarrow M^{+\cdot} + e^-, \quad \Delta H_{PI} = IP(M) - hv \quad (2.15)$$

where M is the neutral analyte molecule, hv is the incident photoionization energy, $M^{+\cdot}$ is the radical cation formed, and e^- is the liberated electron. The change in enthalphy for the photoionization is equal to the ionization potential of the molecule [IP(M)] minus the incident photoionization energy (hv). The second process involves the abstraction of a proton from the surrounding solvent producing a positively charged, protonated analyte species:

$$M^{+\cdot} + S \rightarrow MH^+ + S(-H) \quad \Delta H = IP(H) - IP(M) - PA(M) + D_H(S) \quad (2.16)$$

where PA(M) is the proton affinity of the analyte and D_H is the hydrogen bond energy. Finally, a third process is also observed to take place with photoionization in the form of analyte fragmentation that, when controlled, can give structural information.

2.7.2 APPI VUV Lamps

Most organic molecules have first ionization potentials between 7 and 10 eV, thus a VUV lamp source is needed that can supply this energy. There are three such VUV lamps that can supply energies in this range: the xenon (Xe) lamp that generates light at 8.4 eV, the krypton (Kr) lamp that generates light at 10.0 and 10.6 eV, and the argon (Ar) lamp that generates light at 11.7 eV.[40] The Kr lamp is the most common lamp used between the three VUV lamp choices. Examples of typical VUV lamps are illustrated in Figure 2.19.

2.7.3 APPI Sources

There are two types of APPI sources generally in use at this time and available commercially. This includes the design originally reported by Robb et al.[41] (available

FIGURE 2.19 Examples of common designs of vacuum ultraviolet (VUV) lamp sources.

from Sciex) and is illustrated in Figure 2.20. The second APPI source configuration is based on the original design by Syage et al.[42] and is illustrated in Figure 2.21. In both sources the analytes undergo photoionization using a 10-eV krypton discharge lamp, which can be followed by gas-phase reactions. The gas-phase reaction approach is used in the APPI source shown in Figure 2.20 where we see the addition of a dopant (used to promote ionization of the neutral analyte). Both sources are also similar in that nebulization and high-temperature desolvation are used to evaporate the liquid solution containing the analyte. We can also see that the APPI source in Figure 2.21 contains an open area where the gaseous analyte is ionized while in the APPI source illustrated in Figure 2.20 the ionization takes place within the closed quartz tube.

2.7.4 Comparison of ESI and APPI

In the preceding two sections we looked at the ionization process involving electrospray, both normal electrospray and nanoelectrospray. At this point we are at an advantageous place to compare APPI to electrospray, affording us an opportunity to

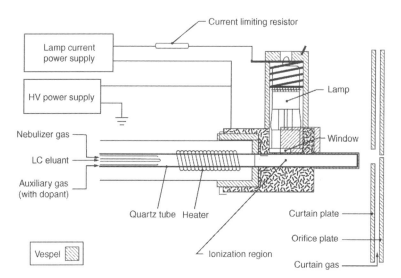

FIGURE 2.20   Schematic of the APPI ion source, including the heated nebulizer probe, photoionization lamp, and lamp mounting bracket. (Reprinted with permission from Robb, D. B., Covey, T. R., and Bruins, A. P. Atmospheric pressure photoionization: An ionization method for liquid chromatography–mass spectrometry. *Anal. Chem.* 2000, *72*, 3653–3659. Copyright 2000 American Chemical Society.)

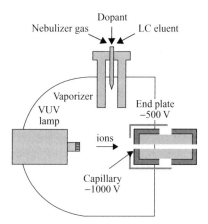

FIGURE 2.21   Schematic of the Agilent/Syagen APPI source (Agilent G1971A APPI source, Agilent Technologies, Palo Alto, CA). (Reprinted with permission. Kauppila, T. J., Bruins, A. P., and Kostianen, R. Effect of solvent flow rate on the ionization efficiency in atmospheric pressure photoionization-mass spectrometry. *J. Am. Soc. Mass Spectrom.* 2005, *16*, 1399–1407. Copyright Elsevier 2005.)

get a closer look at the two ionization techniques and also some examples of the application of APPI to biomolecule analysis. When analyzing neutral biomolecules by electrospray, the mechanism of ionization is primarily dependent upon the proton affinity (PA) (also directly related to the metal cation affinity) of the analyte. For example, the reactions involved during ESI for the production of a proton adduct (acidic solution) or a metal (sodium, Na^+) adduct are:

$$M + H^+ \rightarrow MH^+ \quad \text{(proton adduct)}$$
$$M + Na^+ \rightarrow MNa^+ \quad \text{(sodium adduct)} \tag{2.17}$$

Because the ionization process is dependant upon the proton affinity of the analyte in ESI, often nonpolar species have very low detection. With APPI the ionization process is dependent upon the incident photon possessing greater energy than the ionization potential of the analyte. This in effect can allow the detection of nonpolar species by APPI that are difficult to detect using ESI. Figure 2.22 illustrates the versatility of APPI for the detection of nonpolar compounds by comparing spectra obtained with ESI and APPI.[39] The spectra in Figure 2.22 were collected using a dual source that has the ability to apply APPI in conjunction with ESI. In the top spectrum the APPI lamp source is turned off during the spraying of a 100-ng/μL solution of

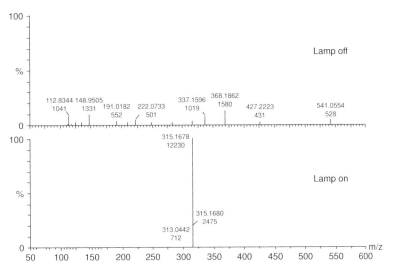

FIGURE 2.22 Positive ion mass spectra of progesterone (100 ng/μL) recorded for the dual ESI/APPI source on the Waters LCT. The top spectrum is in ESI-only mode and the bottom spectrum is in dual mode. Spectra were recorded by flow injection analysis (5 μL sample injection) for mixture of methanol–toluene (95 : 5) solvent. Both mass spectra are on the same absolute intensity scale. (Reprinted with permission. Syage, J. A., Hanold, K. A., Lynn, T. C., Horner, J. A. and Thakur, R. A. Atmospheric pressure photoionization II. Dual source ionization. *J. Chromatogr. A* 2004, *1050*, 137–149. Copyright Elsevier 2004.)

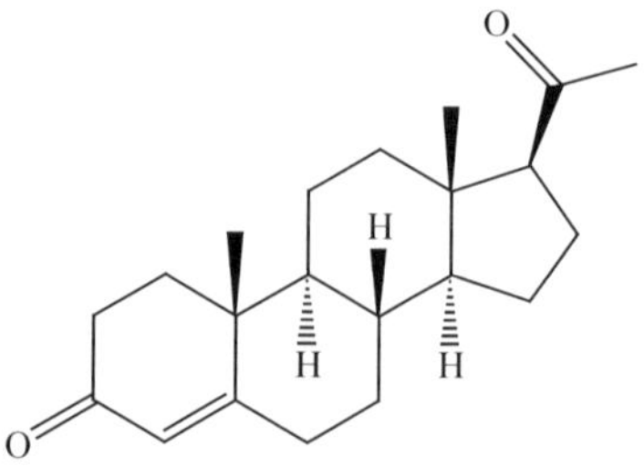

Progesterone

FIGURE 2.23 Structure of neutral progesterone, the biomolecule that is being analyzed by ESI and APPI in Figure 2.22.

progesterone. With the APPI lamp off very little of the analyte is observed. In the bottom spectrum the APPI lamp has been turned on and an abundant $[M + H]^+$ peak for the protonated form of progesterone is observed at m/z 315.1878. Figure 2.23 illustrates the structure of progesterone, the biomolecule that is being analyzed by ESI and APPI in Figure 2.22. The proton adduct that is being observed is formed by the help of the dopant toluene, which is included in the spray liquid.

The ionization process in ESI is a solution-based process while the ionization process in APPI is a gas-phase process. This is where the two ionization processes fundamentally differ. This difference is the case for direct photoionization; however, often the addition of a dopant can also be used in APPI, which brings about a more complicated mechanism for the production of an ionized analyte. The process of APPI ionization with the inclusion of a dopant has similarities to the solution-phase mechanism of ESI where the proton affinity of species now has an influence on the ionization process. The mechanism for dopant-assisted ionization in APPI involves first the ionization of the dopant to a radical cation $(M^{+\cdot})$. Often the dopant is a chemical species that possesses a relatively low ionization potential (IP) that allows its ionization to take place with high efficiency. An example of a common dopant used in APPI is toluene, which has an ionization potential of 8.82 eV. Table 2.3 lists some common solvents and dopants used in APPI along with their respective proton affinities (PA) and ionization potentials (IP).

The next step in the APPI process involving a dopant is the protonation of the solvent molecules by the dopant radical cation. The solvent molecules that are protonated can next transfer a proton to the neutral analyte molecule, producing an ionized species detectable by mass spectrometry. It is also possible for the dopant radical cation to directly ionize the neutral analyte through the process of charge transfer.[43,44]

Other examples of the use of APPI and its comparison to ESI include a study of the analysis of dinitropyrene and aminonitropyrene by LC-MS/MS,[45] the analysis of aflatoxins in cow's milk,[46] and a study of the acylglycerols in edible oils, which we shall take a closer look at.[47] In their comparison of edible oils it was observed that the ESI spectra of the oils primarily gave the spectra of the triacylglycerols.

TABLE 2.3   Compilation of Thermochemical Data for Some Common Solvents and Dopants Used in APPI

Compound	IP	PA
Methanol	10.85	7.89
H_2O	12.61	7.36
Acetonitrile	12.19	8.17
DMSO	9.01	9.16
Naphthaline	8.14	8.44
Benzene	9.25	7.86
Phenol	8.47	8.51
Aniline	7.72	9.09
m-Chloroaniline	8.09	8.98
l-Aminonaphthaline	7.1	9.41
Toluene	8.82	8.23
TNT	10.59	

Source: Redrawn with permission from J. A. Sayage et al. J. Chromatogr. A 2004, 1050, 137–149.

This is illustrated in Figure 2.24 of the ESI spectra of (a) sunflower oil and (b) corn oil. In comparison to this the spectra in Figure 2.25 were obtained for the same oils using APPI where there is observed a much higher abundance of the diacylglycerols and the monoacylglycerols. This was attributed to fragmentation taking place of the triacyglycerols during the APPI process. In Figure 2.24 the majority of biomolecules that are being observed are in the range of m/z 800–900. This is illustrated in the m/z regions that are expanded in the spectra. The m/z 800–900 region is in the range of the molecular weights of the triacylglycerols. In Figure 2.25 the m/z 800–900 region is now much lower in intensity as compared to the m/z 300–700 region. The m/z 250–400 region contains biomolecules such as free fatty acids and monoacylglycerols in these edible oil samples. In the m/z 500–700 region the biomolecules are primarily diacylglycerols. Both the monoacyls and the diacyls are thought to be produced by APPI process from the original triacylglycerols present in the oil samples. Edible oil samples are typically comprised of 80–98% triacylglycerols.

2.8   MATRIX-ASSISTED LASER DESORPTION IONIZATION (MALDI)

Matrix-assisted laser desorption ionization is a process that enables the transfer of compounds in a solid, crystalline phase to the gas phase in an ionized state, thus allowing their measurement by mass spectrometry.[48–50] The process involves mixing the analyte of interest with a strongly UV-absorbing organic compound, applying the mixture to a target surface (MALDI plate) and then allowing it to dry (illustrated in Fig. 2.26). Examples of typical organic matrix compounds used are 2,5-dihydroxybenzoic acid (DHB), 3,5-dimethoxy-4-hydroxy-trans-cinnamic acid (sinapic or sinapinic acid), and α-cyano-4-hydroxy-trans-cinnamic acid (α-CHCA)

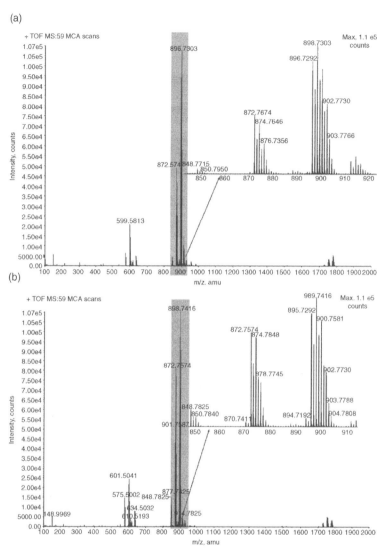

FIGURE 2.24   Full-scan ESI-MS: (*a*) sunflower oil and (*b*) corn oil. The region between *m/z* 800 and 900 is magnified. (Reprinted with permission. Gomez-Ariza, J. L., Arias-Borrego, A., Garcia-Barrera, T., and Beltran, R. Comparative study of electrospray and photospray ionization sources coupled to quadrupole time-of-flight mass spectrometer for olive oil authentication. *Talanta* 2006, *70*, 859–869. Copyright Elsevier 2006.)

(structures illustrated in Fig. 2.27). There are many techniques described for applying the matrix to the target such as the *dried droplet* method[50,51] where the analyte and matrix are mixed usually in an approximate 1 : 1000 ratio and then 0.5–2 μL of the analyte–matrix mixture are spotted onto the target and allowed to dry forming a crystalline solid. The *thin film* method[52] where a thin polycrystalline film is deposited

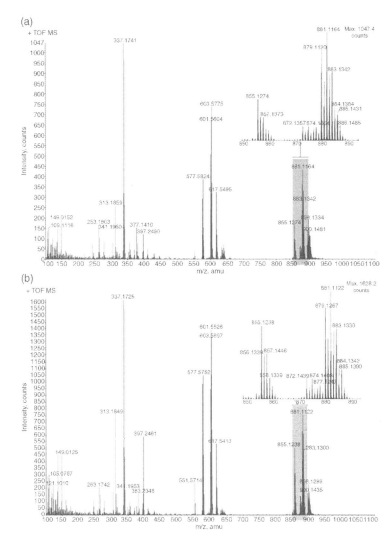

FIGURE 2.25 Full-scan APPI-MS: (*a*) sunflower oil and (*b*) corn oil. The region between *m/z* 800 and 900 is magnified. (Reprinted with permission. Gomez-Ariza, J. L., Arias-Borrego, A., Garcia-Barrera, T., and Beltran, R. Comparative study of electrospray and photospray ionization sources coupled to quadrupole time-of-flight mass spectrometer for olive oil authentication. *Talanta* 2006, *70*, 859–869. Copyright Elsevier 2006.)

of a homogenous mixture of the matrix and analyte. There are layer or sandwich methods where a layer of matrix may be applied and allowed to dry, followed by a layer of the analyte, and then followed by another layer of matrix. Often it is best to try a few approaches to find which method works best for the analytes being measured.

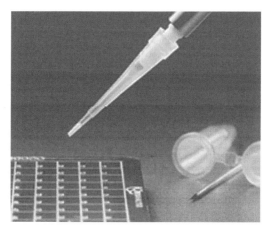

FIGURE 2.26 Spotting of a MALDI target plate. (Courtesy of Millipore Inc., Bedford, MA.)

The target plate after spot drying is placed into the source of the mass spectrometer, and the source is evacuated. The dried crystalline mixture "film" or "spot" is then irradiated with a nitrogen laser (337 nm), or an Nd-YAG laser (266 nm). The strongly UV-absorbing matrix molecules accept energy from the laser and desorbs from the surface, carrying along any analyte that is mixed with it. The desorbed matrix and analyte molecules, ions, and neutrals form a gaseous plume above the target and within the source. The analyte is cationized in the plume above the crystalline surface with, in the positive mode, either a hydrogen proton $[M + H]^+$ transferred from the acidic matrix or a metal cation present such as sodium $[M + Na]^+$. Figure 2.28 illustrates

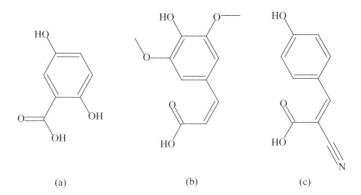

FIGURE 2.27 Examples of typical organic matrix compounds used are (a) 2,5-dihydroxybenzoic acid (DHB), (b) 3,5-dimethoxy-4-hydroxy-trans-cinnamic acid (sinapic or sinapinic acid), and (c) α-cyano-4-hydroxy-trans-cinnamic acid (α-CHCA).

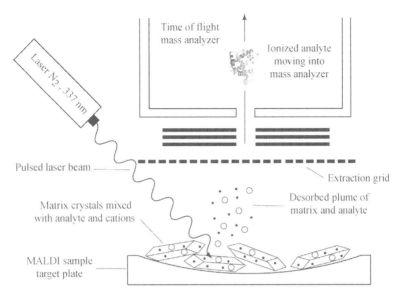

FIGURE 2.28 Matrix-assisted laser desorption ionization, MALDI, process of desorption and ionization.

the MALDI process where a pulsed laser is irradiated upon the crystalline matrix and analyte mixture creating the plume above the target. There is often a delayed extraction period, allowing a decrease in the spatial distribution of the analytes within the plume. There are two known problems with MALDI ionization in the gaseous plume of analytes above the target that cause a decrease in the resolution of the detected analytes. The first resolution decreasing effect is the initial spatial distribution where not all of the desorbed analytes are at the exact same distance from the detector at the start of their flight (placement in crystal structure may attribute to this). The second resolution decreasing effect involves the initial velocity distribution of the analytes within the desorbed gaseous plume above the target. Not all of the analytes may have exactly the same velocity at the start of their flight toward the detector. Together these have attributed to the low-resolution ability of the MALDI-TOF MS technique where spectra are typically generated at a resolution of approximately $m/\Delta m = 500$. We will see later in Chapter 3 that improvements in TOF instrumentation and source design have helped to effectively address these problems [delayed extraction in the source and an electrostatic mirror (reflectron) after the drift tube] and greatly increase spectral resolution up to an $m/\Delta m$ value of 10,000 or higher. The desorbed, ionized compounds from the MALDI process are then introduced into a mass spectrometer for analysis by an extraction grid known as an accelerating voltage or draw-out pulse.

Recently, there has been the addition to the MALDI technique two new types of matrices: liquid ionic matrices[53–57] and solid ionic matrices.[58,59] These new types of MALDI matrices have been introduced to address inherent problems that have been

associated with the traditional solid MALDI matrices. For example, when analyzing biological lipid extracts by MALDI prompt fragmentation involving cleavage
at the head group of phosphorylated lipids is a common problem. There are also
known difficulties when using solid MALDI matrices pertaining to the ability to
obtain quantitative[60] and spatial[61] information. The matrix layer also causes variations in responses due to so-called hot or sweet spots versus the areas that produce
lower level signals that are the results of inhomogeneous deposits of the analyte
within the matrix and nonuniformities in the matrix–analyte spot. There exists between the MALDI spectral signal intensity and the amount of the measured analyte
present in the spot[61] a rather complex relationship. The use of ionic liquid matrices
is demonstrated to have improved shot-to-shot reproducibility of signal intensities
over traditional solid matrices. This has also enabled more accurate quantitative
analysis by MALDI. When using the ionic liquid matrices, there is also observed a
reduction in fragmentation induced by the MALDI ionization technique. Three examples of ionic liquid matrices are illustrated in Figure 2.29: (a) 2,5-dihydroxybenzoic
acid butylamine, (b) 3,5-dimethoxy-4-hydroxycinnamic acid triethylamine, and (c)
α-cyano-4-hydroxycinnamic acid butylamine. Through combining the appropriate
viscous liquid amines with the crystal MALDI matrix, after having dissolved both in
methanol, the ionic liquid matrices are formed. Finally, the methanol and free amine
are removed, producing the ionic pair, which is then mixed with a small amount of
ethanol to reduce the viscosity of the liquid matrix. The ionic liquid matrix enhances
the MALDI technique by more uniformly dissolving the analyte within the liquid
matrix versus the crystalline state that the traditional matrices produce upon drying.
Furthermore, the enhancement may also result from an action that is similar to that

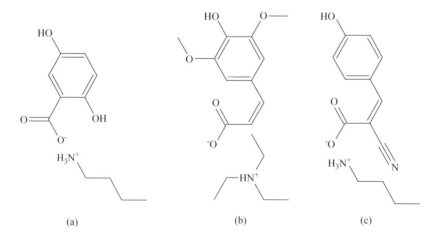

(a)   (b)   (c)

FIGURE 2.29   Ionic liquid matrices used for improved shot-to-shot reproducibility, and a
reduction in fragmentation induced by MALDI. (*a*) 2,5-dihydroxybenzoic acid butylamine,
(*b*) 3,5-dimethoxy-4-hydroxycinnamic acid triethylamine, and (*c*) α-cyano-4-hydroxycinnamic
acid butylamine.

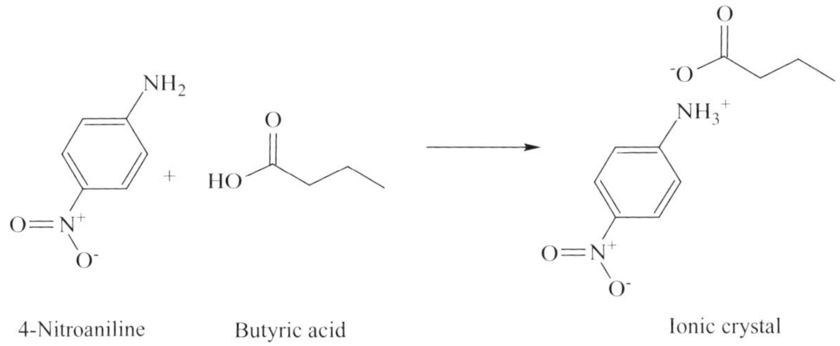

FIGURE 2.30 Synthesis of the solid ionic crystal matrix for MALDI upon the addition of the matrix modifier butyric acid to 4-nitroaniline.

observed with fast atom bombardment (covered in the next section) where the analyte is mixed in glycerol. As the analyte is depleted from the surface of the liquid during the desoprtion/ionization step, the action of the liquid may allow the immediate replenishment of the analyte to the liquid surface, thus greatly enhancing the response. The ionic liquids have been observed to be stable at both atmospheric pressure and reduced pressure within the source. Solid ionic crystal matrices have also recently demonstrated an enhancement in MALDI for certain applications as compared to the traditional crystalline matrices illustrated in Figure 2.27. One example is the ionic pair produced between the addition of the matrix modifier butyric acid to 4-nitroaniline as illustrated in Figure 2.30.

The resultant matrix mixture contrasts with those in Figure 2.29 in that an ionic liquid was not produced, but upon analyte mixing and drying a solid ionic crystal is formed that acts as a powerful gas-phase proton donor and enhances phosphorylated lipid response. This new matrix preparation is crystalline and shows deposition (spotting) behavior similar to other widely used solid MALDI matrices such as sinapinic acid, DHB, or α-CHCA. The new solid crystal ionic matrix was observed to give an enhanced response for phosphorylated lipid analysis by MALDI-TOF mass spectrometry. Primarily protonated analyte ions [phospholipid + H]$^+$ were observed in the mass spectra with a reduced degree of prompt ion fragmentation, which was earlier discussed as a major drawback of the more traditional MALDI matrices illustrated in Figure 2.27. The combination of mostly protonated molecular ions and reduction in the amount of prompt ion fragmentation greatly facilitates the interpretation of unknown spectra.

Difficult applications will be more likely to succeed when a solid ionic crystal matrix is used in preference to an ionic liquid matrix, such as depositing the matrix upon tissue or other surfaces for MALDI analyses, revealing two-dimensional spatial distributions.[62] In contrast to the acidic MALDI matrices DHB or α-CHCA acidification of the analyte environment should not be severe with the solid ionic crystal matrix (where butyric acid is largely ion paired with a p-nitroaniline). This property can facilitate the analysis of certain compounds that are prone to hydrolysis in an acidic environment, such as plasmalogens.

Another recent application of solid ionic MALDI matrices has been in the analysis of peptides.[63] The direct measurement of peptides was obtained by application of solid ionic crystal matrices to tissue samples. It was demonstrated that three solid ionic crystal matrices involving α-CHCA were observed to give an enhanced peptide response for rat brain tissue analysis as compared to α-CHCA only. Three of the CHCA matrix modifiers used in the study were aniline, N,N-dimethylaniline, and 2-amino-4-methyl-5-nitropyridine. In this study the enhanced properties of using a solid ionic MALDI matrix were also observed, including an enhancement in the quality of the spectra obtained in the form of sensitivity, resolution, and high tolerance to contamination. It was also observed that there was a better ability for a homogenous crystallization of the solid ionic matrix upon the tissue to be analyzed. The solid ionic MALDI matrices were also observed to have satisfactory stability in the high-vacuum source, good analyte response in both positive and negative ion mode, and sufficient molecular ion intensities to perform postsource decay (PSD) prompt ion fragmentation studies, thus allowing structural information to be collected and analyzed.

2.9 FAST ATOM BOMBARDMENT (FAB)

A similar ionization technique to MALDI is fast atom bombardment (FAB) where both techniques involve the use of a matrix. In MALDI the matrix is used to transfer energy from the laser to the analyte, inducing desorption of the analyte. The matrix also serves to transfer a charge to the analyte, converting it from the neutral state to the ionized state, thus allowing its measurement by mass spectrometry. In FAB a matrix is also used in the liquid state such as glycerol or m-nitrobenzyl alcohol (NBA). This is in contrast to MALDI, which often uses a matrix in the crystalline state (excluding, however, the more recent room temperature ion pair liquid matrices currently being used with the MALDI technique, e.g., see Section 2.8). In FAB the analyte is dissolved and dispersed within the glycerol liquid and is bombarded with a high-energy beam of atoms typically comprised of neutral argon (Ar) or xenon (Xe) atoms or charged atoms of cesium (Cs^+). As the high-energy atom beam (6 keV) strikes the FAB matrix/analyte mixture, the kinetic energy from the colliding atom is transferred to the matrix and analyte, effectively desorbing them into the gas phase. The analyte can already be in a charged state or may become charged during the desorption process by the surrounding ionized matrix. Figure 2.31 illustrates the ionization mechanism of FAB where Figure 2.31a shows the overall process and Figure 2.31b is a closeup view of the desorption process. FAB is another soft ionization technique where there is the observance of a high yield of cationized analyte species with minimal fragmentation taking place. During the desorption and ionization process, the matrix absorbs the largest amount of the available kinetic energy from the incoming high-energy atom beam, thus sparing the analyte from unwanted decomposition. When decomposition does take place, the product ions derived from the precursor analyte are even electron ions owing to the softer ionization process taking place. The matrix serves to constantly replenish the analyte to the surface of the liquid matrix for

(a)

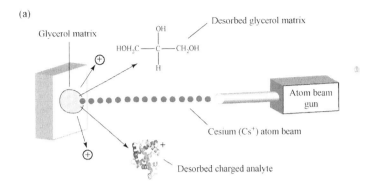

(b)

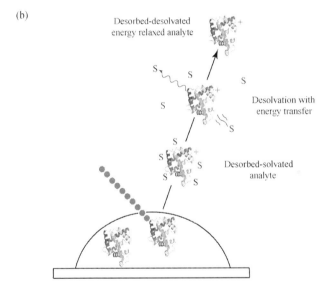

FIGURE 2.31 Fast atom bombardment (FAB) desorption and ionization process. (*a*) A cesium (Cs$^+$) atom beam gun directs a beam at the FAB analyte/matrix mixture desorbing both analyte and matrix. (*b*) Desorbed analyte is solvated with matrix ions. Transference of excess energy takes place between analyte and matrix relaxing the desorbed analyte.

desorption and to limit analyte fragmentation. As shown in Figure 2.31*b* the desorbed analyte goes through a desolvation mechanism where excess energy obtained during the desorption process is transferred to the matrix solvent molecules, thus relaxing the desorbed analyte ion. This effectively prevents excess fragmentation of the desorbed analyte and subsequently increased observation of intact cationized analyte species. This is of course important when the analyst is attempting to determine the molecular weight of an unknown analyte species. If the mass spectrum contains primarily only one major *m/z* peak versus a complicated spectrum of many *m/z* peaks of analyte and fragments, the determination of the molecular weight of the unknown is simple and

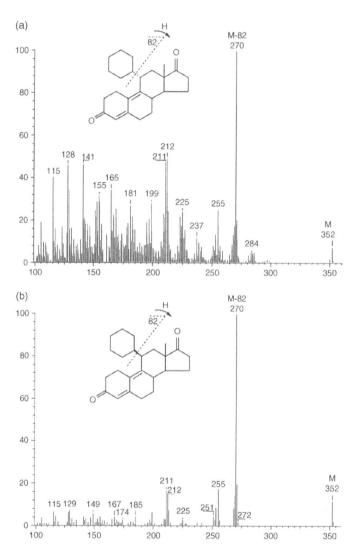

FIGURE 2.32 Electron ionization mass spectra of *1a* and *1b* epimers. (Reprinted with permission. Mak, M., Francsics-Czinege, E., and Tuba, Z. Steroid epimers studied by different mass spectrometric methods. *Steroids* 2004, *69*, 831–840. Copyright Elsevier 2004.)

straightforward. The FAB technique can be useful for the measurement of nonvolatile compounds that can also be thermally labile.

2.9.1 Application of FAB versus EI

As an example of the ionization technique of FAB, we will look at the comparative analysis of steroids by both FAB and by EI. Steroids, especially the unhydroxylated,

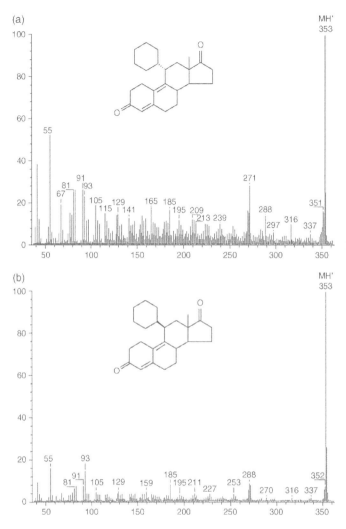

FIGURE 2.33 Fast atom bombardment mass spectra of *1a* and *1b* epimers. (Reprinted with permission. Mak, M., Francsics-Czinege, E., and Tuba, Z. Steroid epimers studied by different mass spectrometric methods. *Steroids* 2004, *69*, 831–840. Copyright Elsevier 2004.)

tend to be very nonpolar and thus do not have a very predominant spectrum when analyzed using ESI. An alternative approach to their measurement has been the application of FAB, which is conducive to nonpolar analytes. Figure 2.32 illustrates the EI mass spectra of two epimers of estrans where Figure 2.32*a* is the 11α-cyclohexyl estran and Figure 2.32*b* is the 11β-cyclohexyl estran.[64] Notice that the product ions, such as the m/z 270 product ion for the loss of the cyclohexyl substituent, are the predominant ions in the spectra versus the precursor molecular ion ($M^{+•}$) at m/z 352. The spectra for the two same steroid species are illustrated in Figure 2.33 collected using FAB. Notice in these spectra that the predominant ion is the m/z 353 proton adduct of the precursor (MH^+) while the peak at m/z 270 and 271 are very minor peaks. This illustrates that the FAB ionization is a soft ionization technique as compared to the EI approach.

2.10 PROBLEMS

2.1. List some source/ionization systems that are typically used in mass spectrometry.

2.2. Why is a source/ionization system important and a necessary component of mass spectrometric instrumentation?

2.3. What parameter is kept constant in the EI source that allows standardization of EI generated mass spectra?

2.4. Small magnets are included as part of the EI source. Are these used to remove neutral analytes from the source? If not, what is their purpose in the EI source?

2.5. Chemical ionization and electron ionization source designs are similar but posses some key differences. What are they?

2.6. What are the most common reagent gasses used in positive chemical ionization? Explain the differences in reactivity between the reagent gasses used.

2.7. If an alcohol had a value of 895.0 $kJ \cdot mol^{-1}$ for its proton affinity (PA), will the use of ammonia as the chemical ionization reagent gas result in the protonated form? What is the change in enthalpy for the reaction?

2.8. Demonstrate by calculating the change in enthalpy what other CI reagent gas may and may not form the protonated alcohols.

2.9. What reagent gasses will protonate ketones? What reagent gasses will not?

2.10. How can the two ionization techniques electron ionization and chemical ionization be used in a complementary fashion?

2.11. In the APCI source a dry nitrogen curtain gas is used. Explain what its inclusion in the source is used to achieve.

2.12. Describe and explain what takes place to produce a Taylor cone and initiate the electrospray process.

2.13. What is the difference between the "charged residue" model and the "ion evaporation" model.

2.14. What are some of the cooling processes taking place during the electrospray process? What do these tend to promote?

2.15. Describe how flow rate enhances nanoelectrospray as compared to normal electrospray.

2.16. Describe the APPI mechanism including the ionization process and the photon source needed.

2.17. Why is it easier at times to detect nonpolar analytes with APPI than with electrospray?

2.18. Though ESI is a solvent-based process and APPI is a gas-phase process, how does the inclusion of a dopant to APPI affect the ionization process? Is APPI still a gas-phase ionization process?

2.19. For the compounds listed in Table 2.3, which could be ionized by the use of a xenon (Xe) lamp? Are there any that could not be analyzed by an argon (Ar) lamp?

2.20. What are some of the physical and chemical characteristics needed in order to make a good MALDI matrix? What are some examples of MALDI matrices used?

2.21. What are three general MALDI plate spotting techniques that are used?

2.22. Briefly describe the desorption/ionization process that takes place with the MALDI technique.

2.23. What are the two known problems with the MALDI technique that affects resolution of the analytes? What instrumental designs have been made to reduce their effect?

2.24. List some differences that are found when using liquid ionic matrices as compared to solid crystal matrices.

2.25. What are some of the advantages of using a solid ionic crystal MALDI matrix?

2.26. Briefly describe the ionization mechanism of fast atom bombardment (FAB).

2.27. What processes take place that tend to relax the analyte and produce a soft ionization technique in FAB.

REFERENCES

1. Whitehouse, C. M., Dryer, R. N., Yamashita, M., and Fenn, J. B. *Anal. Chem.* 1985, *57*, 675.
2. Fenn, J. B. *J. Am. Soc. Mass. Spectrom.* 1993, *4*, 524.
3. Dole, M., Hines, R. L., Mack, R. C., Mobley, R. C., Ferguson, L. D., and Alice, M. B. *J. Chem. Phys.* 1968, *49*, 2240.
4. Kebarle, P., and Ho, Y. In *Electrospray Ionization Mass Spectrometry*, Cole, R. B., Ed. New York: Wiley, 1997, p. 17.
5. Cole, R. B. *J. Mass Spectrom.* 2000, *35*, 763–772.
6. Cech, N. B., and Enke, C. G. *Mass Spectrom. Rev.* 2001, *20*, 362–387.
7. Gaskell, S. J. *J. Mass Spectrom.* 1997, *32*, 677–688.
8. Cech, N. B., and Enke, C. G. *Anal. Chem.* 2000, *72*, 2717–2723.
9. Sterner, J. L., Johnston, M. V., Nicol, G. R., and Ridge, D. P. *J. Mass Spectrom.* 2000, *35*, 385–391.

10. Cech, N. B., and Enke, C. G. *Anal. Chem.* 2001, *73*, 4632–4639.

11. Taylor, G. I. *Proc. R. Soc. London, Ser. A*, 1964, *280*, 383.

12. Gomez, A., and Tang, K. *Phys. Fluids* 1994, *6*, 404.

13. Cao, P., and Stults, J. T. *Rapid Commun. Mass Spectrom.* 2000, *14*, 1600–1606.

14. Ho, Y. P., Huang, P. C., and Deng, K. H. *Rapid Commun. Mass Spectrom.* 2003, *17*, 114–121.

15. Kocher, T., Allmaier, G., and Wilm, M. *J. Mass Spectrom.* 2003, *38*, 131–137.

16. Lorenz, S. A., Maziarz, E. P., and Wood, T. D. *J. Am. Soc. Mass Spectrom.* 2001, *12*, 795–804.

17. Daniel, J. M., Friess, S. D., Rajagopalan, S., Wendt, S., and Zenobi, R. *Int. J. Mass Spectrom.* 2002, *216*, 1–27.

18. Whitehouse, C. M., Dreyer, R. N., Yamashita, M., and Fenn, J. B. *Anal. Chem.* 1985, *57*, 675.

19. Fenn, J. B. *J. Am. Soc. Mass Spectrom.* 1993, *4*, 524.

20. Cole, R. B. *J. Mass Spectrom.* 2000, *35*, 763.

21. Cech, N. B., and Enke, C. G. *Mass Spec. Rev.* 2001, *20*, 362.

22. Wilm, M. S., and Mann, M. *Int. J. Mass Spectrom. Ion Processes* 1994, *136*, 167–180.

23. Wilm, M., and Mann, M. *Anal. Chem.* 1996, *68*, 1–8.

24. Caprioli, R. M., Emmett, M. E., and Andren, P. Proceedings of the 42nd ASMS Conference on Mass Spectrometry and Allied Topics, Chicago, IL, May 29–June 3, 1994, p. 754.

25. Qi, L., and Danielson, N. D. *J. Pharma. Biomed. Anal.* 2005, *37*, 225–230.

26. Fernandez de la Mora, J., and Loscertales, I. G. *J. Fluid Mech.* 1994, *260*, 155–184.

27. Pfiefer, R. J., and Hendricks, C. D., Jr. *AIAA J.* 1968, *6*, 496–502.

28. Juraschek, R., Dulcks, T., and Karas, M. *J. Am. Soc. Mass Spectrom.* 1999, *10*, 300–308.

29. Schmidt, A., and Karas, M. *J. Am. Soc. Mass Spectrom.* 2003, *14*, 492–500.

30. Li, Y., and Cole, R. B. *Anal. Chem.* 2003, *75*, 5739–5746.

31. El-Faramawy, A., Siu, K. W. M., and Thomson, B. A. *J. Am. Soc. Mass Spectrom.* 2005, *16*, 1702–1707.

32. Tang, K., Lin, Y., Matson, D. W., Kim, T., and Smith, R. D. *Anal. Chem.* 2001, *73*, 1658–1663.

33. Driscoll, J. N., and Spaziani, F. F. *Res./Dev.* 1976, 50–54.

34. Langhorst, M. L. *J. Chromatogr. Sci.* 1981, *19*, 98–103.

35. Schermund, J. T., and Locke, D. C. *Anal. Lett.* 1975, *8*, 611–625.

36. Locke, D. C., Dhingra, B. S., and Baker, A. D. *Anal. Chem.* 1982, *54*, 447–450.

37. Driscoll, J. N., Conron, D. W., Ferioli, P., Krull, I. S., and Xie, K. H. *J. Chromatogr.* 1984, *302*, 43–50.

38. De Wit, J. S. M., and Jorgenson, J. W. *J. Chromatogr.* 1987, *411*, 201–212.

39. Syage, J. A., Hanold, K. A., Lynn, T. C., Horner, J. A., and Thakur, R. A. *J. Chromatogr. A* 2004, *1050*, 137–149.

40. Short, L. C., Cai, S. S., and Syage, J. A. *J. Am. Soc. Mass Spectrom.* 2007, *18*, 589–599.

41. Robb, D. B., Covey, T. R., and Bruins, A. P. *Anal. Chem.* 2000, *72*, 3653–3659.

42. Syage, J. A., Evans, M. D., and Hanold, K. A. *Am. Lab.* 2000, *32*, 24–29.

43. Kauppila, T. J., Bruins, A. P., and Kostianen, R. *J. Am. Soc. Mass Spectrom.* 2005, *16*, 1399–1407.

44. Kauppila, T. J., Kotiaho, T., Kostianen, R., and Bruins, A. P. *J. Am. Soc. Mass Spectrom.* 2004, *15*, 203–211.

45. Straube, E. A., Dekant, W., and Volkel, W. *J. Am. Soc. Mass Spectrom.* 2004, *15*, 1853–1862.

46. Cavaliere, C., Foglia, P., Pastorini, E., Samperi, R., and Lagana, A. *J. Chromatogr. A* 2006, *1101*, 69–78.

47. Gomez-Ariza, J. L., Arias-Borrego, A., Garcia-Barrera, T., and Beltran, R. *Talanta*, 2006, *20*, 859–869.

48. Karas, M., Bachmann, D., Bahr, U., and Hillenkamp, F. *Int. J. Mass Spectrom. Ion Processes* 1987, *78*, 53.

49. Tanaka, K., Waki, H., Ido, Y., Akita, S., Yoshida, Y., and Yoshida, T. *Rapid Commun. Mass Spectrom.* 1988, *2*, 151–153.

50. Karas, M., and Hillenkamp, F. *Anal. Chem.* 1988, *60*, 2299.

51. Xiang, F., and Beavis, R. C. *Org. Mass Spectrom.* 1993, *28*, 1424.

52. Xiang, F., and Beavis, R. C. *Rapid Commun. Mass Spectrom.* 1994, *8*, 199–204.

53. Armstrong, D. W., Zhang, L.-K., He, L., and Gross, M. L. *Anal. Chem.* 2001, *73*, 3679–3686.

54. Carda-Broch, S., Berthod, A., and Armstrong, D. W. *Rapid Commun. Mass Spectrom.* 2003, *17*, 553–560.

55. Li, Y. L., and Gross, M. L. *J. Am. Soc. Mass Spectrom.* 2004, *15*, 1833–1837.

56. Mank, M., Stahl, B., and Boehm, G. *Anal. Chem.* 2004, *76*, 2938–2950.

57. Li, Y. L., Gross, M. L., and Hsu, F. F. *J. Am. Soc. Mass Spectrom.* 2005, *16*, 679–682.

58. Ham, B. M., Jacob, J. T., and Cole, R. B. *Anal. Chem.* 2005, *77*, 4439–4447.

59. Lemaire, R., Tabet, J. C., Ducoroy, P., Hendra, J. B., Salzet, M., and Fournier, I. *Anal. Chem.* 2006, *78*, 809–819.

60. Mank, M., Stahl, B., and Boehm, G. *Anal. Chem.* 2004, *76*, 2938–2950.

61. Aebersold, R., and Mann, M. *Nature* 2003, *422*, 198–207.

62. Luxembourg, S. L., McDonnell, L. A., Duursma, M. C., Guo, X., and Heeren, R. M. A. *Anal. Chem.* 2003, *75*, 2333–2341.

63. Lemaire, R., Wisztorski, M., Desmons, A., Tabet, J. C., Day, R., Salzet, M., and Fournier, I. *Anal. Chem.* 2006, *78*, 7145–7153.

64. Mak, M., Francsics-Czinege, E., and Tuba, Z. *Steroids* 2004, *69*, 831–840.

CHAPTER 3

MASS ANALYZERS IN MASS SPECTROMETRY

3.1 MASS ANALYZERS

The mass analyzer is the heart of the mass spectrometric instrumentation used in the separation of molecular ions ($M^{+\cdot}$) and analyte ions (e.g., $[M+H]^+$) in the gas phase. The most fundamental aspect of the mass analyzer is the ability to separate ions according to their mass-to-charge ratio (m/z; see Section 1.4). In this chapter we will take a close look at the most common mass analyzers used today, including their basic design and construction and the theory behind their ability to separate ionized analyte species according to their mass-to-charge ratio. Figure 3.1 illustrates the basic design and layout of most of the mass analyzers that are in use today. Included in Figure 3.1 are (a) an electric and magnetic sector mass analyzer, (b) a time-of-flight mass analyzer (TOF/MS), (c) a time-of-flight/time-of-flight mass analyzer (TOF-TOF/MS), (d) the hybrid (hybrids are mass analyzers that couple together two separate types of mass analyzers) quadrupole time-of-flight mass analyzer (Q-TOF/MS), (e) a triple quadrupole or linear ion trap mass analyzer (QQQ/MS or LIT/MS), (f) a three-dimensional quadrupole ion trap mass analyzer (QIT/MS), (g) a Fourier transform ion cyclotron mass analyzer (FTICR/MS), and finally (h) the linear ion trap–Orbitrap mass analyzer (IT-Orbitrap/MS). Also included later in the chapter are discussions of the two more recently introduced hybrid mass analyzers in use in laboratories today: the linear quadrupole ion trap Fourier transform mass spectrometer (LTQ-FT/MS) and the linear quadrupole ion trap Orbitrap mass spectrometer

Even Electron Mass Spectrometry with Biomolecule Applications   By Bryan M. Ham
Copyright © 2008 John Wiley & Sons, Inc.

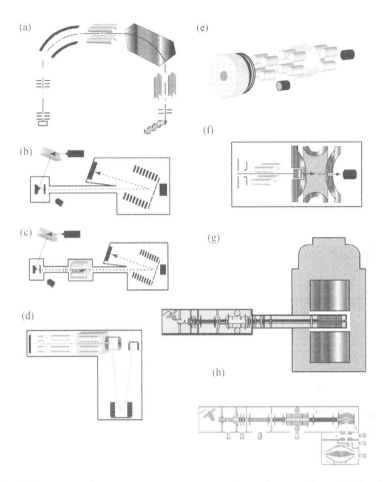

FIGURE 3.1 Some of the most common mass analyzers in use today: (*a*) electric and magnetic sector mass analyzer, (*b*) time-of-flight mass analyzer (TOF/MS), (*c*) time-of-flight/time-of-flight mass analyzer (TOF-TOF/MS), (*d*) quadrupole time-of-flight mass analyzer (Q-TOF/MS), (*e*) triple quadrupole or linear ion trap mass analyzer (QQQ/MS or LIT/MS), (*f*) three-dimensional quadrupole ion trap mass analyzer (QIT/MS), (g) Fourier transform ion cyclotron mass analyzer (FTICR/MS), and (*h*) linear ion trap Orbitrap mass analyzer (IT-Orbitrap/MS).

(LTQ-Orbitrap/MS). As discussed in Section 1.2, the most general configuration for mass spectrometric instrumentation is an ion source, a mass analyzer, and finally a detector (see Fig. 1.1). As we also learned in Chapter 2, two very common techniques of analyte ionization are electrospray ionization (ESI) and matrix-assisted laser desorption ionization (MALDI). We will see in this chapter how these ionization sources are used in conjunction with the common mass analyzers illustrated in Figure 3.1. We will also look at a few newer, state-of-the-art mass analyzers that have come out recently such as the linear ion trap Fourier transform mass analyzer and the Orbitrap mass analyzer. Finally, we will take a look at a few examples of applications of the LTQ-FT/MS and the LTQ-Orbitrap/MS for the measurement of biomolecules at the end of the chapter.

3.2 MAGNETIC AND ELECTRIC SECTOR MASS ANALYZER

When charged particles enter a magnetic field, they possess a circular orbit that is perpendicular to the poles of the magnet. This phenomenon has been applied to a magnetic sector mass analyzer, which is a momentum separator. Figure 3.2 illustrates the basic design of the magnetic sector mass analyzer. Notice that the slits are normally placed collinear with the apex of the magnet. The ions enter a flight tube (first field-free region) from a source through a source exit slit and travel into the magnetic field. The accelerating voltage in the source will determine the kinetic energy (KE) that is imparted to the ions:

$$KE = zeV = \tfrac{1}{2}mv^2 \tag{3.1}$$

where V is the accelerating voltage in the source, e is the fundamental charge of an electron (1.60×10^{-19} C), m is the mass of the ion, v is the velocity of the ion, and z is the number of charges. The magnetic field will deflect the charged particles according to the radius of curvature of the flight path (r), which is directly proportional to the mass-to-charge ratio (m/z) of the ion. The centripetal force that is exerted upon the

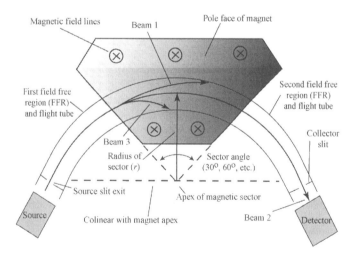

FIGURE 3.2 Basic components of a magnetic sector mass analyzer. The magnetic field lines are directed into the plane of the paper for negative ion mode, which follows the right-hand rule. For positive ion mode the polarity is switched and the magnetic field lines are directed out of the plane of the paper.

ion by the magnetic field is given by the relationship:

$$F_M = Bzev \tag{3.2}$$

where B is the magnetic field strength. The centrifugal force acting upon the ion from the initial momentum is given by:

$$F_c = \frac{mv^2}{r} \tag{3.3}$$

For the ion to reach the detector the centripetal force acting upon the ion from the magnet (down pushing force) must equal the centrifugal force acting upon the ion from the initial momentum of the charged ion (upwardly pushing force). This equality is represented by:

$$F_M = F_C \tag{3.4}$$

$$Bzev = \frac{mv^2}{r} \tag{3.5}$$

where B is the magnetic field strength, e is the electron fundamental charge, m is the mass of the ion, v is the velocity of the ion, z is the number of charges, and r is the radius of the curvature of the sector. Solving Equation (3.5) for v and substituting into Equation (3.1), we can derive the relationship between the mass-to-charge ratio (m/z) of the ion to the force of the magnetic field strength:

$$Bzev - \frac{mv^2}{r} \tag{3.5}$$

$$v = \frac{Bzer}{m} \tag{3.6}$$

$$KE = zeV = \tfrac{1}{2}mv^2 \tag{3.1}$$

$$zeV = \frac{1}{2}m \left(\frac{Bzer}{m} \right)^2 \tag{3.7}$$

$$zeV = \frac{1}{2}m \frac{B^2 z^2 e^2 r^2}{m^2} \tag{3.8}$$

$$V = \frac{1}{2} \frac{B^2 zer^2}{m} \tag{3.9}$$

$$\frac{m}{z} = \frac{B^2 er^2}{2V} \tag{3.10}$$

This relationship is often referred to as the scan law. In normal experimental operation the radius of the sector is fixed, and the accelerating voltage V is held constant while the magnetic field strength B is scanned. From Equation (3.10) it can be seen that a higher magnetic field strength equates to a higher mass-to-charge ratio stability path. The magnetic field strength B is typically scanned from lower strength to higher strength for m/z scanning in approximately 5 s. In exact mass measurements and when calibrating the magnetic sector mass analyzer, the magnetic field strength B of the magnet is kept fixed while the accelerating voltage is scanned. This is done to avoid the slight error that is inherent in the magnet known as historesis where after the magnet has scanned its original strength is slightly changed from its initial condition before the scanning was commenced. Scanning the accelerating voltage does not have this problem. The calibration of the magnetic sector mass analyzer is a nonlinear function; therefore, several points are often required for calibration. In calibrating, a standard mass is scanned by scanning the acceleration voltage and holding the magnetic field strength constant. When the mass is detected, the values are set for that set of acceleration voltage versus magnetic field strength. The resolution of the magnetic sector mass analyzer is directly proportional to the magnet radius and the slit width of the source exit slit and the slit width of the collector slit:

$$\text{Resolution} \propto \frac{r}{S_1 + S_2} \tag{3.11}$$

where $r =$ radius of the magnet
$\quad S_1 =$ slit width of the source exit slit
$\quad S_2 =$ slit width of the collector slit

It can be seen from the relationship in Equation (3.11) that a larger magnet radius and smaller slit widths will equate to a larger resolution value. However, decreasing the slit widths causes a loss in sensitivity, as the amount of ions allowed to pass through will decrease. The typical resolution obtained under normal scan mode for a magnetic sector mass analyzer is ≤ 5000. This means that the magnetic sector mass analyzer can separate an m/z 5000 molecular ion from an m/z 5001 molecular ion, an m/z 500.0 molecular ion from an m/z 500.1 molecular ion, an m/z 50.00 molecular ion from an m/z 50.01 molecular ion, and so on. Two factors that limit the mass resolution of the magnetic sector mass analyzer are angular divergence of the ion beam, and the kinetic energy spread of the ions as they leave the source. Decreasing the source exit slit width will reduce the angular divergence effect upon the resolution; however, this will also decrease the sensitivity of the mass analyzer. The spread in the kinetic energy of the ions as they leave the source is caused by slight differences in the initial velocity of the ions imparted to them by the source. In EI the initial kinetic energy spread is approximately 1–3 eV. To address the kinetic energy spread distribution in a magnetic sector mass analyzer, an electric sector that acts as a kinetic energy separator is used in combination with the magnetic sector. The electric sector shown in Figure 3.3 is used as an energy filter where only one energy of ions will pass through the electric field. The electric sector will only pass ions that have the exact energy match as that with the acceleration voltage in the source. In combination with slits the electric sector can be called an energy focuser. The electric sector is constructed of two cylindrical plates with potentials of $+\frac{1}{2}E$ and $-\frac{1}{2}E$, where the total field $= eE$ $= (+\frac{1}{2}E) + (-\frac{1}{2}E)$. For ions to pass through the electric sector the deflecting force

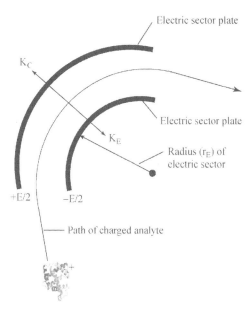

FIGURE 3.3 Path of charged particle through the electric sector energy filter. Forces acting upon the particle include the centrifugal force (K_C) and the deflecting force (K_E).

(K_E) must equal the centrifugal force (K_C). This equality is represented as

$$eE = \frac{mv^2}{r_E} \tag{3.12}$$

where eE is the deflecting force (K_E) and mv^2/r_E is the centrifugal force (K_C). For the ions arriving into the electric sector the kinetic energy (KE) equals

$$KE = zeV = \tfrac{1}{2}mv^2 \tag{3.1}$$

$$v^2 = \frac{2zeV}{m} \tag{3.13}$$

substituting Equation (3.13) into Equation (3.12) we get

$$eE = \frac{m\left(\dfrac{2zeV}{m}\right)}{r_E} \tag{3.14}$$

$$E = \frac{2zV}{r_E} \tag{3.15}$$

$$r_E = \frac{2zV}{E} \tag{3.16}$$

From this relationship we see that the radius of the curvature will equal two times the accelerating voltage divided by the total electric field strength. What ions pass through the electric sector do not depend upon the mass or charge of the ions but

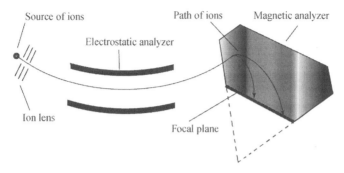

FIGURE 3.4   Mattauch–Herzog double-focusing geometry mass spectrometer.

upon how well the ions' kinetic energy matches the field according to the acceleration voltage within the source. This relationship demonstrates that the electric sector is not a mass analyzer like the magnetic sector but rather is an energy filter.

By combining the magnetic sector with the electric sector a double-focusing effect can be achieved where the magnetic sector does directional focusing while the electric sector does energy focusing. Two types of double-focusing geometries are illustrated in Figures 3.4 and 3.5. The double-focusing geometry illustrated in Figure 3.4 is known as the Mattauch–Herzog double-focusing geometry. The electric sector (electrostatic analyzer) located before the magnetic analyzer in the instrumental

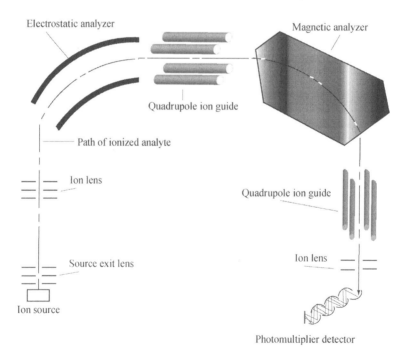

FIGURE 3.5   Nier–Johnson double-focusing geometry mass spectrometer.

design has effectively been added to gain in the resolution of the mass-analyzed ions. The Mattauch–Herzog double-focusing geometry results in a focal plane within the mass analyzer. This is a static analyzer plane where a photoplate or microchannel array detectors can be placed parallel to the focal plane. In this design the different mass-to-charge ratios will be focused at different points on the microchannel plate. The detection of the separated mass-to-charge ratio ions will require no scanning, increasing the sensitivity in the instrumental response. The second design in Figure 3.5 is the Nier–Johnson double-focusing mass spectrometer. Here also the electrostatic analyzer is placed before the magnetic analyzer. In this design ions are focused onto a single point and the magnetic analyzer must be scanned.

3.3 TIME-OF-FLIGHT MASS ANALYZER (TOF/MS)

The most common mass spectrometer that is coupled to the MALDI technique is the time-of-flight mass spectrometer (TOF MS).[1,2] The TOF mass spectrometer separates compounds according to their mass-to-charge ratios (m/z) through a direct relationship between a compound's drift time through a predetermined drift path length and the analyte ion's mass-to-charge ratio (m/z). Initially, all the ions have similar kinetic energies imparted to them from the draw-out pulse (representing time zero), which accelerates them into the flight tube. Because the compounds have different masses, their velocities will be different according to the relationship between kinetic energy and mass represented by $KE = zeV = \frac{1}{2}mv^2$. From this expression, the mass-to-charge ratio is related to the ions flight time by the following expression: $m/z = 2eVt^2/L^2$.

$$KE = zeV = \tfrac{1}{2}mv^2 \tag{3.1}$$

$$v = \left(\frac{2zeV}{m}\right)^{1/2} \tag{3.17}$$

$$t = \frac{L}{v} \qquad L = \text{length of drift tube} \tag{3.18}$$

$$t = L\left(\frac{m}{2zeV}\right)^{1/2} \qquad V \text{ and } L \text{ are fixed} \tag{3.19}$$

Solving for m/z:

$$t^2 = L^2\left(\frac{m}{2zeV}\right) \tag{3.20}$$

$$\frac{m}{z} = \frac{2eVt^2}{L^2} \tag{3.21}$$

There is a high transmission efficiency of the ions into and through the drift tube that equates to very low levels of detection limits, which are in the femtamole

FIGURE 3.6 Example of a commercially available MALDI time-of-flight mass spectrometer (MALDI TOF/MS). An Applied Biosystems Voyager-DE STR time-of-flight mass spectrometer. (Photo provided courtesy of Applied Biosystems.)

(10^{-15}) to atamole (10^{-18}) ranges. Theoretically, the mass range of the time-of-flight mass spectrometer is unlimited due to the relationship of drift time for mass measurement. In practice though, the sensitivity needed to detect a very slow moving large molecular weight compound limits the TOF/MS to 1–2 million Da or so. Figure 3.6 is an example of a commercially available time-of-flight mass spectrometer. Figure 3.7 shows the different parts of a typical time-of-flight mass spectrometer. The first major component of the TOF mass spectrometer is the source where the analytes are ionized and then subsequently transferred into the mass analyzer. When MALDI is used as the ionization technique, the source will be under high vacuum, typically at an approximate pressure of 1×10^{-6} Torr. An atmospheric pressure ionization source such as electrospray is also often used with TOF mass analyzers, which do not require high vacuum but generally require a quadrupole between the source and the drift tube. These quadrupoles can be rf only acting as ion guides or they can be fully functional mass filters used to isolate analyte ions before transfer into the drift tube. This is often done for tandem mass spectrometric analysis of collision-induced dissociation product ion spectra generation (this will be covered in Section 3.8). At this point we will consider a MALDI source. Upon generation of gas-phase ions by the laser and matrix, the ions are drawn into the second component of the TOF mass analyzer by an extraction grid that imparts equal kinetic energy to all of the analyte ions present. The analyte ions then transfer into the second part of the TOF mass analyzer, the drift tube. The analyte ions are separated according to the relationship that is illustrated in Equations (3.17)–(3.21) and are detected typically by electron multipliers.

The early designs of TOF mass spectrometers suffered from poor resolution, which is the ability of the mass spectrometer to separate ions. In earlier work this was expressed as resolution $= m/\Delta m$ where m is the mass of the peak of interest

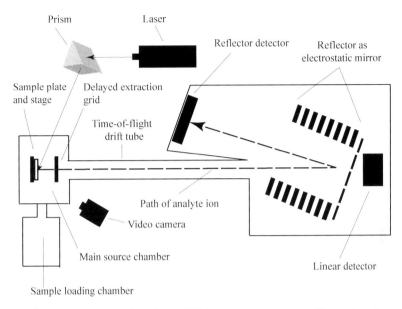

FIGURE 3.7 Components of a time-of-flight mass spectrometer illustrating the major sections including the source, drift tube, reflectron electrostatic mirror, and detector.

and Δm is the difference between this mass and the next closest peak (more of a chromatography approach to resolution). Typically used today the mass spectral peak resolution is calculated as $m/\Delta m$ where m is the mass of the peak of interest and Δm is the full width of the peak at half maximum (FWHM). The poor resolution was due to nonuniform initial spatial and energy distributions of the formed ions in the mass spectrometer's ionization source. Due to the distribution spreads, the mass resolution was directly dependent upon the initial velocity of the formed ions. Early work conducted by Wiley and McLaren[3] reported upon the correction for initial velocity distributions using a technique they described as "time-lag energy focusing" where the ions were produced in a field-free region; then, after a preset time, a pulse was applied to the region to extract the formed ions. They also reported upon a way to correct for initial special distributions through a two-field pulsed ion source. They demonstrated though that correction can only be made for one of the distributions at a time. In MALDI, the spatial distribution spread is not a significant problem, therefore "delayed extraction" of the ions from the field-free source region should correct for the initial energy distribution spread of the formed ions.[4] With this correction, the resolution should be directly dependent upon the total flight time ratio to the error in the time measurement; thus, an increase in the length of the flight path should equate to a higher resolution. The resolution enhancement due to the delayed extraction initial energy focusing and flight path was demonstrated in a study reported by Vestal et al.[4] using a MALDI-TOF mass spectrometer. Early work reported by Brown and Lennon[5] in 1995 gave a good description of a pulsed ion extraction system applied to

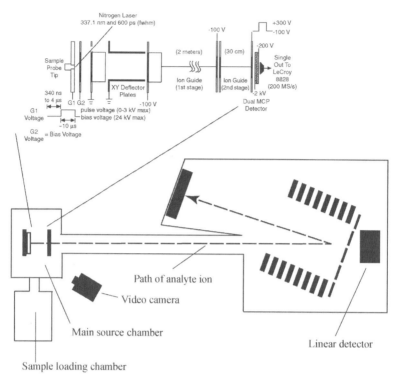

FIGURE 3.8 Expanded view of the delayed extraction setup within the source of the time-of-flight mass spectrometer.[5] (Reprinted with permission from Brown, R. S., and Lennon, J. J. *Anal. Chem.* 1995, *67*, 1998–2003. Copyright 1995 American Chemical Society.)

MALDI-TOF mass spectrometry. Figure 3.8 illustrates a schematic representation of the pulsed ion extraction apparatus they used and the location of it within the source section of the TOF mass spectrometer. It is described as a three-grid system (G1, G2, and a third grounded grid) where the grids are constructed as a mesh of wire for uniform electrostatic field formation. In the delayed extraction mechanism a nitrogen laser is pulsed onto the MALDI matrix/sample mixture, effectively desorbing the matrix and the sample into a gas-phase plume above the G1 and between G2. After a brief delay period (340 ns to 4 μs) for optional plume formation, a short pulse (0–3 kV for ~10 μs) of voltage is applied to G1 while maintaining G2 at the prepulse voltage (bias voltage). This creates a potential field gradient between G1 and G2, highest at G1 and decreasing toward G2. Slower moving ions within the plume will be closer to G1 with respect to faster moving ions and will experience a stronger repelling force from G1. The result will be that the ions that were initially moving slower will effectively catch up to ions with the same m/z value that possessed slightly higher initial velocities. By combing an optimized delay time and extraction pulse, the flight time spread contribution to resolution loss due to different initial velocities can be corrected for.

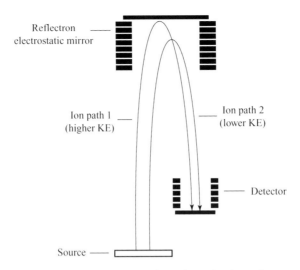

FIGURE 3.9 Electrostatic mirror focusing of two ions that have the same m/z value but slightly different kinetic energies. Ion path 1 possesses slightly higher kinetic energy in relation to ion path 2. Ion path 1 travels slightly farther to match that of ion path 2. The two ions are focused and arrive at the detector at the same time.

A second modification of TOF mass spectrometers involves time focusing of the ions while they are in flight through the use of an electrostatic mirror at the end of the flight tube.[6,7] The electrostatic mirror focuses the ions with the same mass-to-charge ratio but slightly different kinetic energies by allowing a slightly longer path for the higher KE containing ion as compared to a slightly lower KE ion thus allowing the ions to catch up to one another. The electrostatic mirror also increases the flight path length of the mass spectrometer, thus providing a double-focusing effect. All recent TOF mass spectrometers employ both delayed extraction and the reflectron electrostatic mirror for analysis of compounds below 10 kDa. For compounds greater than 10 kDa a linear mode is typically used where the electrostatic mirror is turned off. Figure 3.9 illustrates the focusing of the two ions that have the same m/z value but slightly different kinetic energies. Ion path 1 represents the path of the ion with the slightly higher kinetic energy in relation to ion path 2. Ion path 1 travels slightly farther into the electrostatic mirror field gradient, thus focusing its flight time to match that of ion path 2. After traveling through the second field-free region the two ions are focused and arrive at the detector at the same time. Producing a narrower peak representing its detection and thus increasing the resolution.

3.4 TIME-OF-FLIGHT/TIME-OF-FLIGHT MASS ANALYZER (TOF-TOF/MS)

A modification of the Reflectron time-of-flight (TOF) mass analyzer is the coupling of essentially two TOF mass analyzers together with the chief addition of a central floating collision cell between them. This effectively allows tandem mass spectrometric analyses to be performed that consists of collision-induced dissociation experiments of precursor analyte ions for structural information and identification. However, the design of this system primarily dictates that a MALDI source must be used. The MALDI technique almost exclusively produces singly charged analyte ions that are then mass analyzed. One drawback of singly charged ions, particularly in the mass spectrometric analysis of peptides, is the sometimes observed low efficiency in the number of fragmentation pathways that give useful product ions. Electrospray ionization is a technique that often produces multiply charged analyte ions (peptides and proteins) that tend to have a higher degree in the number of fragmentation pathways that produce product ions. This is due to the higher degree of charging by electrospray, which tends to activate a higher degree of fragmentation in CID experiments. Figure 3.10 illustrates an example of a commercial available time-of-flight (TOF)/time-of-flight (TOF) mass spectrometer. Notice that the flight tubes do not necessarily have to be oriented in a horizontal fashion as is the design of the single flight tube TOF illustrated in Figure 3.6. In the TOF-TOF/MS design in Figure 3.10 the flight tubes are actually oriented vertically. Figure 3.11 illustrates the basic components that make up a TOF-TOF/MS. A tandem TOF-TOF mass analyzer,

FIGURE 3.10 An Applied Biosystems 4800 MALDI time-of-flight/time-of-flight mass spectrometer (4800 MALDI TOF/TOF MS). (Photo provided courtesy of Applied Biosystems.)

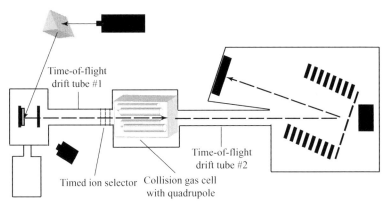

FIGURE 3.11 Basic components that make up a time-of-flight/time-of-flight mass spectrometer (TOF/TOF MS).

such as the one illustrated in Figure 3.11 produces a product ion spectrum by first isolating the gas-phase analyte ion species of interest separated from the other species present by the first drift tube, which is then allowed to pass out of the drift tube and into a central radio frequency (rf) only quadrupole or hexapole collision cell with the use of an ion gate acting as a timed ion selector. The ion gate is comprised of a series of wires that alternating voltages can be applied to usually as ± 1000 V. When the ion gate is switched on, no ions are allowed to pass through the gate, thus preventing the transmission of any ions through the remainder of the instrument. When the gate is switched off, all ions will be able to pass through the gate. By switching off the gate at a predetermined flight time, a specific m/z value will be allowed to pass through the gate. In this way the timed ion selector is turned on and off to allow the passage of the desired m/z species according to its predetermined drift tube flight time. The central rf-only quadrupole or hexapole collision cell is filled with a stationary target gas such as argon. The ion is induced to collide with the gas, thus activating the ion for dissociation. The central rf-only quadrupole or hexapole collision cell is not a mass analyzer itself but functions to refocus the product ions before exiting the cell. The products produced by the collision-induced dissociation are then transferred to the second TOF mass analyzer by a series of ion guides. To this point the series of events include the production of gas-phase ionized analyte species in the source phase, the delayed extraction pulsing of the ions into the first TOF mass analyzer, the separation of the different m/z values according to their drift times in the first flight tube, and finally the selection and allowed passage of a single m/z species by the timed ion selector into the central collision cell. The m/z species that was allowed passage into the collision cell undergoes product ion activating collision with the stationary argon target gas. The refocused product ions are then transferred into the second TOF drift tube where they are separated into different m/z values according to their respective drift times. The TOF-TOF/MS instrumentation also includes the Reflector (electrostatic mirror) used to focus the product ions prior to detection. Finally, the mass-analyzed gas-phase ions are detected and their spectra stored for processing.

3.5 QUADRUPOLE MASS FILTER

A quadrupole mass analyzer is made up of four cylindrical rods that are placed precisely parallel to each other. One set of opposite poles have a dc voltage (U) supply connected while the other set of opposite poles of the quadrupole have a radio frequency (rf) voltage (V) connected. The quadrupole dc and rf voltage configuration is illustrated in Figure 3.12. Ions are accelerated into the quadrupole by a small voltage of 5 eV, and under the influence of the combination of electric fields, the ions follow a complicated trajectory path. If the oscillation of the ions in the quadrupole have finite amplitude, it will be stable and pass through. If the oscillations are infinite, they will be unstable and the ion will collide with the rods. These path descriptions are illustrated in Figure 3.13. In the particular orientation that is illustrated in Figure 3.13, the dc and rf voltages have been selected to give an m/z value of 100 a stable trajectory through the guadrupoles. An m/z value of 10, which is a less massive ion, will have a very unstable trajectory and will collide with the quadrapole rods at an early stage. An m/z 1000 species will be a more massive ion of greater momentum that tend to travel further through the quadrupole field but still possesses an unstable trajectory and will also suffer collision with the rods and thus also be effectively filtered out. The construction of the quadrupoles is based off of the equipotential curves of a quadrupole field, which are illustrated in Figure 3.14.

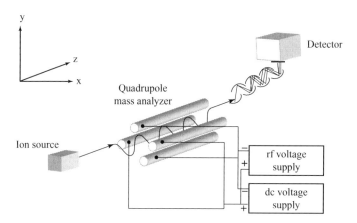

FIGURE 3.12   Quadrupole orientation and the configuration for the connections of the dc voltage (U) and radio frequency (rf) voltage (V). Ions are accelerated into the quadrupole by a small voltage of 5 eV, and under the influence of the combination of electric fields, the ions follow a complicated trajectory path.

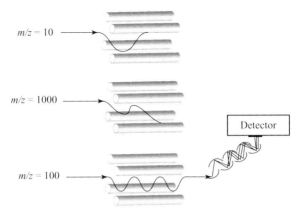

FIGURE 3.13 Stable and unstable trajectories of ions through the quadrupole. The m/z 100 species has been selected for stable path and transmission through the quadrupole for detection.

As Figure 3.14 shows the equipotential curves of the quadrupole field are comprised of rectangular hyperbolas. The field is created in the mass analyzer by selecting one set of the equipotential curve rectangular hyperbolas and placing the quadrupole electrodes in a configuration that follows the dimensions of the rectangular

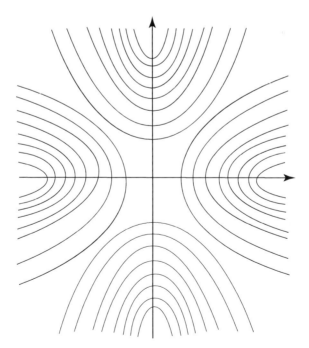

FIGURE 3.14 Equipotential curves of the quadrupole field comprised of rectangular hyperbolas where the inside of the hyperbola curves follow a semicircular path.

hyperbolas. The optimum field and transmission through the quadrupole would be obtained by using hyperbolic rods; however, for ease of manufacturing the rods are often made cylindrical (notice that the inside of the hyperbolas' curves follow a semicircular path). The construction of an ideal hyperbolic field is achieved by orienting the quadrupole electrodes in an imaginary square so that the opposite poles lay 1/1.148 times the electrode diameter away from each other. The inherent faults in the electric field caused by using cylindrical rods instead of perfect hyperbolas can be corrected by placing the electrodes 1/1.148 times the electrode diameter according to the following relationship:

$$r_0 = \frac{r}{1.148} \tag{3.22}$$

where r_0 is the radius of the field and r is the electrode radius.

The ratio of 1/1.148 is theoretically derived to produce an ideal hyperbolic field within the geometric center of the quadrupole. The mathematical derivation of the description of the stable trajectory of a charged particle through the quadrupole has been well characterized and the reader is directed to other references for a more thorough coverage of this subject if interested.[9] However, a brief description will be offered; the derivation of stability starts with the simple $F = ma$ force equation and results in a second-order differential. The canonical form of the Mathieu equation[10] that describes the stability of a charged particle's trajectory through the quadrupole, in a Cartesian coordinate system, is as follows:

$$\frac{d^2u}{d\xi^2} + (a_u - 2q\cos 2\xi)u = 0 \tag{3.23}$$

with u representing either x or y:

$$a_u = a_x = -a_y = \frac{4eU}{m\omega^2 r_0^2} \qquad q = q_x = -q_y = \frac{2eV}{m\omega^2 r_0^2} \tag{3.24}$$

where e is the charge on an electron, U is the applied dc voltage, V is the applied zero-to-peak rf voltage, m is the mass of the ion, ω is the angular frequency, and r is the effective radius between the quadrupole electrodes. Intuitively, the student can instantly recognize that the quadrupole possesses the fundamental dependence of mass-to-charge ratio (m/e) to the effective voltages applied to the electrodes, which is required of a mass analyzer.

Stability diagrams can be constructed from the interdependencies of the stable trajectories of the ions through the quadrupole. The quadrupole field produced by the electrodes focuses the mass-analyzed ion down to the center of the quadrupole by the alternating biases applied to the oppositely aligned electrodes. Usually, the rf amplitude is kept constant while the polarity of the dc amplitude is switched. For a positively charged ion the focusing force will exist in the plane of the positively

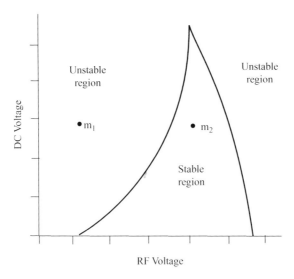

FIGURE 3.15 Stability diagram relating the dc voltage amplitude versus the RF voltage amplitude for masses m_1 and m_2 where m_2 lies within a designated stable area for transmission through the quadrupole.

biased electrodes in the form of $U + V_0 \cos \omega t$. The two rods opposite each other will possess the same exact combination of fields. The two opposed rods to the positively biased electrodes that destabilize the trajectory of the positive ion will posses a negative bias in the form of $-U - V_0 \cos \omega t$. The rapidly changing biases producing rapidly changing fields will cause the ion to oscillate back and forth within the quadrupole. A potential difference applied across the length of the quadrupole will draw the ion through the oscillating fields. Ions with unstable trajectories will collide with the rods. This mass filtering takes place primarily near the beginning of the quadrupole, and often quadrupole mass analyzer designs will include small rods making up a prequadrupole filter. The plot in Figure 3.15 illustrates an ion stability diagram relating the dependence between U and V. At a certain range of dc voltage U and rf potential V, the ion of interest will have a stable region (ion m_1 versus ion m_2). Due to this functionality, quadrupoles are called mass filters; however, they are still dependent upon a mass-to-charge ratio. The plot in Figure 3.16 illustrates the ability to scan mass-to-charge ratios (m/e) with the quadrupole, allowing the measurement of a range of masses, typically up to 4000 Da. The slope of the line is a fixed ratio of U to V. As U and V are varied, the scan follows the scan line and subsequent m/e values are recorded. The k value of the x axis represents an instrument calibration constant. In the plot of Figure 3.16 there are two scan lines labeled I and II. The scan lines demonstrate that the resolution of the quadrupole is inversely proportional to the sensitivity. Scan line I has a higher resolution than scan line II, but the sensitivity is lower in scan line I versus scan line II. Scan line II intersects a greater area under

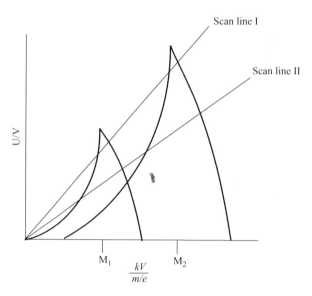

FIGURE 3.16   Scanning mass-to-charge ratios (m/e) with the quadrupole allowing the measurement of a range of masses, typically up to 4000 Da. As the fixed ratio of U and V are varied, the scan follows the scan line, and subsequent m/e values are recorded. The k value of the x axis represents an instrument calibration constant. Scan line I represents higher resolution and lower sensitivity than scan line II.

the curves for masses M_1 and M_2, thus allowing a greater proportion of the ions to transmit through the quadrupole and be detected, increasing the sensitivity of the instrument. The trade-off is that a greater amount of ion transmission coincides with a greater amount of energy spread within the ions that manifests itself as peak broadening, thus lowering the resolution of the quadrupole mass analyzer. It can be seen that the curves in the stability diagrams rise three times slower than they fall. Therefore, as the scan line is lowered, the frontal portion of the peak in the mass spectrum moves toward the lower mass value three times faster than the trailing part of the curve moves toward higher mass values. This results in a shift of the apex of the mass peak to a slightly lower mass value. Figure 3.17 shows the effect of peak shifting and broadening due to scan line I and scan line II.

The quadrupole mass analyzer possesses some unique advantages such as a high acceptance angle capability for the incoming ion particle beam (angle of incidence). As the ion particles enter the quadrupole, they are very effectively focused into a coherent path if they possess the chosen stable trajectory. In this way the quadrupole is highly tolerant to a wider range of incidence angle where the incoming ion particles can be off focus up to a certain critical angle. In relation to this the quadrupole also effectively focuses the position of the incoming ion particle incidence through the field oscillations (position of incidence). The quadrupole is also relatively insensitive

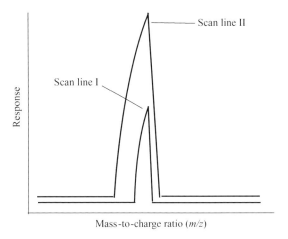

Response

Scan line II

Scan line I

Mass-to-charge ratio (m/z)

FIGURE 3.17 Result of resolution decrease through peak broadening and mass shift for scan line II versus scan line I.

to the spread in the kinetic energy of the incoming ion particle beam (kinetic energy spread). As stated earlier 5 eV is an optimal energy for ions to enter the quadrupole. The ability of the quadrupole to analyze nonideal beams successively increases with the diameter and the rod length of the electrodes. The quadrupole mass analyzer is also relatively easy to operate and calibrate, they are mechanically rugged, and they are relatively inexpensive to produce. Some limitations of the quadrupole are related to the imperfections in the quadrupole field. If the kinetic energy of the ion is too low, it may be lost and not pass through the quadrupole filter even though it is of the correct mass-to-charge ratio. If the kinetic energy of an ion is too high, there is the probability that it may pass through the quadrupole mass filter and be incorrectly detected even though it is not of the proper mass-to-charge ratio. This tends to increase the width of the peak and therefore reduce the resolution of the proper m/z peak. Quadrupoles can also be easily contaminated due to buildup near the front end of the quadrupole. Instruments often possess short rod prefilter quadrupoles directly before the main quadrupoles to decrease this effect and allow ease of cleaning. Quadrupoles are low-resolution mass analyzers typically possessing unit mass resolution at each mass ($1/\Delta m = 1$ for $R = m/\Delta m$). Finally, transmission of the ions that are formed at the source through the quadrupole mass filter is mass dependent, and there exists a slight biased discrimination against high mass ions.

3.6 TRIPLE QUADRUPOLE MASS ANALYZER (QQQ/MS)

In the preceding section a mass analyzer was described comprised of four cylindrical rods placed precisely parallel to one another, making up a quadrupole. One set of opposite poles have a dc voltage (U) supply connected while the other set of opposite poles of the quadrupole have a radio frequency (rf) voltage (V) connected. Ions are accelerated into the quadrupole by a small voltage of 5 eV, and under the influence of the combination of electric fields, the ions follow a complicated trajectory path. The single quadrupole setup can be used as a mass spectrometer scanning a range of m/z values (typically from m/z 1 to 4000) or can be used as a mass filter where all other ions besides the ion of interest will be given unstable trajectories and removed, thus filtering out all except for a single m/z species. Besides the single quadrupole mass filter, a combination of quadrupoles can be constructed in tandem to form a more complex mass spectrometer known as the triple quadrupole mass spectrometer. The triple quadrupole mass spectrometer is quite a versatile instrument in its ability for structural analysis of biomolecules. The alignment of the three quadrupoles in tandem allows a combination of unique processes where the quadrupoles are scanned or held static. Two of the quadrupoles, the first called Q1 and the third called Q3, have individual detectors (MS1 and MS2), while the middle quadrupole, Q2, is surrounded by a gas cell used for collision-induced product ion generation, and is RF only with no detector. Figure 3.18 shows the arrangement in space of the quadrupoles. Having the quadrupoles aligned in tandem allows approaches to compound isolation and subsequent identification by using the quadrupoles as mass filters. The mass spectrometry methods used utilizing the triple quadrupole in biomolecule analyses includes single-stage analysis (ESI-MS) where the first quadrupole (Q1) is scanned and the first detector (MS1) is used. This type of scanning is for the measurement of biomolecule molecular weights as mass-to-charge ratios (m/z). A second method used in biomolecule triple quadrupole mass spectrometry is neutral loss scanning. This is used to determine the initial biomolecule that losses a certain neutral mass upon fragmentation induced within the central quadrupole (Q2) collision cell. In this procedure the first and third quadrupoles are scanned at a differential equal to the neutral loss (NL) value. For example, if the NL value is 50, then when Q1 is

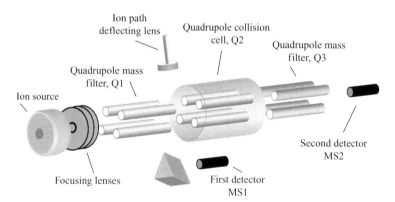

FIGURE 3.18 Illustration of arrangement of tandem quadrupoles in space making up a triple quadrupole mass spectrometer. The central quadrupole is contained within a floating gas cell used for fragmentation studies.

at m/z 200, Q3 is scanning at m/z 150. When m/z 150 is registered by the second detector MS2, the Q1 m/z value at that particular scan time is recorded along with the intensity recorded with detector MS2. A third method used is precursor ion scanning. This is used to determine the initial biomolecule that produces a specific product ion generated in Q2 and detected at MS2. Here Q3 is held static at the ion fragment m/z value of interest and Q1 is scanned over an m/z range. When the ion is detected by Q3, the m/z value in Q1 is recorded in conjunction with the intensity recorded by detector MS2. An example of precursor ion scanning is in the analysis of phosphatidylcholines where the protonated form of the phoshatidylcholine headgroup at m/z 184 is used. The biomolecules are scanned in Q1 and allowed one-by-one to pass through Q2 where collisions take place and the generation of product ions. During the collision activation in Q2 the phosphatidylcholine biomolecule allowed to pass through Q1 will lose the headgroup and generate the m/z 184 product ion. For the specific m/z value that represents the phosphatidylcholine biomolecule and was allowed to pass through Q1 will be associated with the event of the detection of m/z 184 by MS2, thus identifying that this particular species contains the phosphatidylcholine headgroup. Finally, a fourth method used is product ion scanning where a species is filtered and isolated in Q1, allowed to pass through Q2 with collisions generating product ions. The product ions that are generated in Q2 are scanned in Q3 and recorded by MS2. This type of scan is often used for structural information of biomolecules.

3.7 THREE-DIMENSIONAL QUADRUPOLE ION TRAP MASS ANALYZER (QIT/MS)

A second design utilizing the equipotential curves of the quadrupole field is the three-dimensional (3D) Paul quadrupole ion trap. This was a second mass analyzer invented and described by Wolfgang Paul[11] in addition to the quadrupole mass filter discussed in the preceding section (see Fig. 3.12 for design of quadrupole). Initially the ion trap mass analyzer was not utilized to the extent the quadrupole mass filter was due to its complexity in design and operation. The early quadrupole ion traps were operated in stability scan mode the same as the quadrupole mass filters are. This is where the amplitude of the dc and RF components of the ring electrode are ramped at a certain ratio, which would subsequently stabilize higher and higher mass-to-charge ratio ions for detection. In the design of the quadrupole ion trap a potential well is formed by wrapping a hyperboloidal electrode into a ring that has two electrodes on each side, creating a trapping space that is also of hyberboloidal shape. The basic design and construction of the quadrupole ion trap mass analyzer is illustrated in Figure 3.19. The two opposite electrodes, known as end caps, are characterized by passages or slits used for ion introduction into the trap (shown on the left) and for detection of ions exiting the trap (shown on the right). Probably the most striking ability that distinguishes the quadrupole ion trap from the quadrupole mass analyzer illustrated in Figure 3.12 is the capability of the quadrupole ion trap to perform multiple stages of product ion fragmentation as MS^{n}. The quadrupole ion trap has been able to achieve up to 12 stages of product ion fragmentation, MS^{12}.

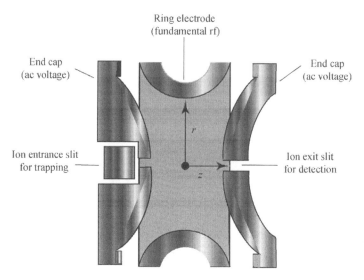

FIGURE 3.19 Basic construction of a quadrupole ion trap mass analyzer including the end cap electrodes and the ring electrode.

This can allow the investigation of structural aspects of compounds to a very exact and specific level.

An ideal quadrupole field is characterized according to the following relationship between the radius of the ring electrode and the distance to the end cap from the center of the trap:

$$r_0^2 = 2z_0^2 \tag{3.25}$$

where r_0 is the ring electrode radius and z_0 is the end cap electrodes' axial distance from the trap's center. Using this relationship when r_0 has been set then the distance of the placing of the end caps is predetermined. However, recent work has demonstrated that moving the end caps to a distance slightly greater than the relationship of Equation (3.25) dictates allows access to greater multiple fields, thus increasing the controlling of the efficiency of ion ejection and excitation while also increasing resolution. Typically, the size of a quadrupole ion trap used in most mass analyzers employs an r_0 value of 1.00 or 0.707 cm equating to a z_0 value of 0.707 or 0.595 cm, respectively.

The derivation of the useful form of the Mathieu equation for quadrupole ion trap mass analyzers is similar to the quadrupole discussed in the preceding section except that for the three-dimensional quadrupole ion trap the derivations for a_z and q_z are multiplied by a factor of 2, giving the following expressions:

$$a_z = \frac{-8eU}{m\omega^2 r_0^2} \qquad q_z = \frac{4eV}{m\omega^2 r_0^2} \tag{3.26}$$

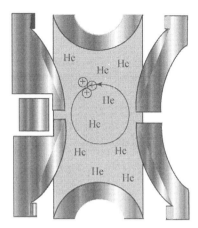

FIGURE 3.20 Simplified trajectory of ions trapped within the ion trap in the presence of helium bath gas.

The dimensionless parameter q_z describes the dependency of the motion of the trapped ion upon the parameters of the quadrupole ion trap, which includes the mass (m) and charge (e) of the ion, the ion trap radial size (r_0), the fundamental RF oscillating frequency (ω), and the voltage amplitude on the ring electrode (V). Related to q_z the parameter a_z in Equation (3.26) describes the dc potential effect upon the motion of the ions trapped within the trap when applied to the ring electrode. Ions are introduced into the trap through slits in one of the end caps. The operation of the trap usually includes the presence of a bath or dampening gas such as helium at a pressure of approximately 1 mTorr. The purpose of the bath gas is to induce low-energy collisions with the ions to thermally cool the ions. These dampening collisions will decrease the kinetic energy of the trapped ions. This has the effect of focusing the ion packets into tighter ion trajectories near the center of the trap. Figure 3.20 illustrates a simplified ion trajectory within the ion trap in the presence of helium bath gas. A second focusing of the ions toward the center of the trap is achieved through the fundamental rf voltage that is applied to the ring electrode. This is an oscillating voltage whose amplitude will determine the mass-to-charge range of ions that are trapped. The terminology of "fundamental" implies that there are other higher order frequencies, but they are not of practical use. In trapping operation and scanning for masses, the end caps are held at ground potential; thus the field that confines the ions is oscillatory only from the frequency of the fundamental rf ring electrode voltage. The method for scanning the ion trap for the mass-to-charge m/z species present is known as *mass-selective axial instability mode*. The amplitude of the radio frequency applied to the ring electrode is ramped in a linear fashion called an analytical scan. This will cause increasingly unstable trajectories of the ions from lower m/z values to higher m/z values, in other words lower m/z values that have trajectories closer to the center of the trap will be ejected first before larger m/z values, which have larger orbital trajectories. To store ions in the ion trap their trajectories must be stable in both the r and the z directions.

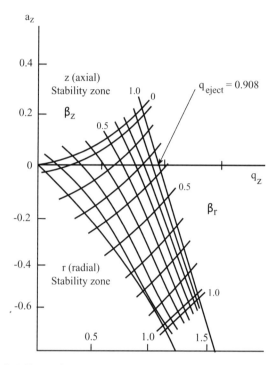

FIGURE 3.21   Stability region diagram for the radial and axial trajectories as a function of a_z and q_z.

The theoretical region of radial (r direction) and axial (z direction) stability overlap is illustrated in the stability diagram of Figure 3.21. Quadrupole ion trap mass analyzers are typically operated where the a_z parameter is equal to zero in the stability diagram of Figure 3.21. The region in the stability diagram where the z-axial stability zone and the r-radial stability zone intersect is known as the A region and is closest to the origin. There is a second z-stable/r-stable intersection zone known as region B, but we will not discuss this second one further. In the stability diagram of Figure 3.21 an ion will have a q_z value according to the amplitude of the rf voltage (V) applied to the ring electrode. If the ion's q_z value is less than the q_z value of 0.908, then the ion will have a stable secular motion. The motion of the ions in the trap is described by two secular frequencies that are axial (ω_z) and radial (ω_r). The ion trajectory within the quadrupole ion trap follows a figure 8 "saddle" path as illustrated in Figure 3.22. The sloping path is due to the quadrupole potential field surface, which resembles the form of a saddle. The small curving oscillations along the path are due to the oscillating quadrupole potential surface that has the effect of spinning the surface of the saddle. In Figure 3.21 the value of 0.908 for q_z at $a_z = 0$ is known as the low-mass cutoff (LMCO) and is the mass-to-charge lowest value that can be stored within the trap for a given rf voltage (V) applied to the ring electrode. In most quadrupole ion trap mass analyzers with $r = 1$ cm, the working applied rf voltage to the ring electrode

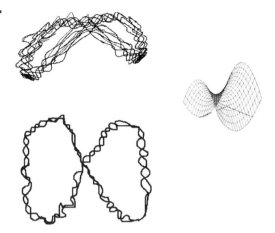

FIGURE 3.22 Trajectory path of the secular motion of an ion trapped within the quadrupole ion trap. The ion follows a figure 8 saddle path.

ranges from 0 to 7500 V due to limitations of high voltages and circuitry. This has a specific consequence, namely at 7500 V a mass-to-charge value of m/z 1500 will only have a q_z value of 0.404 (unitless) and thus cannot be ejected from the ion trap. This severely limits the mass range that can be trapped and subsequently measured by the quadrupole ion trap mass analyzer. A technique known as resonance ejection has been developed that allows the extension of the mass-to-charge range that can be measured by the quadrupole ion trap. Resonance ejection is obtained through the application of an ac voltage applied to the end caps of the quadrupole ion trap. Resonance of the ion is induced by applying an ac voltage (usually a few hundred millivolts) across the end caps and then adjusting the q_z value to match the secular frequency of the ion to the frequency of the applied ac voltage. This effectively uses the axial secular frequencies of the ions to induce resonant excitation. To scan the quadrupole ion trap for the mass-to-charge species trapped within the ion trap, the resonant excitation is set at a low applied resonance voltage and then ramped up to high resonance voltage, subsequently ejecting the ions from the trap. If a sufficiently large enough amplitude resonance signal is applied, the ions will be lost or ejected from the ion trap. Smaller amplitude resonance signals can be used to excite the ions to greater kinetic energy values that will produce more energetic collisions with the bath gas. This can be used to induce unimolecular dissociation reactions within the quadrupole ion trap for structural elucidation. Another advantage to using resonance excitation is the inversely proportional relationship between q_z and mass, which allows the ejection of larger molecular weight ions trapped within the quadrupole ion trap leaving lower molecular weight ions still trapped within the ion trap. This is known as resonance ejection and is used to isolate a specific mass-to-charge ion within the quadrupole ion trap. Forward and reverse voltage sweeps are performed on both sides of the ion of interest in order to isolate the ion by ejecting all other mass-to-charge species present. When ions are ejected and subsequently detected, one-half of the ions are collected through the ion exit slit in the end cap, allowing one-half of the ions trapped within the cell to be lost and not detected. This has the deleterious effect of reducing the sensitivity of the three-dimensional ion trap, a problem that has been eliminated with the introduction of the linear quadrupole ion trap covered in the next section.

3.8 LINEAR QUADRUPOLE ION TRAP MASS ANALYZER (LTQ/MS)

The linear ion trap is comprised of rod-shaped quadrupole electrodes similar to those used in the quadrupole mass analyzer and often placed with the same geometry in space as four oppositely opposed rods. Again, as discussed previously, the design of cylindrical rods is easier to manufacture as compared to rods with a truer hyperbolic shape. Ions in a linear quadrupole are confined by a two-dimensional (2D) rf field derived from the four electrodes and stopping potentials induced from two end caps. The rf field defines the radial motion of the trapped ions while the end caps define the axial motion. In contrast to the three-dimensional quadrupole ion traps, linear ion traps can store a greater amount of ions and the introduction of ions into the trap is more efficient. The ion density in the linear ion trap can be increased by increasing the length of the quadrupoles, making a longer ion trap, thus allowing the storage of a greater number of ions. In two-dimensional quadrupole fields the motion of ions has the same form of the Mathieu equation as the three-dimensional quadrupole ion trap [Equation (3.26)] except the directions are in the x-y plane:

$$a_x = -a_y = \frac{8eU}{m\omega^2 r_0^2} \qquad q_x = -q_y = \frac{4eV_{rf}}{m\omega^2 r_0^2} \qquad \xi = \frac{\omega t}{2} \qquad (3.27)$$

where m is the mass of the ion, e is the charge of the ion, r_0 is the ion trap radial size, ω is the RF oscillating frequency, and V_{rf} is the radio frequency voltage amplitude on the electrodes. Figure 3.23 shows the construction of the four quadrupole electrode rods aligned in space. The axial direction through the quadrupole is along the z axis, which is perpendicular to the plane of the page. The ions trapped within the quadrupole will have oscillations in the x-y direction as shown in the direction of ion movement in the graph. At the back of the poles there is a stopping plate aperture used to trap the ions axially. Usually, a second aperture is also placed at the front end of the quadrupole system acting as a stopping plate. In quadrupole fields the motion in the x direction is independent of the motion in the y direction. In higher multipoles such as hexapoles and octapoles the two directional movements are not independent of one another, and, therefore, it is not possible to construct stability diagrams. In a description of the construction and application of a two-dimensional quadrupole ion trap mass spectrometer, Schwartz et al.[12] show the use of quadrupole rods that have a hyperbolic rod design of the two-dimensional linear ion trap illustrated in the bottom of Figure 3.23. The radius of the field (r_0) has been set to 4 mm and each rod has been cut into 12-, 37-, and 12-mm-length axial sections. This helps to avoid problems with fringe field distortions to the resonance excitation and trapping fields. This quadrupole ion trap design also contains end plates on the front and back

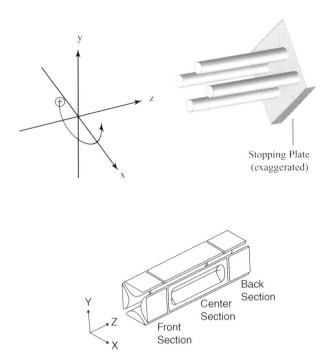

FIGURE 3.23 (*Top*) 3D arrangement of quadrupole electrodes with end-plate aperture for applying stopping potential. Plot shows basic motion of ions in the *x-y* plane. (*Bottom*) Illustration of the hyperbolic rod-shaped quadrupole used in the two-dimensional ion trap shown in Figure 3.24.

sections used for axial trapping (not showing in Fig. 3.23). For ion ejection from the trap a 0.25-mm-high slot was cut into the middle electrode.

A complete view of the two-dimensional ion trap mass spectrometer is illustrated in Figure 3.24. In the top of the figure we see the successive decrease in pressures moving from the electrospray source at ambient pressure (760 Torr) to the ion trap that is at 2.0×10^{-5} Torr. The ion beam produced from the electrospray process passes through a series of ion lenses and RF-only quadrupole and octopole systems to be finally introduced into the ion trap axially. The detector, located radially to the ion trap, is comprised of a conversion dynode and a channeltron electron multiplier.

The Mathieu equations of motion for the ions in the 2D quadrupole field are

$$\frac{d^2x}{d\xi^2} + (a_x - 2q_x \cos 2\xi)x = 0 \qquad (3.28)$$

$$\frac{d^2y}{d\xi^2} + (a_y - 2q_y \cos 2\xi)y = 0 \qquad (3.29)$$

The solutions to these equations give zones of stability and zones of instability, allowing the construction of a stability diagram as illustrated in Figure 3.25. The

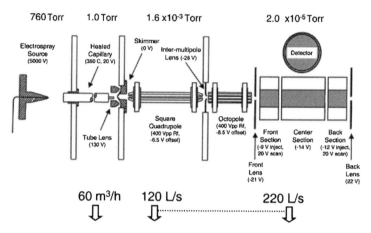

FIGURE 3.24   Configuration of two-dimensional ion trap including the associated pressures and voltages used. (Reprinted with permission. Schwartz J. C., Senko, M. W., and Syka, J. E. P. A two-dimensional quadrupole ion trap mass spectrometer. *J. Am. Soc. Mass Spectrom.* 2002, *13*, 659–669. Copyright Elsevier 2002.)

frequencies with which the ions will oscillate within the 2D linear quadrupole ion trap are given by

$$\omega_n = (2n + \beta)\frac{\omega}{2} \qquad 0 \le \beta \le 1 \qquad n = 0, \pm 1, \pm 2, \ldots \qquad (3.30)$$

where β is described as a function of a and q.

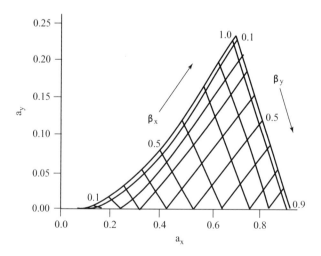

FIGURE 3.25   Stability diagram for the motion of ions within a 2D quadrupole ion trap.

As was observed in the 3D quadrupole ion trap, the presence of a bath gas within the linear quadrupole ion trap improved performance by way of enhancing the transmission of the ions axially through the end trapping aperture. This is due to the cooling of the ions through low-energy collisions with the neutral bath gas where the ion's kinetic energy is slightly dissipated through the bath gas collisions. The ions with lowered kinetic energy will focus closer to the middle of the quadrupole trap where the quadrupole field effective energy is lower. Since the ions are more focused at the center of the quadrupole, they can be transmitted through the end stopping plate aperture more efficiently for detection.

There are three methods that can be used to excite ions within the linear ion trap, including resonant excitation, dipole excitation, and quadrupole excitation. When using dipole excitation, an auxiliary voltage is applied to two opposite rods that is at a frequency that matches the motion of the ion to be excited. The amplitude of the ion's oscillation will increase with the application of the dipole excitation, which can be used for ion ejection or ion fragmentation studies using multiple collisions with the bath gas. The frequency used in dipole excitation is obtained using Equation (3.30) where n is usually set at zero ($n = 0$). With quadrupole excitation there is an excitation waveform that is applied to the electrodes. To scan a range of mass-to-charge values for ejection, the dipole or quadrupole excitation is set and the trapping rf voltage is changed to coincide with different ion frequencies for ejection. There are other approaches for ion excitation and ejection such as applying multiple frequencies as a frequency scan or chirp or using broadband waveforms. Finally, there are two designs for ion ejection, axial and radial. Axial ejection is obtained through an exit aperture leading to an electron multiplier for signal detection. Radial ejection is obtained through modified rod electrodes. Designs have included two detectors enabling the collection of all of the ejected ions greatly increasing sensitivity as compared to 3D quadrupole ion traps.

3.9 QUADRUPOLE TIME-OF-FLIGHT MASS ANALYZER (Q-TOF/MS)

A second variation of the TOF mass analyzer is the quadrupole time-of-flight mass analyzer that is a hybrid instrument. Mass spectrometers that are a coupling of two types of mass analyzers are known as hybrid mass spectrometers. This is different from the TOF instrument that contains an initial rf-only quadrupole that acts as an ion guide and not a mass filter. This type of instrumentation is often used with an electrospray ionization source for exact mass measurements of analyte ions. An internal standard is used to obtain a high level of accuracy in the determination of the mass-to-charge ratio of other ions within the mass spectrum, often with errors well below 10 ppm. This type of analysis is used for compound identification and molecular formula confirmation. Another source that has been successfully used for the rf-only quadrupole TOF mass analyzer is the atmospheric pressure MALDI source. A chief advancement with the design in this instrumentation lies in the orthogonal arrangement of the TOF mass analyzer to the source. Following the quadrupole or hexapole ion guide is a pulsed ion storage chamber where the ions momentarily reside before being introduced into the TOF mass analyzer. A pulsed extraction grid will draw the ions out of the ion storage chamber, imparting equal kinetic energy to all of the ions present. The ions will then be separated according to their mass-to-charge ratio drift times within the TOF mass analyzer. This configuration has made it possible to couple a quadrupole to a TOF mass analyzer. The kinetic energy of the ions can be controlled by the pulsed ion storage chamber and extraction grid design, allowing introduction into the TOF mass analyzer. A second benefit of this design is the ability of the orthogonal acceleration to effectively eliminate the nonuniform spatial distribution of the ions from the source. When the ions are pulsed out of the ion storage chamber into the TOF mass analyzer, the drift time is set to the zero starting point. The initial kinetic energy distribution imparted to the ions from the source is greatly reduced by the pre-TOF ion optics, skimmers, and the quadrupole. The spatial distribution produced by the ion storage chamber is the result of collimation. The TOF mass analyzer's reflectron will compensate for the kinetic energy distribution that was formed from the spatial distribution in the storage chamber when the ions were extracted. These instrumental reductions in spatial and kinetic energy distributions coupled with internal standard calibration allows the orthogonal acceleration design to achieve high resolutions that are used in high mass accuracy exact mass measurements (usually well below 10 ppm error).

In the quadrupole TOF hybrid mass analyzer (Q-TOF/MS) instrumental design, a fully functional quadrupole mass analyzer is placed in tandem to a TOF mass analyzer. This type of configuration allows sources at atmospheric pressure to be used such as an electrospray ionization source in place of a source that requires a vacuum such as a MALDI source. Second, a floating collision cell is also incorporated into the path of the ion particle beam that can be used for collision-induced product ion formation. The fully functional quadrupole TOF mass analyzer also incorporates the orthogonal acceleration design described above, allowing the reduction in spatial and kinetic energy distribution spread before introduction into the TOF mass analyzer, increasing resolution and mass accuracy.

A commercially available Q-TOF mass spectrometer is illustrated in Figure 3.26. The design of this mass spectrometer includes an electrospray source inlet that can be used to couple the instrument to liquid eluant, containing samples from high-performance liquid chromatography, capillary electrophoresis, or direct infusion from syringe pumps. The instrument includes a source inlet, which is most often electrospray (but can also be MALDI). The ions that are introduced into the Q-TOF will pass through an rf-dc quadrupole that is a fully functional mass analyzer. Notice that prior to the quadrupole mass analyzer there is also included a small quadrupole prefilter. The RF-dc quadrupole mass analyzer can be used to filter out all ions except an ion with a single m/z value that is allowed to pass through. The isolated m/z species can then be accelerated into the rf-only quadrupole with the floating gas collision cell for product ion formation. Recent Q-TOF designs also offer functionality that was primarily associated with triple quadrupoles such as neutral loss scanning by employing the abilities of the rf-dc quadrupole. A drawing of the inner components of the Q-TOF/MS and their arrangement in space is illustrated in

FIGURE 3.26 Waters Micromass Q-TOF Premier. The instrumental setup illustrated in the picture includes an electrospray source inlet used for coupling to liquid eluants such as from high-performance liquid chromatography, capillary electrophoresis, or direct infusion using a syringe pump. For a more detailed description of the inner workings of the Q-TOF see Figure 3.27. (Photo provided courtesy of Waters.)

Figure 3.27. Following the floating gas collision cell is the orthogonal ion deflector. The deflector is used to pulse the ions into the TOF drift tube where the ions are allowed to separate according to their m/z associated drift times. A reflection electrostatic mirror is also included in the path of the ion, helping to increase measured resolutions. Typical detectors used are conversion dynodes and channel electron multipliers.

3.10 FOURIER TRANSFORM ION CYCLOTRON RESONANCE MASS ANALYZER (FTICR/MS)

3.10.1 Introduction

Out of all the mass analyzers the Fourier transform ion cyclotron resonance mass analyzer (FTICR/MS) has the ability to achieve mass resolutions and mass accuracies that are unsurpassed by all other mass analyzers. Routinely, mass accuracies of <5 ppm are obtained with resolutions as high as 1 million, using Fourier transform ion cyclotron resonance mass spectrometry. FTICR mass spectrometers are trapping mass analyzers that use the phenomenon of ion cyclotron resonance in the presence of a homogenous, static magnetic field. When an ionized particle enters a strong magnetic field, it will undergo a circular motion that is perpendicular to the magnetic field lines known as cyclotron motion. The cyclotron motion that the ions exhibit

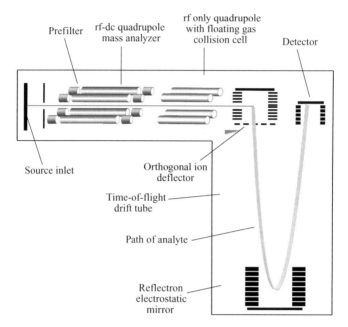

FIGURE 3.27 Schematic diagram of the internal components of a Q-TOF mass spectrometer. Important components include the quadrupole mass filter, the floating collision cell, the orthogonal ion deflector, the time-of-flight drift tube, the reflectron electrostatic mirror, and the detector.

has a resonance frequency that is specific to the ions' mass-to-charge ratio (m/z). Therefore, mass analysis is achieved in FTICR mass analyzers by detecting the cyclotron frequencies of the trapped ions, which are specifically unique to each m/z value. Unlike the quadrupole, time-of-flight, and magnetic sector mass analyzers, which perform a separate spatial ion formation, separation, and detection according to their mass-to-charge ratio, the FTICR mass analyzer can perform a temporal ion formation, separation, and subsequent ion detection all within the ICR cell. All of the ions trapped within the cell undergo a simultaneous time-domain cyclotron motion that allows the measurement of the signals produced by each mass-to-charge ratio present without the need for scanning. The need for scanning a mass analyzer reduces the amount of signal that can be measured. In the FTICR mass analyzer the entire signal produced by the resonating ions is measured at the same time, thus increasing the signal detection and therefore increasing the sensitivity. By keeping the amount of charge in the ion cyclotron cell at an optimum, usually less than 10^7 to ensure that spatial charge interactions are minimized, the ability to measure the signal without scanning allows the FTICR mass analyzer to achieve a high degree of sensitivity. Permanent magnets and electromagnets are typically not used due to the low field strengths obtainable, usually less than 2 teslas (T). In general superconducting magnets are used that range in magnetic field strengths from 3.5, 7, 9.4, to 11.5 T where a 20-T magnet has also been constructed, but such high strengths are not often used. It will be shown later that performance of the FTICR mass analyzer is directly proportional to the magnetic field strength.

3.10.2 FTICR Mass Analyzer

The basic components of an FTICR mass analyzer include: (1) a magnet capable of producing a stable and uniform magnetic field that is constant in space distribution and in time, (2) a trapping cell located within the middle of the magnet where the ions are measured (can be cubicle, cylindrical, or other shapes), (3) an extremely high vacuum system capable of levels down to 10^{-9} or 10^{-10} Torr (required for high-resolution measurements as will be discussed later), and a data system that is capable of collecting and storing large amounts of signal responses required for high-resolution mass measurements. Figure 3.28 illustrates the basic components of an FTICR mass spectrometer. Though ions can be formed within the FTICR cell itself by processes such as electron ionization (EI) or photoionization (PI), the standard technique is to produce the ions using an external source such as MALDI or ESI prior to injection into the cell. Often there is an ion trap in tandem with the FTICR cell located between the cell and the source where the source-generated ions can be

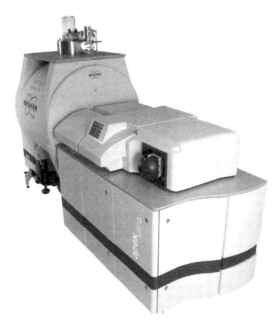

FIGURE 3.28 Bruker Daltonics Apex Ultra™ Fourier transform ion cyclotron mass spectrometer (FTICR-MS). The large circular portion of the mass spectrometer is the shielded magnet, which comes in 7.0, 9.4, 12, or even 15 tesla magnets that are actively shielded. The ion cyclotron cell is located within the central bore of the magnet. The front, right portion of the mass spectrometer houses an inlet source and a trapping quadrupole for ion accumulation prior to entry into the ICR cell. (Photo provided courtesy of Bruker Daltonics.)

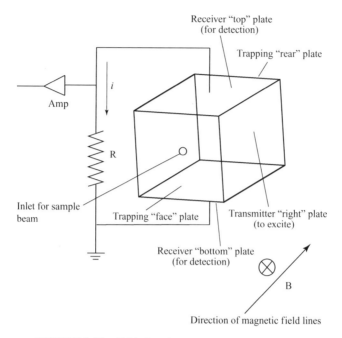

FIGURE 3.29   Cubic Penning trap design of an ICR cell.

collected and accumulated. Once a sufficient amount of ions have been accumulated, they are guided into the FTICR cell by ion optics for mass analysis. Figure 3.29 illustrates the basic design of a cubic Penning trap FTICR cell, which is useful in understanding the components of the cell used for mass analysis. Cubic cells fit well within the shape of the lower strength permanent and electromagnetic magnets; however, for the higher field strength superconducting magnets a cylindrical shaped cell fits more optimally within the central bore of the magnet. To trap the ions within the cell an electrostatic field is produced by two opposing plates where the direction of the electrostatic field is parallel to the magnetic field. The cylindrical cell functions similarly to the cubic cell where two outer cylinders are used for trapping the ions while the inner cylinder is comprised of four electrodes used for excitation and signal measurement.

3.10.3   FTICR Trapped Ion Behavior

The trapped ions within the cell have a force imposed upon them by the strong magnetic field. This is described by Newton's force equation:

$$\text{Force} = \text{mass} \times \text{acceleration} \tag{3.31}$$

$$\mathbf{F} = m \frac{d\mathbf{v}}{dt} \tag{3.32}$$

$$= q\mathbf{v} \otimes \mathbf{B} \tag{3.33}$$

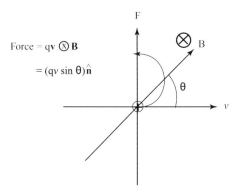

FIGURE 3.30   Circular path of an ion within a strong magnetic field.

where q is the charge of the ion, v is the velocity of the ion, and B is the magnetic field strength. The magnetic component of the Lorentz force (using the right-hand rule, the force on the ion is both perpendicular to the magnetic field and the direction of its velocity) will cause the circular path of the ion shown in Figure 3.30 (positive ion). The circular path is in the x, y plane and perpendicular to the z plane. The magnetic field lines are toward the negative direction (opposed to) of the z axis (into the plane of the paper). Note that the path of a negative ion will be in the opposite direction but will experience the exact same force magnitude.

For the magnetic force equation:

$$\text{Force} = q\mathbf{v} \otimes \mathbf{B} \tag{3.34}$$

due to the definition of the cross product the magnetic force is in the direction given by

$$q\mathbf{v} \otimes \mathbf{B} = (qvB\sin\theta)\hat{\mathbf{n}} \tag{3.35}$$

where $\hat{\mathbf{n}}$ is a unit vector that is perpendicular to both B and v. If $\theta = 90°$ (the velocity vector is perpendicular to the magnetic field lines), then magnetic force is

$$F = qvB \tag{3.36}$$

From this relationship we see that if the direction of the velocity vector is parallel with the magnetic field, then $\theta = 0$ and $\sin\theta = 0$. This demonstrates that there is no magnetic force exerted upon a charged particle that is moving axially with the magnetic field and will therefore be lost from the ICR cell. This is why a trapping electric field is also applied in conjunction with the imposed magnetic field to the ICR cells whose combination produces a three-dimensional ion trap. The electric and magnetic forces from the applied field can be added as vectors; therefore, the net force exerted upon the ion is described by the Lorentz force law:

$$\mathbf{F} = q(\mathbf{E} + \mathbf{v} \otimes \mathbf{B}) \tag{3.37}$$

The imposed strong magnetic field induces the ions to precess or move in a circular orbit at cyclotron frequencies that are uniquely characteristic to each mass-to-charge (m/z) ratio. The cyclotron frequency may be derived from Newton's second law:

$$\sum F = ma \tag{3.38}$$

$$qvB = m\left(\frac{v^2}{R}\right) \tag{3.39}$$

where circular motion velocity can be described as

$$v = 2\pi\omega_c R \tag{3.40}$$

and substituting into Equation (3.39):

$$q(2\pi\omega_c R)B = m\left[\frac{(2\pi\omega_c R)^2}{R}\right] \tag{3.41}$$

$$qB = 2\pi m\omega_c \tag{3.42}$$

$$\omega_c = \frac{qB}{2\pi m} \tag{3.43}$$

Equation (3.43) demonstrates that smaller ions are moving with a greater cyclotron velocity as compared to larger ions, which precess with slower cyclotron velocity. Also demonstrated by Equation (3.43) is that the cyclotron frequency for ions of the same mass-to-charge ratio is independent from the initial position or velocity of the ions. Because the cyclotron frequency is independent from the initial velocity, it is independent from the initial kinetic energy of the ion. This is one direct reason why FTICR mass spectrometry is able to achieve such high mass resolutions where other mass analyzers are required to focus the translational energy of the ions to achieve resolution of the m/z values. The FTICR independence from initial position is also in contrast to the MALDI-TOF mass spectrometer where initial spatial distribution affects resolution, and the quadrupole mass analyzer where a critical angle exists for introduction and transmission of the ion through the quadrupole.

The ions are generally trapped with low kinetic energies so their cyclotron radius are low, resulting in orbits near the middle or center of the ICR cell, thus making the detection of their cyclotron frequencies unavailable. As illustrated in Figure 3.29, the ICR cell design includes two opposing plates that are used as rf transmitter plates, which excite the m/z ion packets as they move in their circular orbits. The excitation of the ion packets is performed by applying an oscillating electric field that has the same frequency as the motion of the m/z ion packet. By keeping the applied rf oscillating electric field in resonance with the ion packet, the m/z ions in resonance with the imposed field will absorb energy in the form of kinetic energy. The frequency of motion will not change as illustrated in Equation (3.43); however, the velocity of the ions will increase according to the resonant rf electric field. Taking Equation (3.39) and rearranging, we obtain

$$qvB = m\left(\frac{v^2}{R}\right) \tag{3.39}$$

$$R = \frac{mv}{qB} \tag{3.44}$$

and by substituting in the relationship between kinetic energy (eV) and mv:

$$KE = eV = \tfrac{1}{2}mv^2 \tag{3.45}$$

$$mv = (2eVm)^{1/2} \tag{3.46}$$

we obtain

$$R = \frac{(2eVm)^{1/2}}{qB} \tag{3.47}$$

The relationship illustrated in Equation (3.47) demonstrates that the radius (R) of the ion's cyclotron rotation will increase with increasing kinetic energy (eV) as imposed by the resonant rf electric field. However, the cyclotron frequency of resonant motion will remain the same, also indicating that the velocity of motion has increased. Finally, Equation (3.47) can be presented as

$$R = \frac{V_P t}{2dB} \tag{3.48}$$

where R is the cyclotron radius, V_P is the amplitude of the rf excitation signal, t is the time duration of the rf pulse, and d is the distance between the transmitter plates. Equation (3.48) demonstrates that the radius of the ions' cyclotron motion after excitation is independent of the mass-to-charge ratios. This means that there is no mass discrimination in the excitation of the ions into greater orbital radius.

Excitation of the m/z ion packet is performed to (1) drive the ions away from the center of the ICR cell and subsequently closer to the plate boundary of the ICR cell for signal measurement, (2) to drive ions to a larger cyclotron radius than the constrained dimensions of the ICR cell to remove (eject) them from the cell, and (3) to add enough internal energy in the form of translational kinetic energy for dissociative collision product ion generation and ion–molecule reactions. In Figure 3.29 there are two opposed plates that act as signal receiver plates used for the detection of the ion packets trapped within the ICR cell. Figure 3.31 illustrates the processes of ion excitation resulting in increased cyclotron rotation radius and ion detection. As the ion packet moves through its cyclotron rotation, it will approach one of the receiver

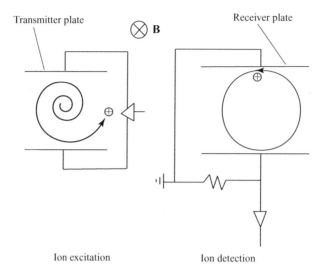

Transmitter plate   Receiver plate

Ion excitation   Ion detection

FIGURE 3.31   (*Left*) Ions are excited to a larger cyclotron rotation radius through a resonant rf electric field induced by the transmitter plates. (*Right*) Ions are detected by the receiver plates acting as electrodes.

plates and induce a current in the form of a flow of electrons through the resistor to the plate (this is in positive ion mode where the ions in the ion packet are positively charged). Subsequently, as the ion packet moves toward the opposing receiver plate, it will induce a flow of electrons through the resistor, producing a current in the form of electron buildup. The excited ions coherent motion will produce a transient signal, which is called an image current in the two opposing receiver plates. This in turn produces a time-dependent waveform by the excited ions, which are resonating at a specific cyclotron frequency where the ion abundances are directly proportional to the magnitude of the frequency components.

A graphical plot of an image (transient) current for a group of ions present in the ICR cell is illustrated in the bottom of Figure 3.32. The top graph of Figure 3.32 depicts the rf burst used to excite the ions trapped within the ICR cell to larger radius cyclotron orbits (unchanged cyclotron frequencies). The top of Figure 3.32 shows a burst of 7 ms where the excitation frequency is scanned from 70 kHz to 3.6 MHz. The bottom of Figure 3.32 shows the transient image current produced by the excited ion packets within the ICR cell. In the excitation process a single frequency can also be applied in the rf excitation process where only a single m/q ion packet, which is in resonant with the rf frequency will be excited to a larger radius cyclotron orbit. A range of frequencies can also be swept to enable broadband detection of many different m/q ions present in the ICR cell by using a rapid frequency sweep known as an rf chirp. This is what is illustrated in the bottom plot of Figure 3.32, a composite of different sinusoids of varying frequencies and amplitudes. With time the magnitude of the cyclotron frequency amplitudes are observed to steadily decrease as is illustrated

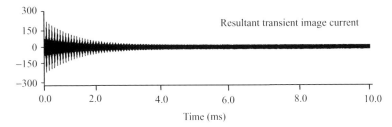

FIGURE 3.32 (*Top*) Excitation rf burst to excite trapped ions within the ICR cell to larger radius cyclotron orbits. (*Bottom*) Resultant transient image current produced by resonant cyclotron frequencies of ion packets trapped and excited within the ICR cell.

in the composite transient signal in the bottom of Figure 3.32. This is due to collisions taking place between the trapped ions and between the trapped ions with neutrals. This is a process called collision-mediated radial diffusion where the coherence of the ion packets is reduced. Collisions will drive the ions toward the outer dimensions of the cell where they will be lost to neutralizing collisions with the cell walls. Optimal conditions require high vacuum in the order of 10^{-9} Torr where collisions are greatly reduced. To better understand what the transient image represents and therefore what the transient image contains, we will look at a single-frequency rf excitation of the ICR cell followed by the addition of many rf frequencies through an rf chirp. Figure 3.33 shows the resultant transient plot for a single m/z ion trapped within the ICR cell after coherent rf excitation by the transmitter plate. The first plot shows the sinusoidal behavior of the cyclotron frequency (ω_c) of the trapped m/z ion. The magnitude of the ω_c amplitude equates to the charge density of the orbiting ion packet that is proportional to the number of ions present and therefore proportional to the ion concentration within the cell. Using the technique of Fourier transform, the time-domain transient signal can be converted to a frequency-domain plot as illustrated in the right half of Figure 3.33. With calibration of the mass analyzer the frequency domain can be converted to a mass spectrum as illustrated in the bottom of Figure 3.33.

In continuation of the demonstration in the construction of the composite transient signal illustrated in Figure 3.32, Figure 3.34 represents the measurement of two orbiting ion packets within the ICR cell that have undergone a coherent rf excitation from the cell's transmitter plates. The transient signal for M_1 indicates an ion packet with a greater (higher) cyclotron frequency equating to a lower mass as compared to M_2. The ion packet represented by M_2 has a lower (less) cyclotron frequency value indicating a higher mass ion packet as compared to M_1. The right transient

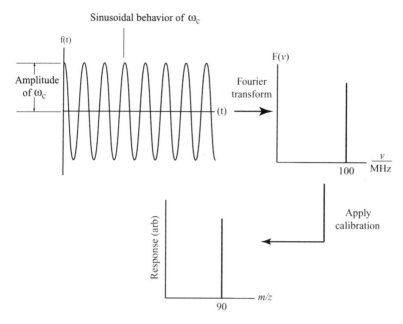

FIGURE 3.33 Progression of the production of a mass spectrum from transient signal measurement. Top left is the transient signal recorded for a single ion trapped within the cell. Fourier transform is applied to change from a time domain to a frequency domain as plotted to the right. A calibration is applied that allows the conversion of the frequency domain plot to be converted to a mass spectrum as illustrated by the bottom mass spectrum.

plot in Figure 3.34 illustrates the effect of the combination of the simultaneous detection of the cyclotron frequencies of M_1 and M_2. We now begin to get a feel for the complexity that is illustrated in the composite transient signal of Figure 3.32. Figure 3.35 ties together all of the aspects discussed in Figures 3.32–3.34. In the top of Figure 3.35 there is the illustration of the construction of the complex composite transient signal plot of the cyclotron frequencies for a large number of orbiting ion packets $M_1 + M_2 + \cdots + M_n$. Next Fourier transform is applied to the time-domain transient signal converting it to a frequency-domain signal containing multiple signals at specific frequencies. A calibration is applied to the frequency-domain signals, converting them to their respective mass-to-charge ratio values, which are used to construct the multiple peak mass spectrum illustrated in the bottom of Figure 3.35. The mass spectral representation of the orbiting ion packets within the ICR cell is the typical type of plot used when performing experiments using the FTICR mass analyzer.

3.10.4 Cyclotron and Magnetron Ion Motion

Within the ICR cell the ions undergo three types of motion that consist of cyclotron motion due to the magnetic field, trapping motion due to the electric field produced

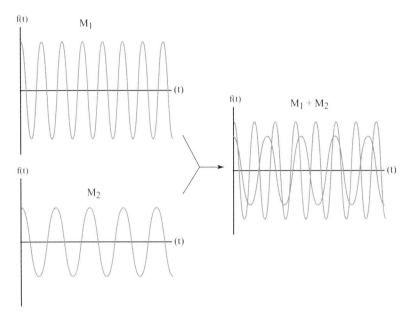

FIGURE 3.34 Illustration of the transient signal composite of the two orbiting ion packets M_1 and M_2. As illustrated by the cyclotron frequencies the transient signal for M_1 indicates an ion packet with a greater (higher) cyclotron frequency equating to a lower mass as compared to M_2. The ion packet represented by M_2 has a lower (less) cyclotron frequency value indicating a higher mass ion packet as compared to M_1.

by the trapping plates, and magnetron motion, which is attributed to a combination of the magnetic and electric fields acting upon the trapped ions. This is illustrated in the Lorentz force law of Equation (3.37) where the total force acting upon the ions is the electric field (E) times the charge (q) upon the ion plus the cross-product relationship between the magnetic field (B) and the ion's velocity. As presented earlier the magnetic field induces a circular motion of the ions in an orbit that is perpendicular to the direction of the magnetic field. In the z-axis direction (direction of the magnetic field) there is no force acting upon the ion by the magnetic field. Therefore, an electric field is produced by two opposite trapping plates located perpendicular to the magnetic field to confine the ions within the ICR cell. A positive trapping potential is applied to the plates to trap positive ions, and a negative potential is applied to trap negative ions. This produces an oscillation motion for the ions back and forth between the trapping plates that follows simple harmonic motion. The combination of the two fields, however, produces a third magnetron motion that is a circular motion that effectively follows the contours of one of the isopotential curves of the electric field. The combination of cyclotron and magnetron motion is illustrated in Figure 3.36 where the larger overall motion is due to the cyclotron orbit with the smaller tube producing motion due to the magnetron orbit. A closer look at the ion motion within the trap reveals that the ion's circular oscillation is actually a complex composition of

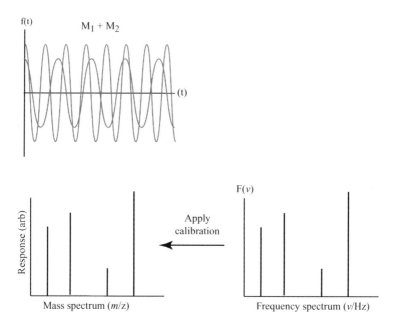

FIGURE 3.35 Construction of the complex composite transient signal plot of the cyclotron frequencies for a large number of orbiting ion packets $M_1 + M_2 + \cdots + M_n$. Fourier transform is applied to the time-domain transient signal converting it to a frequency domain signal containing multiple signals at specific frequencies. A calibration is applied to the frequency-domain signals converting them to their respective mass-to-charge ratio values used to construct the multiple peak mass spectrum.

three difficult motions whose frequencies we can describe and compare. These three motions are comprised of an axial motion (z), and two radial motions of magnetron ($-$) and cyclotron ($+$). The frequencies of these oscillatory motions are

$$\omega_z^2 = \frac{qV_0}{md^2} \quad \text{(axial)} \tag{3.49}$$

$$\omega_\pm = \frac{\omega_c}{2} \pm \sqrt{\frac{\omega_c^2}{4} - \frac{\omega_z^2}{2}} \quad [\text{radial}, (+)\,\text{cyclotron}, (-)\,\text{magnetron}] \tag{3.50}$$

where ω_c is the cyclotron frequency in Equation (3.43), V_0 is the depth of the ICR trap, m is the mass of the ion, q is the charge, and d is a trap characteristic parameter. Figure 3.37 illustrates the three motions that make up the total cyclotron motion of the ion within the ICR cell. Space charge effects can also influence the motion of the ions within the ICR cell. When ions are accumulated above the trapping optimization of the ICR cell space, charge effects can broaden the ion packet, producing a wider peak shape of the signal. At higher ion charge densities peak coalescence can occur where the coherent oscillation of closely spaced ion packets becomes joined into

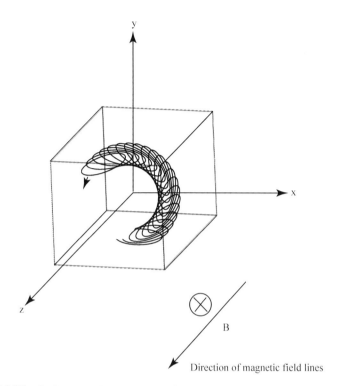

FIGURE 3.36   Cyclotron and magnetron motion of the ions within the ICR cell. The larger overall circular motion is due to the cyclotron orbit and the smaller circular tube producing motion is due to the magnetron orbit.

a single cyclotron orbit. To obtain high resolution a transient signal often must be collected for relatively long periods of time, thus allowing the peak coalescence to occur within this time frame. Peak coalescence can often be controlled by reducing the charge density within the ICR cell along with adjustment of other parameters such as the trapping voltage. A breakdown of the three components of the cyclotron and magnetron motion of the ions within the ICR cell is illustrated in Figure 3.37.

3.10.5   Basic Experimental Sequence

The basic experimental sequence using FTICR mass spectrometry is different from other mass spectrometers where the sequence of events happen in time instead of in space. The sequence is basically comprised of cleaning out the ICR cell by removing all of the ions present, an ionization process that can be either internal or external, excitation of the ions and then subsequent detection of the ions. The purging of the ICR cell in preparation for introduction of new ions into the cell is usually accomplished by applying an asymmetric potential to the opposed trapping plates. One trapping plate will be given a positive potential while the opposite trapping plate possesses

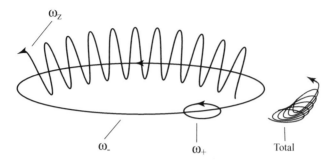

FIGURE 3.37 Breakdown of the three components of the cyclotron and magnetron motion of the ions within the ICR cell. The three motions are comprised of an axial motion (z), and two radial motions of magnetron ($-$) and cyclotron ($+$).

a negative potential. The asymmetric potential within the ICR cell causes the ions to exit along the z axis, thus ejecting all the ions within the ICR cell within a very short period of time. Once the ICR cell has been purged of ions, a new set of ions is introduced into the cell. In earlier work, electron ionization (EI) was used to form ions directly within the cell by passing a stream of electrons through the cell, producing odd electron molecular ions. Recently, the application of external ionization sources such as electrospray ionization (ESI) or matrix-assisted laser desorption ionization (MALDI) have been widely used in FTICR mass spectrometry experiments. Often, preceding the ICR cell will be an ion trap that allows accumulation of ions prior to introduction into the ICR cell when using ionization techniques such as electrospray. When MALDI is used, the source is usually also under high vacuum, and the generated ions are guided into the ICR cell with lenses and ion guides. Figure 3.38 illustrates the steps involved in the experimental sequence. On the left the ions are purged from the cell, essentially leaving the cell empty of ions and neutrals, which have been removed by high vacuum. The cell is then filled with ions that are either generated within the cell or externally. The ions are then excited to higher orbit radius and detected as shown in the right of Figure 3.38.

The next process to take place is the input of kinetic energy into the orbiting ion packets to increase the cyclotron orbit radius of the ions. Initially, the ions within the ICR cell usually have low kinetic energy and low cyclotron orbital radius. This is the case for both externally ionized species and internally ionized species. The ions are orbiting near the center of the ICR cell, which is to far away from the receiver plates to generate the transient signal used for mass detection and analysis. To excite the ions, a sinusoidal voltage is applied to the transmitter plates whose frequency is in resonance with the cyclotron frequency of an orbiting ion packet. By scanning a frequency range, all ions present within a certain mass-to-charge ratio range that is constrained by the complete FTICR mass analyzer system will be excited to a greater orbital radius, thus allowing detection of the ion packets present within the ICR cell.

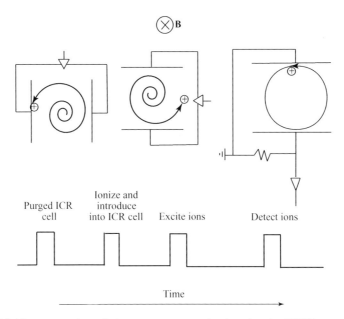

FIGURE 3.38   Progression of the event sequence in time for the FTICR mass analyzer. Purging of the ICR cell in preparation for introduction of new ions into the cell, ionization of species, input of kinetic energy into the orbiting ion packets to increase the cyclotron orbit radius of the ions, and lastly the orbiting ions are detected.

Finally, as shown in the right cell of Figure 3.38 the orbiting ions are detected by the transient electric current induced in the receiver plates. The mass range that can be detected using FTICR mass spectrometry, typically up to 10,000 m/z, is directly dependent upon the magnetic field strength and constrained by the dimensions of the ICR cell. The size of the ICR cell will physically allow only so great of a cyclotron radius diameter, however, in practical use the largest radius diameter is smaller than the upper limit of the cell dimension. Typically, ICR cells are usually a few centimeters in length upon each side.

As a consequence of the relationships derived from Equation (3.43) we see that many aspects of the functionality of the FTICR mass spectrometer are directly influenced by the magnetic field strength B. For example, the maximum amount of ions that can be trapped within the ICR cell is proportional to the square of the magnetic field (B^2). Increasing the concentration of ions within the ICR cell without deleterious space charge effects will directly increase the sensitivity of the mass analyzer. Another important relationship is the increase in maximum ion kinetic energy with increasing magnetic field strength (B^2). In the relationship of increasing energy with increasing magnetic field strength, the experimenter can put more energy into the orbiting ions before they are removed by cell wall neutralization. This gives access to higher collision energies that can be used to utilize fragmentation pathways that require larger activation energies. Finally, when using a higher magnetic field strength the radius of the orbiting ion packet will be smaller for the same kinetic energy as compared to a lower field strength. This allows the trapping of a larger mass-to-charge range and increases the resolution and dynamic range of the mass analyzer.

3.11   LINEAR ION TRAP FOURIER TRANSFORM MASS ANALYZER (LTQ-FT/MS)

Increasingly in today's research, industrial method development, and pharmaceutical drug discovery, there is a need to couple separation science as a front-end analytical methodology to mass spectrometric instrumentation that is subsequently used as an extremely versatile detector. The front-end separation science used today is often high-performance liquid chromatography (HPLC) but can also be gas chromatography (GC) or capillary zone electrophoresis (CZE). The HPLC is used to separate complex mixtures of analytes prior to their introduction into the mass spectrometer. A recent and important example of this is in the field of proteomics where a set of proteins isolated from a biological system (the proteome) such as a culture of cells, tissue, or bodily fluids is digested into peptides, creating an extremely complex mixture of molecules. For example, a cell's proteome may contain as a hypothetical estimate 3000 proteins that when digested to the peptide level turn into 10,000 individual peptides. It is utterly impossible to infuse this complex mixture directly into the mass spectrometric instrumentation for analysis to any reasonable extent of identification of many of the individual species, often only the most abundant species present will be observed obscuring most of the other peptides present. Therefore, HPLC is used in proteomic studies to chromatographically separate the peptides into less complex mixtures that can be detected and identified using mass spectrometry. There has also grown an ever-increasing need for the mass spectrometer to possess the ability to scan the incoming analytes in a rate that is comparable to the complexity and speed of the analytes eluting from the HPLC column. A mass spectrometer that samples an ion beam such as a triple quadrupole mass analyzer must use the level of the analyte within the ion beam for its mass measurement and product ion experiments. The ion trap on the other hand does have the ability to trap and accumulate the incoming analytes from the inlet ion stream, thus allowing the enrichment of the sample (increased sensitivity and detection). The relatively recent introduction of the linear ion trap has also allowed a much faster scan rate of the trapped ions as compared to the three-dimensional quadrupole. While the dynamic range and fragmentation efficiency of these instrumentation are quite good, these instruments possess a very low resolution power, often confined to unit mass resolution with no better than a 20-ppm mass accuracy. The advent and introduction of the orthogonal quadrupole time-of-flight hybrid mass analyzer did allow much higher resolution and mass accuracy, but the instrumentation is still sampling an ion beam where analytes are mass filtered in the first-stage quadrupole and measured in the second-stage time-of-flight mass analyzer. This arrangement suffers from low transmission of the ions and a limited dynamic range for mass accuracy measurements.

FIGURE 3.39 Linear ion trap Fourier transform ion cyclotron resonance mass spectrometer (LTQ-FTICR/MS) developed and designed by Syka et al. at the University of Virginia. (Reprinted with permission from Syka et al., *J. Proteome Res.* 2004, *3*, 621–626. Copyright 2004 American Chemical Society.)

The next hybrid mass spectrometer that has been constructed is the combination of the high scan rate and trapping ability of the linear ion trap mass spectrometer (LTQ-IT/MS) with the ultrahigh resolution Fourier transform ion cyclotron resonance mass analyzer (FTICR/MS). With this hybrid the linear ion trap accumulates ions externally to the ICR cell and can isolate and introduce a single analyte to the ICR cell or fragment an analyte and transmit the product ions into the ICR cell. The linear ion trap also has the ability for MS^n fragmentation studies. The FTICR mass spectrometer possesses the highest obtainable mass resolution (routinely up to 1 million), thus allowing routine mass accuracies in the 1–2 ppm range with a relatively broad dynamic range.

The first hybrid LTQ-FTICR mass spectrometer was designed and reported by Syka et al.[13] at the University of Virginia and is illustrated in Figures 3.39 and 3.40. Figure 3.39 is a picture of the LTQ-FTICR mass spectrometer where the linear ion trap is located in the front of the system and the superconducting magnet is the large cylindrically enclosed apparatus in the back of the picture. In this system the superconducting magnet has a 3-T magnetic field strength. An illustration of the ion optics and trapping cells used in the construction of the mass spectrometer is illustrated in Figure 3.40. The sample is introduced into the mass spectrometric instrumentation through the inlet located to the far left of the drawing. The analytes will pass through an inlet orifice into a lower pressure region under vacuum and guided into the linear quadrupole ion trap by ion guide lenses and an rf-only quadrupole. The analytes enter the linear quadrupole ion trap where they are trapped and accumulated. Here the ions can be isolated and fragmented for structural information and elucidation. Beyond the linear quadrupole ion trap are a series of ion guide lenses and RF-only quadrupoles used to guide the ions from the linear quadrupole ion trap into the Penning trap of the FTICR mass spectrometer.

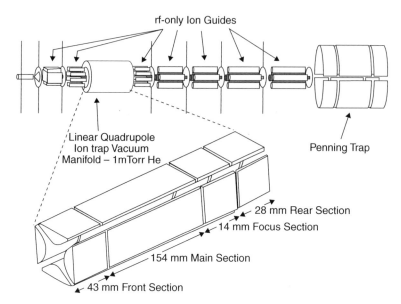

FIGURE 3.40 Ion optics and trapping cell used in the construction of the mass spectrometer. The superconducting magnet has a 3-T magnetic field strength. (Reprinted with permission from Syka et al., *J. Proteome Res.* 2004, *3*, 621–626. Copyright 2004 American Chemical Society.)

The sensitivity of the system was demonstrated at 550 zmol (10^{-21}) of angiotensin I and is illustrated in Figure 3.41. The sample was analyzed by nano-HPLC and both a single stage and a product ion spectrum were recorded. The dynamic range of the system was demonstrated at 4000 : 1 with routine mass accuracies of 1–2 ppm.

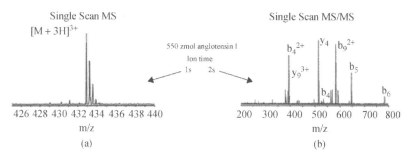

FIGURE 3.41 Sample levels for the detection and sequence analysis of peptides on the prototype QLT/FTMS instrument: (*a*) MS spectrum recorded on 550 zmol (550×10^{-21} mol) of angiotensin 1, (*b*) MS/MS spectrum recorded on $(M + 3H)^{+3}$ ions generated from angiotensin 1 at the 550-zmol sample level. (Reprinted with permission from Syka et al., *J. Proteome Res.* 2004, *3*, 621–626. Copyright 2004 American Chemical Society.)

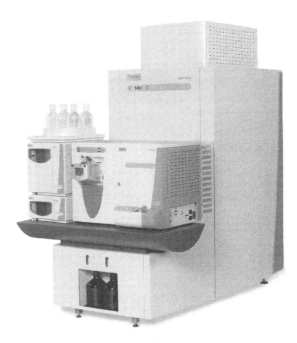

FIGURE 3.42 Picture of the LTQ FT Ultra™ mass spectrometer. (Photo provided courtesy of Thermo Fisher Scientific.)

Figure 3.42 is a picture of a commercially available hybrid mass spectrometer that combines a linear ion trap mass spectrometer with a Fourier transform ion cyclotron resonance mass spectrometer. Also included in the figure is an HPLC system for sample separation prior to the introduction into the ion trap via electrospray ionization. A schematic diagram of the instrument is illustrated in Figure 3.43 showing the linear ion trap, ion optics, rf-only multipoles, and the ICR cell located in the middle of a 7-T superconducting magnet. The ECD assembly (electron capture dissociation), and IRMPB laser assembly (infrared multiphoton dissociation), are features added to the instrument designed to enhance fragmentation of ions within the ICR cell.

3.12 LINEAR ION TRAP ORBITRAP MASS ANALYZER (LTQ-ORBITRAP/MS)

The most recent addition to the mass analyzer family is the Orbitrap mass analyzer developed and reported by Alexander Makarov.[14] As we will see, this mass analyzer achieves very high mass resolution (up to 150,000) without the need of a supercooled superconducting magnet with magnetic field strengths starting at 7 T and going higher, as we saw in Section 3.10 for FTICR mass spectrometry. This eliminates the need for the manufacturing of a complicated magnetic system that requires both the refilling

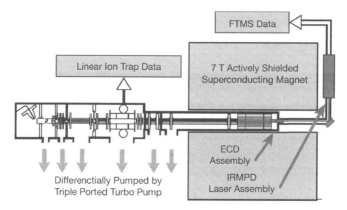

FIGURE 3.43 Schematic diagram of 7-T LTQ FTICR mass spectrometer showing the linear ion trap, ion optics, rf-only multipoles, and the ICR cell located in the middle of a 7-T superconducting magnet. The ECD assembly (electron capture dissociation) and IRMPB laser assembly (infrared multiphoton dissociation) are features added to the instrument designed to enhance fragmentation of ions within the ICR cell. (Figure provided courtesy of Thermo Fisher Scientific.)

of liquid nitrogen and liquid helium reservoirs to maintain the low temperatures needed for superconductivity and the active shielding required to reduce the strong magnetic field surrounding the superconducting magnet. The Orbitrap mass analyzer has the ability to achieve high mass resolution because its detection system is based off of an orbiting packet of ions very similar to those detected in FTICR where an oscillating transient current is produced and can be fast Fourier transformed (fast FT) from a transient time-domain signal into an m/z versus intensity mass spectrum. However, there are significant differences between the construction, motion, and mass detection in the Orbitrap mass analyzer as compared to the FTICR mass analyzer. The physicist Kingdon first introduced orbital trapping of ions in the gas phase in 1923.[15] His design, which was subsequently named the Kingdon trap, consisted of an outer open cylinder that contained a central wire running axial to the cylinder and end flanges that were used to enclose the trapping volume. The orbiting of the ions was achieved by applying a voltage between the outer cylinder and the central axial wire. The ions were attracted by the central wire and would move toward it. If the ions possessed enough kinetic energy (tangential velocity), they would not be stopped by the central wire but would begin to orbit around the wire. The ions' axially movement in relation to the wire was constrained by the fields induced from the flanges that restrained the ions from escaping the Kingdon trap axially. Figure 3.44 shows the basic design and principle of the Kingdon trap where a centrally located wire runs axial to the outer cylinder.

The Orbitrap does not use the simple design illustrated in Figure 3.44 for the Kingdon trap but rather uses outer and inner electrodes that are curviture in nature. The basic design of the Orbitrap is illustrated in Figure 3.45. Both the outer and inner

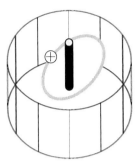

FIGURE 3.44 Simple Kingdon trap illustrating the central axial electrode (wire) surrounded by the outer cylindrical electrode.

electrodes are curved and the electrostatic field produces ion movement both in a circular path around the inner electrode and an axial oscillation. The axial oscillation back and forth along the central electrode was what was finally used for the ion detection instead of the frequency of the ion rotation around the central electrode. Remember in the FTICR cell the frequency of the ion rotation is measured, in the form of a transiently induced oscillating current, as the ions orbit in their cyclotron radius within the cell induced by the strong surrounding magnetic field, by opposing

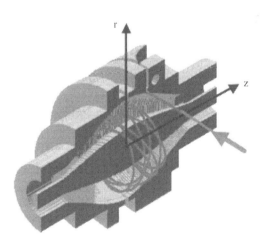

FIGURE 3.45 Cutaway view of the Orbitrap mass analyzer. Ions are injected into the Orbitrap at the point indicated by the perpendicular arrow. The ions are injected with a velocity perpendicular to the long axis of the Orbitrap (the z axis). Injection at a point displaced from $z = 0$ gives the ions potential energy in the z direction. Ion injection at this point on the z potential is analogous to pulling back a pendulum bob and then releasing it to oscillate. (From Hu, Q., Noll, R. J., Li, H., Makarov, A., Hardman, M., and Cooks, R. G. The Orbitrap: A new mass spectrometer. *J. Mass Spectrom.* 2005, *40*, 430–443. Copyright John Wiley & Sons, Inc. 2005. Reproduced with permission.)

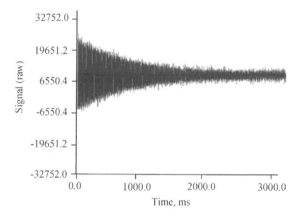

FIGURE 3.46 Image current transient from ions of doxepin (280.1696 Da). (Reprinted with permission from Makarov et al. *Anal. Chem.* 2003, *75*, 1699–1705. Copyright 2003 American Chemical Society.)

detector plates. It was observed for the Orbitrap that the frequency of ion rotation was greatly dependent upon the initial radius of the ion rotation and upon the ion velocity. This type of dependency will result in poor mass resolution (e.g., in FTICR mass spectrometry the frequency of the orbiting ion packet is independent of the ions' initial kinetic energy or orbital radius affording extremely high levels of mass resolution). In the application of the Orbitrap the harmonic ion oscillation frequencies along the field of axis are used to derive mass-to-charge ratios (m/z). While mass selective instability can be used in the Orbitrap to eject and subsequently detect the trapped ions using a secondary electron multiplier, the method of choice is to use image current detection of the axial frequency and fast FT algorithms. The time-domain transient image current for the detection of doxepin (280.1696 Da) is illustrated in Figure 3.46. Notice that it is very similar to the time-domain transient image currents that were presented in the FTICR MS section (see Fig. 3.32). The time-domain transient image currents measured in an Orbitrap also suffer from dissipation of the signal as was observed for the time-domain transient image currents measured in FTICR cells. If you recall, with time the magnitude of the cyclotron frequency amplitudes measured in an FTICR time-domain transient signal are also observed to steadily decrease as is illustrated in the Orbitrap composite transient signal in Figure 3.46. In FTICR this is due to collisions taking place between the trapped ions and between the trapped ions with neutrals. This is a process called collision-mediated radial diffusion where the coherence of the ion packets is reduced. Collisions will drive the ions toward the outer dimensions of the cell where they will be lost to neutralizing collisions with the cell walls. Optimal conditions require high vacuum in the order of 10^{-9} Torr where collisions are greatly reduced. This is also the same case within the Orbitrap where ultrahigh vacuum pressures are incorporated in the Orbitrap mass spectrometer in excess of 2×10^{-10} mbar. The high resolving power achievable by the Orbitrap mass analyzer is demonstrated in Figure 3.47, which illustrates an expanded view

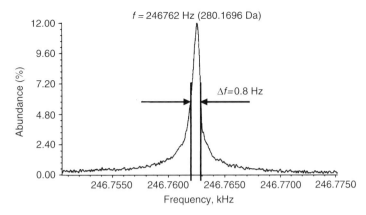

FIGURE 3.47 Expansion of the frequency spectrum demonstrating frequency resolving power of ~300,000 and mass resolving power of 150,000. The spectrum was obtained from the transient in Figure 3.46 by fast Fourier transform with double zero filling. Due to high magnification, only the main isotopic peak of doxepin fits into the window. (Reprinted with permission from Makarov et al. *Anal. Chem.* 2003, *75*, 1699–1705. Copyright 2003 American Chemical Society.)

of the frequency spectrum of the doxepin time-domain transient signal shown in Figure 3.46. The frequency resolving power is in excess of 300,000, which equates to a mass resolving power of 150,000. This high mass resolving power clearly rivals the FTICR in the fact of the relative simplicity of the construction of the Orbitrap mass analyzer, which does not require the incorporation of a superconducting magnet.

The description of the Orbitrap's potential field has similarities to a three-dimensional quadrupole; however, the design of the Orbitrap requires an additional potential description. The Orbitrap induces trapping within static electrostatic fields while the quadrupole ion trap uses an electrostatic field that is dynamic with an oscillation that is approximately 1 MHz. The Orbitrap obtains an electrostatic field orbital trapping with a potential distribution that is described by the following function[16]:

$$U(r, z) = \frac{k}{2}\left(z^2 - \frac{r^2}{2}\right) + \frac{k}{2}(R_m)^2 \ln\left(\frac{r}{R_m}\right) + C \qquad (3.51)$$

where r and z are cylindrical coordinates ($z = 0$ as the plane of the field symmetry), R_m is the characteristic radius, k is field curvature, and C is a constant. The electrostatic field of the Orbitrap comprises two independent fields that include the ion trap's quadrupole field and the cylindrical capacitor's logarithmic field. The field for the Orbitrap has been described by combining the two into what is known as a quadro-logarithmic field.

The physical design and dimensions of the trap as illustrated in Figure 3.45 include a central electrode with a spindle-like-shape that runs through the axis of the outer barrel-like-shaped electrode. In the illustration R_1 is the radius of the inner spindle-like electrode, R_2 is the radius of the outer barrel-like electrode, and r is the cylindrical coordinate with $z = 0$ denoting the plane of symmetry. The shape of the two electrodes has been derived by Makarov from Equation (3.51) as

$$z_{1,2}(r) = \sqrt{\frac{r^2}{2} - \frac{(R_{1,2})^2}{2} + (R_m)^2 \ln\left(\frac{R_{1,2}}{r}\right)} \qquad (3.52)$$

This equation allows the determination of the maximum radius of the inner (R_1) electrode, the outer (R_2) electrode, the characteristic radius (R_m), and the corresponding r polar coordinate with respect to the $z = 0$ plane of symmetry. Makarov has also described that the stable trajectories of trapped ions within the Orbitrap posses both oscillations around the central axis combined with rotation around the central electrode. This results in movement that is a repeating intricate spiral back and forth within the trap. The polar coordinates (r, ϕ, z) equation of motion has shown to (including mass-to-charge ratio m/z) be

$$\ddot{r} - r\dot{\varphi}^2 = -\frac{q}{m}\frac{k}{2}\left[\frac{(R_m)^2}{r} - r\right] \qquad (3.53)$$

$$\frac{d}{dt}(r^2\dot{\varphi}) = 0 \qquad (3.54)$$

$$\ddot{z} = -\frac{q}{m}kz \qquad (3.55)$$

It is deduced from these relationships [Equation (3.53)] that if $r < R_m$ there is an attraction of ions by the electric field to the central axis. If $r > R_m$, then there is a repulsion of the ions. Therefore, in the physical design of the Orbitrap only radii below the R_m are useful. Also deduced from these equations [Equations (3.53)–(3.55)] is that the equation of motion along z is a simple harmonic oscillator with an exact solution of:

$$z(t) = z_0 \cos(\omega t) + \sqrt{\left(\frac{2E_z}{k}\right)} \sin(\omega t) \qquad (3.56)$$

From this relationship the frequency of axial oscillations (in rad/s) can be derived as

$$\omega = \sqrt{\left(\frac{q}{m}\right)k} \qquad (3.57)$$

Within the Orbitrap's quadro-logarithmic field each respective packet of m/q ions will posses a distinctive frequency of axial oscillation (ω). This relationship allows the transient time-domain signal measurement of the oscillating ions according to their mass-to-charge ratios (m/z), which is in effect a mass analyzer. Again, for the same reasons that an FTICR mass analyzer can achieve such high resolution values, the Orbitrap mass analyzer can also obtain resolution values in the order of 100,000–200,000. Like the cyclotron frequency in FTICR, the Orbitrap's frequency of axial oscillations for ions of the same mass-to-charge ratio is independent from the initial position or velocity of the ions. Because the frequency is independent from the initial velocity, it is independent from the initial kinetic energy of the ion. This is one very important reason why the FTICR and Orbitrap mass spectrometers are able to achieve such high mass resolutions where other mass analyzers are required to focus the translational energy of the ions to achieve resolution of the m/z values. The FTICR and Orbitrap's independence from initial position are also again in contrast to the MALDI-TOF mass spectrometer where initial spatial distribution affects resolution, and the quadrupole mass analyzer where a critical angle exists for introduction and transmission of the ion through the quadrupole that directly affects resolution. As a consequence of the design of the Orbitrap based on the optimal radii dimensions, the Orbitrap has a larger trapping volume than the FTICR traps and the quadrupole Paul traps. Because the Orbitrap's field can be defined to a very high accuracy, high mass resolution is achievable routinely to 150,000. Also, due to the independence of the mass-to-charge ratio to the trapping potential, an increased space charge capacity at higher masses is possible.

Figure 3.48 is a picture of a commercially available hybrid linear ion trap/Orbitrap mass spectrometer. The linear ion trap is located in the front while the Orbitrap is

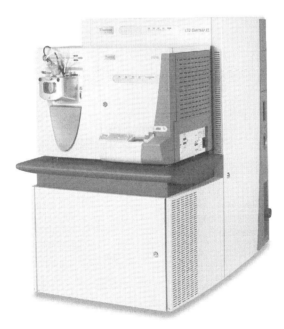

FIGURE 3.48 Picture of the LTQ Orbitrap™ mass spectrometer. (Photo provided courtesy of Thermo Fisher Scientific.)

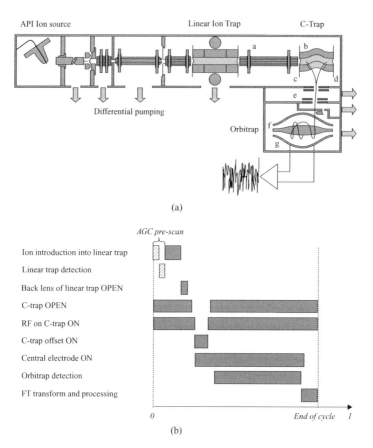

(a)

(b)

FIGURE 3.49 (*a*) Schematic layout of the LTQ Orbitrap mass spectrometer: (a) Transfer octapole, (b) curved rf-only quadrupole (C-trap), (c) gate electrode, (d) trap electrode, (e) ion optics, (f) inner orbitrap electrode, and (g) outer Orbitrap electrode. (*b*) Simplest operation sequence of the LTQ Orbitrap mass spectrometer (not shown are the following: optional additional injection of internal calibrant; additional MS or MS^n scans of linear trap during the Orbitrap detection). (Reprinted with permission from Horning et al. *Anal. Chem.* 2006, *78*, 2113–2120. Copyright 2006 American Chemical Society.)

located in the back portion of the instrument. The major components of the LTQ Orbitrap hybrid mass spectrometer (Thermo Fisher Scientific) are illustrated in the schematic diagram of Figure 3.49. Also illustrated is an experimental sequence that is performed with the Orbitrap for ion mass analysis. For experimental work that has been performed thus far with the Orbitrap, an electrospray ionization source has been used and is depicted at the far left of Figure 3.49. Ions are formed by the electrospray process and transferred into the first stages of differential vacuum pumping through an inlet orifice. The desolvated ions introduced into the instrument are now in the gas phase and are subsequently guided by a series of ion lenses and rf-only multipoles into the linear quadrupole ion trap (LTQ) mass analyzer. The LTQ in itself is also a fully functional ion trap mass spectrometer, and this is why the LTQ-Orbitrap mass spectrometer is a hybrid mass analyzer (as you remember the hybrid mass analyzer is the coupling of two different, fully functional mass analyzers into one instrument). The recent models of LTQ mass analyzers also now possess an electronically controlled procedure known as automatic gain control (AGC) where the ion current within the predefined mass range trapped within the LTQ is scanned using a prescan prior to the full analytical scan. This allows the storage of a targeted number of ions within the trap, producing an enhanced signal due to an optimal signal without the degradation of signal from space charge effects of trap overpopulation. The design of the Orbitrap has many similarities to the LTQ-FT mass spectrometer where the ability to analyze ions in the second mass analyzer is possible (at very high mass resolutions). Following the LTQ mass analyzer is a transfer octapole [dimensions include 300 mm long and an inscribed diameter of 5.7 mm; see (a) in Fig. 3.49] that guides the ions ejected from the LTQ into a curved RF-only C-trap. The C-trap is a recently added feature of the instrumentation as compared to the first Orbitrap mass analyzer that was reported in the literature.[17] The first Orbitrap mass analyzer was a developmental mass spectrometer that was not commercially available and differed from the present commercially available one. The first system coupled a storage quadrupole with transfer lenses to introduce ions into the Orbitrap. The storage quadrupole is necessary because the Orbitrap works on a pulsed set of ions introduced into the trap and not upon a constant beam of ions like a triple quadrupole mass spectrometer. The recent Orbitrap instrumental design includes a C-trap after the LTQ mass analyzer and before the Orbitrap mass analyzer. The axis of the C-trap follows an arc that is C-shaped composed of rods with hyperbolic surfaces. On the two ends of the C-trap are plates used for ion introduction (plate located between the octapole and the C-trap) and trapping (plate located at other end of C-trap). The C-trap is also filled with nitrogen bath gas at a pressure of approximately 1 mTorr. The bath gas is used for collisional dampening of the ions trapped within the C-trap. The collisions between the nitrogen bath gas and the trapped ions are at energies low enough not to activate fragmentation of the trapped ions. The ions experience enough collisional cooling in the C-trap to where they form a stable, thin thread along the curved axis of the C-trap. The trapping plates at the two ends of the C-trap are given a positive potential of 200 V that compresses the trapped ions axially. Using a combination of pullout potentials, the ions are removed from the C-trap and introduced into the Orbitrap. This is achieved in a pulsed fashion where large ion populations are transferred from the C-trap into the Orbitrap in a fast and uniform fashion. As illustrated in Figure 3.49 the ions pass through slightly curved ion optics from the C-trap to the Orbitrap. The packet of ions will also pass through three different stages of vacuum upon arrival into the Orbitrap. This is done because to achieve the high mass resolution within the Orbitrap, there must be a very low incidence of collisions between the trapped ions and any form of ambient gas. As you remember, it is collisions within the Orbitrap that decay the inherent oscillations, thus decaying the transient time-domain signal over time. The ions that are trapped within the C-trap are at a relatively high background pressure (~ 1 mTorr) as compared to the optimal operating pressure of the Orbitrap ($\sim 10^{-9}$ Torr). The distance between the C-trap and the Orbitrap is very small, thus any time-of-flight separation is kept to a negligibly small unwanted negative effect.

Upon exiting the C-trap, the ions are moved through a series of curved ion optics (as depicted in Fig. 3.49) and are also simultaneously accelerated to high kinetic energies. This has the effect of compressing the ion packet into a tight cloud that is able to transverse an entrance aperture that is relatively small and is offset tangentially to the center of the Orbitrap. The ions are injected into the trap at a position that is offset from the center of the Orbitrap at a distance of 7.5 mm from the equatorial center of the trap. In this manner the ions will start coherent axial oscillations due to the present fields in the trap, thus not requiring any type of initial excitation. The capturing of the ions within the trap is obtained by applying a rapidly increasing electric field that squeezes or contracts the radius of the ion cloud to a trajectory closer to the axis. The ion clouds follow a path similar to that shown in Figure 3.45 where clouds with higher m/z values will have a radius of orbit with respect to the central electrode axis that is larger as compared to ion clouds of smaller m/z values. The detection of the oscillating clouds is achieved using the outer electrodes as the ion clouds move back and forth axially within the Orbitrap.

The LTQ-Orbitrap hybrid mass spectrometer also, like the LTQ-FT hybrid mass spectrometer, possesses multifunctional capabilities including high mass resolution and accuracy of precursor ions and precursor ion dissociation through collision-induced dissociation (CID) for structural elucidation. The Orbitrap mass analyzer has been demonstrated to possess a wide dynamic range for mass accuracy determinations such as that illustrated in Figure 3.50 where the measurement of impurities and propanolol in a propanolol sample demonstrate mass accuracy. The impurity species

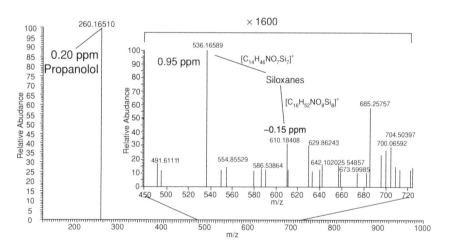

FIGURE 3.50 Illustration of dynamic range of mass accuracy of the LTQ Orbitrap (siloxane impurities in propanolol sample) in a single 1-s scan ($R = 60,000$, $N = 2 \times 10^6$, external mass calibration, reduced profile mode). (Reprinted with permission from Horning et al. *Anal. Chem.* 2006, *78*, 2113–2120. Copyright 2006 American Chemical Society.)

of siloxanes being present in the sample enabled the comparison of mass accuracies for both the most abundant species present being propanolol at m/z 260.16510 with a mass accuracy of 0.20 ppm and siloxane impurities at m/z 536.16589 with a mass accuracy of 0.95 ppm and the m/z 610.18408 siloxane species with a mass accuracy of –0.15 ppm.

The LTQ-Orbitrap mass analyzer also has the capability of performing data-dependent scanning for precursor fragmentation studies of species as they elute from an HPLC column. In this type of experiment species are eluting from an HPLC column and are converted to gas-phase ions using the electrospray ionization source (ESI-LTQ-Orbitrap/MS). The Orbitrap mass analyzer is used to scan the precursor species at a given point in time as they are eluting from the HPLC column. The m/z value for these species are then recorded at a high mass resolution (typically from $R = 50,000$ to 100,000). The controlling software of the LTQ mass analyzer is set to pick a predetermined number of the most abundant species measured by the Orbitrap and subsequently isolates them within the linear trap and fragments them. The product ions can then either be detected using the linear trap or the Orbitrap. The linear trap will collect the spectra at a much lower resolution than the Orbitrap, but the use of the linear trap for product ion spectral measurements is often used when a large number of data-dependent spectra are to be collected (such as a setting of the collection of the top 10 most intense precursor ions, a setting that is often used when analyzing a complex peptide mixture by nano-HPLC LTQ-Orbitrap/MS). Figure 3.51 illustrates

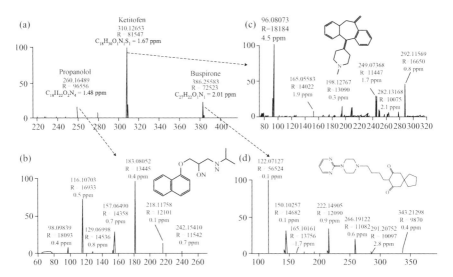

FIGURE 3.51 Example of data-dependent acquisition with external mass calibration for a sample containing small molecules, with one high-resolution mass spectrum recorded of the precursors at $R = 60,000$ and $n = 500,000$: (a) followed by three data-dependent MS/MS spectra at $R = 7500$, $N = 30,000$, (b) for precursor at $m/z = 260$, (c) for precursor at m/z 310, and (d) for precursor at $m/z = 386$. (Reprinted with permission from Horning et al. *Anal. Chem.* 2006, *78*, 2113–2120. Copyright 2006 American Chemical Society.)

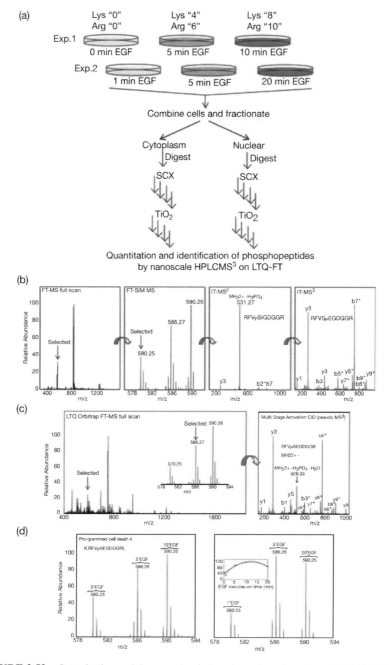

FIGURE 3.52 Quantitative and time-resolved phosphoproteomics using SILAC. (*a*) Three-cell populations are SILAC encoded with normal and stable isotope-substituted arginine and lysine amino acids, creating three stages distinguished by mass. Each population is stimulated for a different length of time with EGF, and the experiment is repeated to yield five time points.

the use of data-dependent scans for a limited amount of precursors (top 3) where the Orbitrap was used for product ion spectral collection at a resolution of $R = 7500$, which is still much higher than that achievable by the LTQ mass analyzer (typical $R = 4000$). Notice, however, in Figure 3.51a the resolutions actually range from $R = 72,923$ (with a mass accuracy of 2.01 ppm) for buspirone at m/z 386.25583 to $R = 90,556$ (with a mass accuracy of 1.48 ppm) for propanolol at m/z 260.16489. The spectra in Figure 3.51b, c, and d are the product ion spectra of the precursors in Figure 3.51a, all demonstrating quite high mass accuracies themselves, which can be used in confirmation of structural elucidation studies.

Lastly, there is one more very interesting feature that has been recently introduced with the LTQ-Orbitrap mass analyzer that allows the collection of MS^3 spectra into a single spectral result. This capability is known as multistage activation (MSA) and is now actually available on the most recently introduced LTQ mass analyzers. Before the introduction of multistage activation, MS^3 spectra were actually separately collected from the second-stage MS^2 product ion spectra. The traditional MS^n spectra collected by an ion trap was performed by first isolating the precursor ion within the trap and then activating it, thus inducing fragmentation and product ion production. Isolating the product ion of interest by excluding all other ions from the ion trap and then activating it, inducing fragmentation and product ion production would then perform the next stage of MS^3. This would be a separately collected spectrum from the MS and the MS^2 spectra. With multistage activation the spectral result contains the product ions from the second stage MS^2 and the third stage MS^3 experiments. In multistage activation instead of removing all of the product ions from the second-stage activation of the precursor ion, the product ion of interest is activated while retaining the entire previously produced product ions and subsequently fragmented creating a product ion spectrum that contains the second- and third-stage product ions. The capability of multistage activation is very useful when studying the product ion spectra of posttranslational modifications of proteins such as phosphorylation. Often

---

Cells are combined, lysed, and enzymatically digested, and phosphopeptides are enriched and analyzed by mass spectrometry. (b) Mass spectra of eluting peptides reveal SILAC triplets (same peptide from the three-cell populations), and these triplets are remeasured in selected ion monitoring (SIM) scans for accurate mass determination. Phosphopeptides are identified by loss of the phospho-group in a first fragmentation step followed by sequence-related information from a second fragmentation step. (c) Same peptides as in (b) but measured on the LTQ-Orbitrap. Inset shows a magnification of the SILAC peptide selected for fragmentation. Right-hand panel shows the result of multistage activation of the peptide. (d) Raw data of a phosphopeptide from the protein programmed cell death 4. The three peptide intensities in the two experiments are combined using the 5-min time point, resulting in the quantitative profile shown in the inset. (Reprinted with permission. *Cell*, Olsen, J. V., Blagoev, B., Gnad, F., Macek, B., Kumar, C., Mortensen, P., and Mann, M. Global, in vivo, and site-specific phosphorylation dynamics in signaling networks. 2006, *127*, 635–648. Copyright Elsevier 2006.)

when a product ion spectrum is collected of phosphorylated peptides, the predominant product ion observed in the spectrum is a peak formed through a neutral loss of the phosphate group, with very little other information in the spectrum in the form of product ions. A typical approach for studying the product ion spectra of phosphorylated peptides was to look for this predominant phosphate neutral loss product ion and then isolate it for MS^3 spectral collection. The MS^3 spectra obtained could either be interpreted manually by inspecting it in conjunction with the MS^2 product ion spectrum of the precursor or by offline combining the spectra into a single spectrum. With multistage activation the composited spectrum is obtained in real time during the spectral acquisition. Presented in Figure 3.52 is an excellent example of the use of multistage activation in the study of posttranslational modification of proteins by phosphorylation. The experiments were collected on an LTQ-FT mass spectrometer using the approach described above where separate spectra are obtained for the MS^2 and the MS^3 product ion fragmentation studies (see Fig 3.52b). The same capability is available on the LTQ-Orbitrap mass spectrometer along with the capability of collecting MS^3 spectra from multistage activation studies (see Fig. 3.52c). Figure 3.52 also affords the opportunity to see what type of biological experiments can be studied using mass spectrometry.

3.13 PROBLEMS

3.1. List some common mass spectrometers in use today.

3.2. With a source exit slit width of 4 μm and a collector slit width of 6 μm, what magnet radius is needed to obtain a resolution of 10,000?

3.3. Is the electric sector a mass analyzer? If not, explain what it is.

3.4. In the TOF-TOF mass analyzer what is the ion gate and for what is it used?

3.5. By what energy are ions accelerated into the quadrupole?

3.6. How do the quadrupoles induce stable and unstable trajectories?

3.7. What is it in the Mathieu equations of (3.24) that demonstrates the quadrupole to be a mass analyzer?

3.8. Explain the relationship between sensitivity and resolution for the quadrupole mass analyzer.

3.9. What are some advantages of quadrupole mass analyzers?

3.10. What are some limitations of quadrupole mass analyzers?

3.11. What are the four methods used in scanning with the triple quadrupole mass analyzer? Briefly describe their usefulness.

3.12. What is the purpose of the bath gas used in ion trap mass analyzers?

3.13. What is mass-selective axial instability mode?

3.14. How is resonance ejection obtained in ion traps?

3.15. What are the three methods to excite ions within the ion trap?

3.16. What is the chief difference between a time-of-flight mass analyzer and a quadrupole/time-of-flight mass analyzer?

3.17. What kind of spatial configuration is used in quadrupole/time-of-flight mass analyzers?

3.18. What two analyses routinely done sets FTICR mass analyzers apart from all others?

3.19. What are the strengths of permanent electromagnets and superconducting magnets?

3.20. What vacuum pressures are needed in the FTICR cell and why?

3.21. What are three reasons why excitation of the m/z ion packet is performed in FTICR mass spectrometry?

3.22. What is collision-mediated radial diffusion. How is it observed?

3.23. What functionalities of the FTICR mass spectrometer are influenced by the magnetic field strength?

3.24. What advantages are there for the Orbitrap mass analyzer not requiring a strong magnet?

REFERENCES

1. Hillenkamp, F., Unsold, E., Kaufmann, R., and Nitsche, R. *Appl. Phys.* 1975, *8*, 341–348.

2. Van Breemen, R. B., Snow, M., and Cotter, R. J. *Int. J. Mass Spectrom. Ion Phys.* 1983, *49*, 35–50.

3. Wiley, W. C., and McLaren, I. H. *Rev. Sci. Instrum.* 1955, *26*, 1150–1156.

4. Vestal, M. L., Juhasz, P., and Martic, S. A. *Rapid Commun. Mass Spectrom.* 1995, *9*, 1044–1050.

5. Brown, R. S., and Lennon, J. J. *Anal. Chem.* 1995, *67*, 1998–2003.

6. Mamyrin, B. A., Karateev, V. I., Shmikk, D. V., and Zagulin, V. A. *Sov. Phys. JETP* 1973, *37*, 45.

7. Della Negra, S., and Le Beyec, Y. *Int. J. Mass Spectrom. Ion Processes* 1984, *61*, 21.

8. Johnstone, R .A. W., and Rose, M. E. *Mass Spectrometry for Chemists and Biochemists*, 2nd ed. New York: Cambridge University Press, 1996, Chapter 2.

9. March, R. E., and Hughes, R. J. In *Quadrupole Storage Mass Spectrometry*, New York: Wiley, Interscience, 1989, Chapter 2, pp. 31–110.

10. White, F. A. *Mass Spectrometry in Science and Technology*. New York: Wiley, 1968.

11. Paul, W. *Angew. Chem. Int. Ed. Engl.* 1990, *29*, 739.

12. Schwartz J. C., Senko, M. W., and Syka, J. E. P. *J. Am. Soc. Mass Spectrom.* 2002, *13*, 659–669.
13. Syka, J. E. P., Marto J. A., Bai, D. L., Horning, S., Senko, M. W., Schwartz, J. C., Ueberheide, B., Garcia, B., Busby, S., Muratore, T., and Shabanowitz, J. *J. Proteome Res.* 2004, *3*, 621–626.
14. Makarov, A. *Anal. Chem.* 2000, *72*, 1156–1162.
15. Kingdon K. H. *Phys. Rev.* 1923, *21*, 408–418.
16. Hardman, M., and Makarov, A. A. *Anal. Chem.* 2003, *75*, 1699–1705.
17. Hu, Q., Noll, R. J., Li., H., Makarov, A., Hardman, M., and Cooks, G. R. *J. Mass Spectrom.* 2005, *40*, 430–443.

CHAPTER 4

COLLISION AND UNIMOLECULAR REACTION RATE THEORY

4.1 INTRODUCTION TO COLLISION THEORY

Product ion spectra collected in mass spectrometric experiments are often obtained by colliding the precursor ion species with a stationary target gas that is contained within a collision cell within the mass spectrometer. This is the predominant mode of dissociation used in triple quadrupole mass spectrometers (QQQ/MS), in ion trap (three-dimensional and linear) mass spectrometers (LCQ/MS and LTQ/MS), in hybrid quadrupole time-of-flight mass spectrometers (Q-TOF/MS), in hybrid linear ion trap Fourier transform mass spectrometers (LTQ-FT/MS), in Fourier transform ion cyclotron mass spectrometers (FTICR/MS), and in the hybrid linear ion trap Orbitrap mass spectrometers (LTQ-Orbitrap/MS). Typically, the precursor analyte ion is provided with increased internal energy in the form of kinetic energy and allowed to collide with a stationary target gas molecule such as nitrogen, helium, or argon. The energy of the collision is dissipated within the molecule through vibrational modes of the collisionally activated molecule. Some of the vibrational modes are energetic enough to induce dissociation of bonds that have the lowest activation barrier as compared to other bonds within the molecule associated with the localized vibrational modes. As a result of the collisional activation, the molecule will fragment, producing product ions that are subsequently measured by the detector of the mass spectrometer. The events within the activated molecule leading up to the production of the product ions are described by fragmentation pathways also known as product ion channels. Figure 4.1 illustrates a simple collision process within the mass spectrometer where the precursor analyte is accelerated into the collision cell

Even Electron Mass Spectrometry with Biomolecule Applications   By Bryan M. Ham
Copyright © 2008 John Wiley & Sons, Inc.

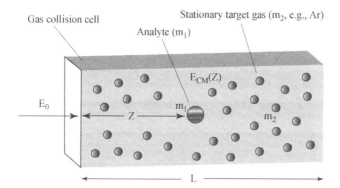

FIGURE 4.1   Simple collision cell within the mass spectrometer. The boxed-in area depicts the region where the stationary target gas is contained at a certain maintained pressure.[2]

from the left of the figure. The boxed-in region is the enclosed target stationary gas that is maintained at a controlled and measurable number gas density (n). The number gas density, n, is derived according to the ideal gas law[1]:

$$n \equiv \frac{N}{V} \cong \frac{PL}{RT} \tag{4.1}$$

$$n = \frac{P}{k_B \alpha T} \tag{4.2}$$

where P is the pressure in atm, k_B is the Boltzmann constant, α is a conversion factor of 9.869 cm^{3}atm/J, and T is the temperature in kelvin. The accelerated analyte enters the collision cell from the left and transverses the cell, colliding with the stationary target gas. In collision reactions there are two types of collisions that take place: (1) collisions that result in product ion formation and (2) collisions that do not result in product ion formation. The probability that a collision will take place in the vicinity of the species involved in the collision process is described by the collisional cross section.

These two processes are depicted pictorially in Figure 4.2. In the left half of Figure 4.2a the collision does not result in product ion formation but only in the deflection of the colliding species. In this process the gas-phase ions and/or neutrals have come in close enough distance for contact to take place, but the collision may not have been of sufficient energy to induce fragmentation and subsequent product ion formation. In the right half of Figure 4.2b the collision was of sufficient energy and has resulted in the formation of product ions.

These two processes have resulted in the formation of two collisional probabilities that are described as collision total cross sections (σ_t) or collision product cross sections (σ_p). The collision total cross section is the physical measured cross section of the species involved in the collision. In a hard sphere collision involving an atom, the total cross section is the diameter of the atom. The collision product cross section

(a)   (b)

FIGURE 4.2 Collisions resulting in either (*a*) no product ion formation or (*b*) sufficient collisional energy resulting in labile bond breakage and product ion formation.

is the area probability that a collision will take place resulting in products. In general, the product cross section is smaller than the total cross section. The two types of cross sections are illustrated in Figure 4.3 where the total cross section is represented by the area σ and the product cross section is represented by the area σ^*. In the collision where no product ions are formed, the reactant molecule is simply deflected. In this collision there has not been enough energy that has been converted from energy of motion to internal vibrational energy for activation of the breakage of energy labile bonds. For a collision to take place one of the molecules must be contained within a cylinder that has a radius that is a sum of the two radii of the colliding species:

$$r_{\text{collision}} = r_1 + r_2 \tag{4.3}$$

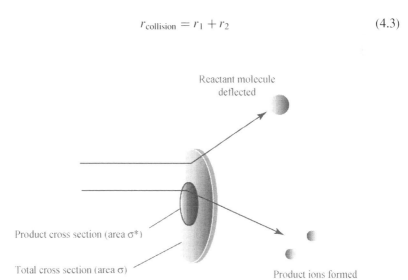

Reactant molecule
deflected

Product cross section (area σ^*)

Total cross section (area σ)

Product ions formed

FIGURE 4.3 Description of total cross section (area σ) versus product cross section (area σ^*). The total cross section is larger in area than the product cross section and results in a deflected reactant molecule only. The collision described by the product cross section comprises a smaller area and results in the formation of product ions.

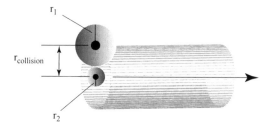

FIGURE 4.4 Cylindrical space that is swept out where, if a collision is to take place, one of the molecules must be contained within this cylinder that has a radius that is a sum of the two radii of the colliding species. In this simplified illustration the two colliding species are hard-body spheres with a clearly defined radius.

This is depicted in Figure 4.4 where the radius of the collision cylinder is the sum of the two species colliding. In this simplified illustration the two colliding species are hard body spheres with a clearly defined radius. With large molecules, however, this is not as simple and straightforward as two hard body spheres. The area of the total collision cross section will depend upon the spatial conformation of the species. If a species can exist in multiple conformations that are stable, then each conformation will possess its own total collision cross section. When a molecule is moving through space, it will cover a cylindrical volume that has a *radius r* and a cross-sectional *area of* πr^2. The volume that is swept out by the moving molecule with an average speed of

$$v = \int_0^\infty v F(v)\, dv = \sqrt{\frac{8RT}{\pi M}} \tag{4.4}$$

in a time of Δt is

$$v(t) = \pi r^2 v\, \Delta t \tag{4.5}$$

This is the volume of space that describes the probability of a collision in that any molecule whose center is contained within this volume will collide with the moving species. Using the number gas density, n, we can depict the frequency of collisions as

$$\text{Frequency of collision, } z = \sqrt{2}\pi r^2 v n \tag{4.6}$$

Normally, when describing reactions based off of collisions, we are working with average relative velocities because both species involved in the collision are moving. However, for simplicity when describing collisions that take place within the floating collision gas cell of a mass spectrometer, we describe the target collision gas as stationary and thus not moving with its own velocity. In this way we consider only the velocity and hence the kinetic energy of the charged analyte within the mass spectrometer for gas-phase thermochemical calculations and analysis. Another important consideration to take in collisions in the gas phase within the mass spectrometer is the mean free path (ℓ) of the analyte as the average distance that the molecule has traveled between collisions. For a given time interval a molecule will travel

$$\Delta x = v \, \Delta t \tag{4.7}$$

distance. The number of collisions that the molecule will encounter is

$$z \, \Delta t \tag{4.8}$$

The mean free path can therefore be described as

$$l = \frac{v}{z} \tag{4.9}$$

Substituting in Equation (4.6) for the frequency of collisions (z) we obtain

$$l = \frac{1}{\sqrt{2}\pi r^2 n} \tag{4.10}$$

and finally with the insertion of the number gas density (n) from Equation (4.2) we have as the mean free path of collision:

$$l = \frac{k_B \alpha T}{\sqrt{2}\pi r^2 P} \tag{4.11}$$

The study of thermodynamic properties of complexes such as bond dissociation energies by collision-induced dissociation (CID) tandem mass spectrometry has been a useful approach in the measurement of the dissociation of gas-phase complexes. For the study of the bond dissociation energy (BDE) of a cobalt carbene ion Armentrout and Beauchamp[3] introduced an ion beam apparatus, and they have since refined the measurement of noncovalent, collision-induced threshold bond energies of metal–ligand complexes by mass spectrometry.[4–11] For the measurement of thermodynamic properties of gas-phase complexes, the array of uses of CID tandem mass spectrometry[12] includes investigations of bond dissociation energies,[13] noncovalent interactions,[12,14–16] gas-phase dissociative electron-transfer reactions,[17] bond dissociation energies using guided ion beam tandem mass spectrometry,[4–11] critical energies for ion dissociation using a quadrupole ion trap,[14] or a flowing afterglow triple quadrupole,[15,16] or an electrospray triple quadrupole,[18,19] and gas-phase equilibria.[20–22]

To produce intact gas-phase ions with minimal fragmentation during the ionization process, electrospray[23–26] is an ionization method that is well known. Thus, studies are quite accessible by electrospray mass spectrometry of gas-phase thermochemical properties of species originating from solution. Another "soft" ionization approach to gas-phase ionic complex analysis is MALDI; however, the complex's character in the solution phase can be lost during crystallization with the matrix. Experimental cross sections for a set of alkali metal ion cationized poly(ethylene terephthalate) (PET) oligomers were effectively derived by Gidden et al.[27] using a MALDI ion source coupled to a hybrid mass spectrometer.

The collision cell path length, or the mean free path of reaction [Equation (4.11)], is often unknown when calculating thermochemical values such as the bond dissociation energy for a gas-phase ionic complex by CID tandem mass spectrometry. Interjecting guesses into data reduction equations to iteratively test which value results in a best fit is one approach to estimating this collision path length in gas-phase dissociation studies. Describing the collision cell path length as the average distance that ions travel before dissociating is one approach that Chen et al.[28] presented for a new model for determining bond energies for holomyoglobin noncovalent gas-phase complexes. Until a realistic fit was obtained, estimated values for the path length were iteratively used in their method. Essentially, the method used by Ervin and Armentrout[4] was to employ the actual measured dimensions of the gas cell, for example, the central hexapole, where the dimensions of their gas cell are well defined, but they also take into account the pressure profile of the collision gas within the system to derive an effective cell length. A third experimental approach would be to measure the dissociation of standards with known BDE values to obtain I_P/I_0 values, which then can be related to the reaction path length according to the thin target limit equation[4]:

$$\frac{I_P}{I_0} = \frac{I_P}{I_R + \sum I_P} = \sigma_P nl \tag{4.12}$$

where I_P is the transmitted product ion signal intensity, I_R is the transmitted reactant ion signal intensity, I_0 is the incident ion signal intensity, σ_P is the product cross section (cm^2), n is the number gas density (cm^{-3}), and ℓ is the collision cell path length (cm). It is assumed that the sum of the transmitted product peak intensities is equal to the incident reactant ion peak intensity in this approach; in actuality the latter incident reactant ion peak intensity is not readily measurable. Therefore, from the measurement of known standards a "derived effective reaction path length" is being obtained. The effective path length represents the average distance that the complex travels before dissociation takes place (or the average distance of travel before one collision resulting in dissociation occurs) at the thin target limit. According to the thin target limit [Equation (4.12)], the slope of a plot of the $I_P/(I_R + \sum I_P)$ ratio, versus number gas density, is related to the collision cross section.[4]

4.2   NONCOVALENT BOND DISSOCIATION ENERGY

As an example of the measurement of noncovalent dissociation energies using mass spectrometry, we will consider a set of experiments involving metal and halide adducts

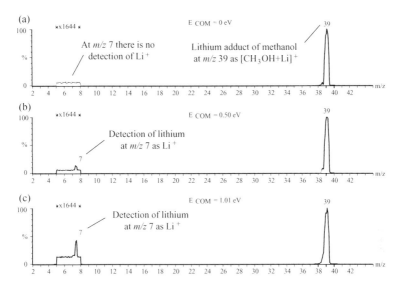

FIGURE 4.5 Methanol lithium adduct (m/z 39) collision-induced dissociation product ion mass spectra illustrating the appearance of the dissociation of the lithium cation (m/z 7, BDE = 1.596 eV[12]). (*a*) At 0 eV with no detection of lithium, (*b*) at 0.50 eV (E_{COM}) there is the emergence of the dissociated lithium cation, and (*c*) detection of lithium cation at 1.01 eV (E_{COM}).

of some selected lipid species. Used as a standard for effective reaction path length determination, lithium adducts of methanol were subjected to collision studies. Mass spectra obtained from these collision studies are illustrated in Figure 4.5, depicting the emergence of the dissociation of the lithium cation from the precursor adduct. For these types of bond dissociation energy calculations from mass spectrometry collision studies, the supplied energy for the collision is converted from the laboratory frame of reference (E_{LAB}) to the center-of-mass frame of reference (E_{COM}) according to

$$E_{COM} = E_{LAB} \left(\frac{m}{m + M} \right) \qquad (4.13)$$

where E_{LAB} is the selected value of the offset for the central hexapole, m is the mass of the stationary target Ar atom, and M is the mass of the precursor ion. At 0-eV offset value for the central hexapole, Figure 4.5*a* shows the measurement of the incident lithium adduct ion without the detection of the lithium cation. The emergence of the dissociated lithium cation is illustrated in Figure 4.5*b* at m/z 7 at an offset value = 0.50 eV (center-of-mass frame of reference, E_{COM}) for the central hexapole from the precursor methanol/lithium adduct. The spectra indicate an earlier threshold energy value than what is expected from the known BDE of 1.596 eV.[9] Next illustrated in Figure 4.5*c* is the product ion mass spectrum acquired

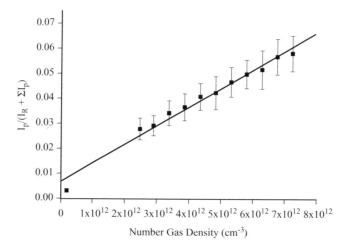

FIGURE 4.6 Ratio of product ion abundances to incident ion abundances [the latter is approximated by $(I_R + \sum I_P)$] versus number gas density for the lithium adduct of DMSO at constant collision energy ($E_{LAB} = 10$ eV). The slope of the curve is related to the product cross section (σ_P) of the lithium–DMSO cationic complex according to the thin target limit as $\sigma_P =$ slope/ℓ. (Reprinted with permission from Ham, B. M., and Cole, R. B. *Anal. Chem.* 2005, *77*, 4148–4159. Copyright 2005 American Chemical Society.)

at a 1.01 eV E_{COM} offset for the central hexapole. The ratio of the intensities of the lithium cation peak to the precursor peak at 0.5 eV is 8.0×10^{-5}, while at 1.0 eV the ratio is 2.5×10^{-4}.

Figure 4.6 is a graph of a dimethyl sulfoxide (DMSO) lithium adduct standard for the $I_P/(I_R + \sum I_P)$ ratio [Equation (4.12)] versus number gas density (n). The CID of the DMSO lithium adduct at fixed collision energy ($E_{LAB} = 10$ eV) but increasing gas cell pressures was performed to obtain the values in the graph. A "zero-point" collision cell gas pressure can be included and corresponds to the pressure measured in the analyzer in the absence of argon, that is, $7–8 \times 10^{-6}$ mbar ($1.7–1.9 \times 10^{11}$ cm^{-3} expressed as number gas density). However, this represents a very small probability for collisions with nitrogen or other ambient gas. The slope of the curve in Figure 4.6 is related to the product cross section according to the thin target limit [Equation (4.12)][29] as slope $= \sigma_P \ell$. This relationship can be used to calculate the cross sections of the DMSO lithium adducts, and estimate ℓ of the QhQ mass spectrometer. We assign $\ell = 1$ cm for the initial cross-section calculations using the thin target limit ratio relationship $\sigma_P =$ slope/ℓ. This assignment necessitates a subsequent correction of the derived reaction path length (i.e., actual length $\ell = 1$ cm + correction) but also allows the simplifying approximation that $\sigma_P =$ slope. At a fixed collision gas pressure of 1.3×10^{-4} mbar (number gas density of 3.14×10^{12} cm^{-3}), Figure 4.7 is a plot of the $I_P/(I_R + \sum I_P)$ ratio versus E_{COM} for CID of the DMSO lithium adduct. According to the thin target limit [Equation (4.12)] the $I_P/(I_R + \sum I_P)$ ratio is directly related to the cross section. For cross-sectional calculations it is assumed

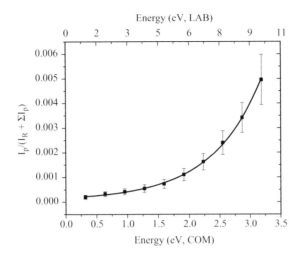

FIGURE 4.7 Plotted results of $I_P/(I_R + \sum I_P)$ ratio versus collision energy (E_{LAB}, top) (E_{COM}, bottom) for the dissociation of the DMSO–lithium adduct. The best fit gives an $I_P/(I_R + \sum I_P)$ ratio of 1.879×10^{-3}, at 2.36 eV,[29] and a number gas density of 3.14×10^{12}, used in calculating a derived effective reaction path length. (Reprinted with permission from Ham, B. M., and Cole, R. B. *Anal. Chem.* 2005, *77*, 4148–4159. Copyright 2005 American Chemical Society.)

that at constant n and ℓ values, the change in the $I_P/(I_R + \sum I_P)$ ratio will be directly proportional to the change in the cross section. From the initial number gas density determined cross section, the magnitude in the change of the $I_P/(I_R + \sum I_P)$ ratio can be used to predict the subsequent decreasing cross sections, for decreasing energies in the center-of-mass frame of reference.

The derived cross sections for the DMSO lithium adduct are illustrated in Figure 4.8. By first evaluating the cross section from the number gas density plot of Figure 4.6 obtained at the highest collision energy ($E_{LAB} = 10$ eV) where the S/N of the product ion is maximized, Figure 4.8 can be constructed. According to the magnitude of the change in the ratios given in Figure 4.7, each subsequent energy cross section at progressively lower energies is calculated relative to this initial point. For the DMSO lithium adduct the reference BDE value is 2.36 eV.[29] Using this liter-ature BDE value, at a number gas density of 3.14×10^{12}, the best fit of Figure 4.7 gives an $I_P/(I_R + \sum I_P)$ ratio of 1.879×10^{-3}. At 2.36 eV, and number gas density of 3.14×10^{12}, the best fit of Figure 4.8 gives a cross-section value of 2.878×10^{-16} cm^2. Using these three values, the $I_P/(I_R + \sum I_P)$ ratio of 1.879×10^{-3}, the number gas density of 3.14×10^{12}, and the cross section of 2.878×10^{-16} cm^2, can subsequently be used to calculate an experimentally derived corrected effective reaction path length (initially set at 1 cm) using the thin target limit relationship. For example, if we take for the lithium adduct of DMSO that the number gas density, $n = 3.14 \times 10^{12}$ cm^{-3}, the $I_P/(I_R + \sum I_P)$ ratio $= 1.879 \times 10^{-3}$, and the product cross section, $\sigma_P = 2.878 \times 10^{-16}$ cm^2, the unit cell path length can be calculated using

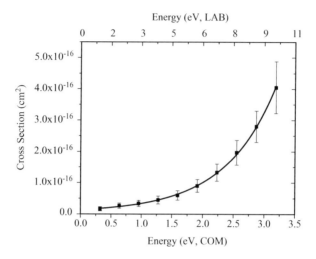

FIGURE 4.8 Calculated cross sections for the DMSO–lithium cationic adduct employing the $\sigma_P = $ slope/ℓ relationship from Figure 4.6 and the proportional changes of the $I_P/(I_R + \sum I_P)$ ratios with collision energy in Figure 4.7. Using the 2.36 eV BDE value, the best fit gives a cross section value of 2.878×10^{-16} cm^2 and, by the thin target limit relationship, a path length of $\ell = 2.08$ cm.

the thin target limit relationship:

$$\frac{I_P}{I_0} = \frac{I_P}{I_R + \sum I_P} = \sigma_P n l \qquad (4.14)$$

rearranged into the form

$$l = \frac{\dfrac{I_P}{\left(I_R + \sum I_P\right)}}{\sigma_P n} \qquad (4.15a)$$

$$l = \frac{I_P}{\left(I_R + \sum I_P\right)\sigma_P n} \qquad (4.15b)$$

and substituting into Equation (4.15b) the values for the number gas density, $n = 3.14 \times 10^{12}$ cm^{-3}, the $I_P/(I_R + \sum I_P)$ ratio $= 1.879 \times 10^{-3}$, and the product cross section, $\sigma_P = 2.878 \times 10^{-16}$ cm^2, we obtain a path length of

$$l = \frac{1.879 \times 10^{-3}}{\left(2.878 \times 10^{-16}\,\text{cm}^2\right)\left(3.14 \times 10^{12}\,\text{cm}^{-3}\right)} \qquad (4.16a)$$

$$\ell = 2.08\,\text{cm} \qquad (4.16b)$$

An average effective reaction path length of 1.95 ± 0.56 cm (as AVG $\pm$ STDEV using seven individual determinations) was calculated using the lithium adducts of

MeOH, DMF, and ACN to obtain an experimentally derived effective reaction path length value that can then be used in all subsequent cross-section calculations and ensuing BDE evaluations.

The rate of a unimolecular decomposition is described by the Rice–Ramsperger–Kassel–Marcus (RRKM) theory. The reaction's rate constant is dependent upon the activated complex's energy state population with respect to the decomposing ion's populations in all other energy states. The mathematical representation of the unimolecular decomposition rate constant is given by

$$k(E) = \frac{1}{h} \frac{z^{\pm}}{Z^*} \frac{\sum P^{\pm}(E - E_0)}{\rho^*(E)} \tag{4.17}$$

where h is Planck's constant, Z is the adiabatic degrees of freedom partition function, the superscript $\pm$ represents the activated complex, the superscript $*$ represents the ionic active molecule, $\rho(E)$ is the density of states, and $P^{\pm}(E - E_0)$ is the number of states contained within the energy range $E - E_0$. The rate of a unimolecular decomposition is approximated by the simplified version of RRKM theory[30]:

$$k(E) = v \left(\frac{E - E_0}{E} \right)^{n-1} \tag{4.18}$$

where n is the number of vibrational degrees of freedom, v is the frequency factor, E is the kinetic energy, and E_0 is the threshold energy for reaction. Equation (4.18) is directly related to the activated complex's populated energy states. According to the transient energy states of the decomposing ion at the time of the dissociating collision, the theoretically derived mathematical function used to obtain a best fit of the experimentally determined cross sections is[3,11]

$$\sigma(E) = \sigma_0 \sum g_i \frac{(E + E_i - E_0)^n}{E^m} \tag{4.19}$$

where E and E_0 remain as above, E_i represents the individual reactant states (electronic, rotational, and vibrational) with populations g_i, σ_0 is a scaling factor, and n and m are adjustable parameters. According to the thin target limit relationship of the reaction cross sections (σ_p) to the measured intensities in Equation (4.12), the cross sections for the determination of the threshold energy for Equation (4.20) can be calculated for the lithium adducts of the solvents such as the methanol lithium adduct (m/z 39) in Figure 4.5. A general expression for the endothermic CID reaction of the dissociation of the lithium adduct from the solvent is

$$M^+S \xrightarrow{\text{Ar (gas)}} M^+ + S \tag{4.20}$$

where M is the metal cation, S is the solvent (e.g., MeOH, DMSO, ACN, etc.), and Ar is the stationary target gas. The free metal cation M^+ (lithium, Li^+, m/z 7) and the precursor ion M^+S signals are measured at increasing collision energies, as are all other products formed via CID.

Figure 4.8 shows the results of these cross-section calculations at increasing collision energies, covering an activation energy range approximately three times that of the threshold energy region, for decompositions of the DMSO–lithium cationic adduct. The cross sections in Figure 4.8 (left axis) are related to the ratio of abundances of the product ions over all ions in Figure 4.7 (left axis) through Equation (4.12). Consistent with the mathematical forms of Equations (4.18) and (4.19), curve fitting of the cross sections obtained from the mass spectrometric measurements of the BDEs for the lithium–monopentadecanoin electrostatic adduct, as well as all the other adducts measured, was empirically observed to fit an exponential growth equation of the following form:

$$\sigma_p(E) = \sigma_0 \exp\left(\frac{E_{COM}}{\alpha}\right) + y_0 \qquad (4.21)$$

where $\sigma_p(E)$ is the experimentally determined cross section in centimeters squared, $\sigma_0 + y_0$ is the y intercept of the fit in centimeters squared, and E_{COM} is the energy of the system in the center-of-mass frame of reference. This equation effectively relates the product cross section to the system energy (supplied by the central hexapole offset) in the center-of-mass frame of reference.

4.3 LOW-MOLECULAR-WEIGHT BDE PREDICTIVE MODEL

To evaluate bond dissociation energies of adducts of the relatively low-molecular-weight solvent standards, the general multivariate growth curve model (GCM–polynomial regression)[31] was used. The GCM–polynomial regression was applied to the electrospray triple quadrupole mass spectrometer cross-section results of the lithium cation solvent adducts. Single exponential cross section versus energy curves, derived as outlined in the previous section for each lithium adduct of the solvents, were used to create a multivariate curve polynomial expression that can be used to calculate BDE values for unknown complexes. To create the polynomial expression, the independent parameters of Equation (4.21), σ_0 and α, are regressed with the dependent variable represented by literature BDE values for each experimentally evaluated complex that had generated a single exponential cross section versus energy curve. Analysis of the variance gave an adjusted regression factor of 0.9, and a probability factor of < 0.0001, demonstrating the statistical significance of the multivariate growth curve model. The model resulted in the following general second-order polynomial descriptive equation that was subsequently used to predict the bond dissociation energies of the lithium–solvent adducts:

$$BDE\,(eV) = -4.22589 - 9.83647 \times 10^{15*}\sigma_0 + 14.3319^*\alpha$$
$$+ 2.20399 \times 10^{31*}\sigma_0^{\,2} - 7.43328^*\alpha^2 \qquad (4.22)$$

TABLE 4.1   Comparison of Model-Fitted Results for Low-Molecular-Weight Lithium Adduct Standards vs. the Measured BDE Calculated from Growth Curve Multivariate Regression Models from the Single Exponential Cross Section vs. Energy Method, Obtained by ES Triple Quadrupole Tandem Mass Spectrometry[a]

Adduct	Reference BDE (eV)	Model Fit BDE (eV)
CH_3OHLi^+	1.596 $(0.083)^{11}$	1.57 ± 0.08
$ACNLi^+$	1.468^{29}	1.60 ± 0.15
$DMSOLi^+$	2.36^{29}	2.44 ± 0.12
$DMFLi^+$	2.43^{29}	2.20 ± 0.09
$C_2H_5OHLi^+$	1.70 $(0.08)^{11}$	1.72 ± 0.06
$CH_3COCH_3Li^+$	1.71 $(0.11)^{11}$	1.68 ± 0.18

[a] Values in parentheses represent uncertainties in the BDE values.

To use this equation to predict a BDE for an unknown complex, it is first necessary to generate a single exponential cross section versus energy curve that will provide values for the σ_0 and α parameters. Table 4.1 is a compilation of the results obtained from the newly developed single exponential cross section versus energy method using Equation (4.22) to predict the BDEs. Overall, excellent agreement between the literature BDEs and the single exponential cross section versus energy method is observed for the low-molecular-weight lithium–solvent complexes studied.

4.4   COMPUTER MODELING OF BDE VALUES

In this section we will see that computer modeling can be used to help aid in thermochemistry values such as the noncovalent bond dissociation energy values that we are currently measuring and calculating in the gas phase using mass spectrometry. For many small compounds that are adducted to lithium ion (ranging anywhere from 4 up to 30 atoms), such as the solvent adducts listed in Table 4.1, there are now numerous bond dissociation energy values in the literature. However, this is often not the case with larger compounds such as lipid metal adducts (ranging from 40 to 119 atoms). Theoretical bond energy values for various lithium–lipid adducts were calculated using Becke-style 3-parameter density functional theory (DFT) computational methods (B3LYP, using the Lee–Yang–Parr correlation functional, at the 6-31G* level). This is a computer modeling experiment that consists of minimizing the potential energy of a lithium–lipid adduct structure, and then sequentially altering the lithium ion distance to the lipid (i.e., assigning incrementally larger distances to the nearest neighboring atom via increases in the "R" parameter of Gaussian 98^{32}) while performing potential energy surface scans of the adduct complex at each step. The computer-modeled theoretical potential energies are then plotted versus distance (in angstroms) of the lithium cation from the lipid. From the difference between the potential energy obtained at the minimum and the potential energy at complete dissociation (i.e., where the charge on the lithium has reached +1 and does not change),

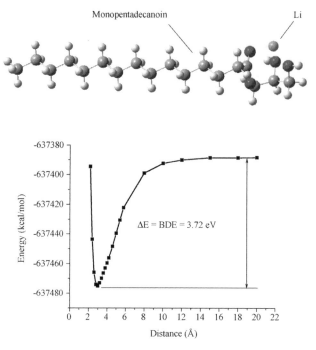

FIGURE 4.9   Optimized structures and computer modeling results for the theoretical determination of the electrostatic bond dissociation energies for the monopentadecanoin lithium adduct. The lithium cation is initially located between the carbonyl oxygen at a distance of 1.86 Å, and the 1-position hydroxyl oxygen at 1.96 Å. The distance to the 2-position hydroxyl oxygen is 1.92 Å. The bond dissociation energy value is extracted as the difference in the energy from the minimum of the potential well to the electrostatic bond distance where the charge has reached unity for the adducted ion. The BDE was calculated to be 3.72 eV.

the electrostatic bond dissociation energy can be calculated. Figure 4.9 illustrates an optimized structure, and a plot of the potential energy surface scans versus distance for the lithium adduct of monopentadecanoin. With the computer modeling it is found that the lithium cation has orientated itself between the carbonyl oxygen and the oxygen atoms of the two hydroxyl groups. The distance to the 1-position hydroxyl oxygen is 1.96 Å, the distance from the lithium cation to the carbonyl oxygen is 1.86 Å, and the distance to the 2-position hydroxyl oxygen is 1.92 Å. For the lithium adduct of 1,3-dipentadecanoin, a similar preferred orientation is found where the lithium cation has oriented itself between the two carbonyl oxygen and the oxygen atom of the hydroxyl group. By computer modeling it was found that the distance between the lithium cation and the two carbonyl oxygen is 1.88 and 2.00 Å, and the distance between the lithium cation and the hydroxyl oxygen is 2.01 Å. For the lithium adduct of 1-stearin-2-palmitin diacylglycerol, the lithium cation was also

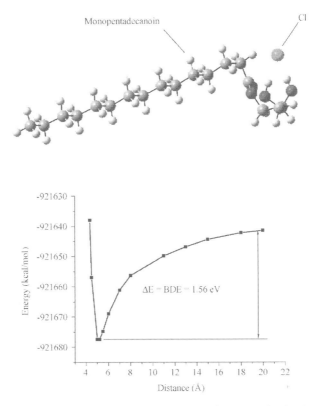

FIGURE 4.10   Optimized structures and computer modeling results for the theoretical determination of the electrostatic bond dissociation energies for the monopentadecanoin chloride adduct. The chloride anion is between the 1-position and 2-position hydroxyl hydrogen, each at a distance of 2.18 Å. The BDE was calculated to be 1.56 eV.

found to orient itself between the two carbonyl oxygen of the 1- and 2-position fatty acyl chains and the oxygen of the hydroxyl group. The distance between the lithium cation and the hydroxyl oxygen is 1.97 Å, whereas the distance between the lithium cation and the two carbonyl oxygen is 1.88 and 2.04 Å.

For chloride adducts of the lipids the same approach was used for calculating the bond dissociation energies using computer modeling. Figure 4.10 illustrates a plot of the potential energy surface scans versus bond distance for the chloride adduct of monopentadecanoin and an optimized structure of the complex. As would be expected due to their higher acidities (more electropositive hydrogen) as compared to hydrocarbon hydrogen atoms, the chloride anion is situated nearest to the 1-position and 2-position hydroxyl hydrogen at distances of 2.18 Å and 2.18 Å, respectively. Also found for the optimized structures of the 1,2- and 1,3-diacylglycerols, orientation of the chloride ion is in close proximity to the lone hydroxyl hydrogen.

TABLE 4.2   Becke-Style 3-Parameter Density Functional Theory Values

Adduct	Theoretical BDE (eV)	Charge on Li^+ or Cl^-	Electrostatic Bond Distance of Li^+ or Cl^- (angstrom)
Li^+ monopentadecanoin	3.72	+0.628	1.9
Cl^- monopentadecanoin	1.56	−0.777	2.2
Li^+ monopentadecanoin dimer	—	+0.527	—
Li^+ 1,3-dipentadecanoin	3.52	+0.669	1.9
Cl^- 1,3-dipentadecanoin	1.48	−0.845	2.1
Li^+ 1,3-dipentadecanoin dimer	—	+0.595	—
Li^+ 1-stearin-2-palmitin glycerol	3.98	+0.579	1.9
Cl^- 1-stearin-2-palmitin glycerol	1.37	−0.836	2.1

The distance between the hydroxyl hydrogen and the chloride ion is 2.09 Å for the 1,2-diacylglycerol-chloride adduct, while for the 1,3-diacylglycerol, the distance is 2.07 Å. Table 4.2 is a compilation of the modeling results for the lithium and chloride adducts of monopentadecanoin, 1,3-dipentadecanoin, 1-stearin-2-palmitin glycerol, and associated dimers.

4.5   HIGH-MOLECULAR-WEIGHT BDE PREDICTIVE MODEL

Using the cross-section results obtained from the electrospray triple quadrupole mass spectrometer, the general multivariate growth curve model (GCM–polynomial regression)[31] was used to create a predictive model for bond dissociation energies involving the relatively high molecular weight lipid standards. Single exponential cross section versus energy curves, derived as outlined in the previous section for each lithium and chloride adduct of the monopentadecanoin, dipentadecanoin, and 1-stearin,2-palmitin diacylglycerol standards, were used to create a multivariate curve polynomial expression that can be used to calculate BDE values for unknown complexes. To create the polynomial expression, the independent parameters of Equation (4.21), σ_0 and α, are regressed with the dependent variable represented by the BDE values (obtained from the DFT computational methods above) for each experimentally evaluated complex that had generated a single exponential cross section versus energy curve. Analysis of the variance gave an adjusted regression factor of 0.93 and a probability factor of <0.0001, demonstrating the statistical significance of the multivariate growth curve model. The model resulted in the following general second-order polynomial descriptive equation that was subsequently used to predict the bond dissociation energies of the higher molecular weight (e.g., m/z 300–700) adducts:

$$\text{BDE (eV)} = 0.78051 - 1.7180 \times 10^{13}\sigma_0 + 1.4627\alpha - 0.16327\alpha^2 \qquad (4.23)$$

It is first necessary to generate a single exponential cross section versus energy curve that will provide values for the σ_0 and α parameters to use this equation to predict a BDE for an unknown complex.

4.6 NONCOVALENT BDE OF Li$^+$ ADDUCT OF MONOPENTADECANOIN

The electrostatic complex of the cationic lithium adduct of monopentadecanoin was measured using the derived effective path length approach. To determine the thresholds for the endothermic CID reaction:

$$M^+L \xrightarrow{\text{Ar (gas)}} M^+ + L \qquad (4.24)$$

where M is the metal cation, L is the lipid, and Ar is the stationary target gas, the free metal cation M^+ (lithium, Li$^+$, m/z 7) and the precursor ion M^+L (m/z 323) signals are measured at increasing collision energies, as are all other products formed via CID. Figure 4.11 shows the emergence of the lithium cation (m/z 7) near the threshold energy for Equation (4.24). Figure 4.11a is the CID product ion mass spectrum obtained at 2.75 eV (E_{COM}), where the lithium metal cation is present only at noise level (S/N = 2). In Figure 4.11b, a significant peak at m/z 7 has emerged for the lithium cation (S/N = 6) at 3.30 eV (E_{COM}). This indicates that the threshold energy for Equation (4.24) has been achieved. From the computer modeling results illustrated in Figure 4.9 the bond dissociation energy value for the lithium–monopentadecanoin adduct was calculated at 3.72 eV, thus giving a threshold energy value that is just below the E_{COM} value of 3.85 eV that was used to obtain the CID product ion spectrum shown in Figure 4.11c.

A combination of factors that contribute to error in the direct measurement of a threshold energy value for Equation (4.24) can be the result of the appearance of decomposition products at an energy lower than the threshold energy for dissociation as calculated by computer modeling. The computational method does not significantly contribute to this error as the magnitude of uncertainty in potential energy calculations at the B3LYP 6-31G* level has been estimated to be 3.9 kcal·mol^{-1}, or 0.17 eV[33] (mean absolute deviation). The uncertainty in using mass spectrometry for estimating dissociation energies has been attributed to (1) collisions outside of the reaction cell (e.g., in Q_1)[16,18] (underestimation of BDEs), (2) non-neglible translational energy at 0 eV (E_{LAB}) collision energy (underestimation of BDEs),[16,18] (3) initial internal energy content, which may lead to metastable decompositions (resulting in underestimation of BDEs), (4) instrumental bias against observation of "slow kinetics" reactions (overestimation of BDEs),[34] and (5) deposition of energy in competitive vibrational excitation modes (overestimation of BDEs).[35] Also known to contribute to this error are competitive fragmentation reactions that are also observed in the product ion mass spectra of the lithium adduct of monopentadecanoin, as illustrated in

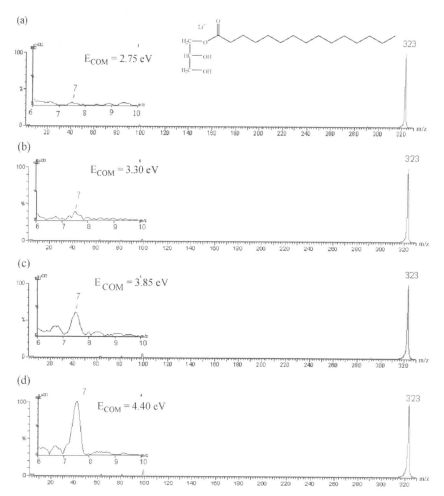

FIGURE 4.11   Product ion mass spectra illustrating the appearance of the lithium cation upon CID of the lithium–monopentadecanoin adduct near the threshold energy for dissociation. E_{COM} energies: (*a*) 2.75 eV, (*b*) 3.30 eV, (*c*) 3.85 eV, threshold area, and (*d*) 4.40 eV.

Figure 4.12. Figure 4.12 is the CID spectrum obtained for the decomposition of the lithium adduct of monopentadecanoin in the reaction threshold range at 3.85 eV (E_{COM}), illustrating a small contribution to the relative error in the measurement of the BDE if only the intensity of the lithium cation is used as $\sum I_p$ instead of using the actual sum of the products as $\sum I_P$ in calculating $I_0 = I_R + \sum I_P$. The product ion intensities that are not being monitored can be neglected when applying Equation (4.12) if the intensities of the "extraneous" products are much smaller than the precursor.[4] However, in this study it was observed that many of the lipids yielded

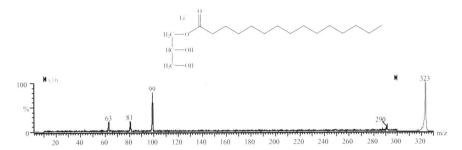

FIGURE 4.12 Competitive fragmentation reactions in CID of product ion mass spectra of lithium–lipid adducts illustrating a contribution to error in the measurement of BDEs. Lithium adduct of monopentadecanoin is in the reaction threshold range (3.85 eV E_{COM}).

other CID fragment ions, in addition to the loss of the lithium cation, at abundances that could not be considered much smaller than the precursor. As can be seen in the product ion spectrum in Figure 4.12, there are product ions at m/z 290, 99, 81, and 63. Thus, it is concluded that in evaluating cross sections for the precursor adduct displayed in Figure 4.12, product ion abundances for all fragmentation pathways must be summed along with the transmitted precursor abundance when applying Equation (4.12).

Next we will look at the plotted results of the BDE study of the lithium adduct of monopentadecanoin. In Figure 4.13a the solid line shows the fit of the experimentally determined cross sections in the threshold region of the dissociation of the lithium–monopentadecanoin adduct, calculated using Equation (4.21). The results of single exponential curve fitting of the cross section versus energy for the chloride adduct of monopentadecanoin is illustrated in Figure 4.13b.

4.7 PRACTICE PROBLEMS

4.7.1 Problem 1

A CID dissociation experiment for the sodium adduct of water, $[NaH_2O]^+$, was collected with a gas cell pressure of 1.1×10^{-4} mbar at a temperature of 27°C. The intensity of the sodium ion was measured at 15.4 abu (arbitrary units) while the total incident ion beam was measured at 10,000 abu. The product cross section (α_p) for the dissociation of the sodium–water adduct is 3.267×10^{-15} cm². Calculate the relative reaction cell path length associated with this experiment.

4.7.2 Problem 2

Convert the following collision-supplied energies from the laboratory frame of reference (E_{LAB}) to the center-of-mass frame of reference (E_{COM}) where m is

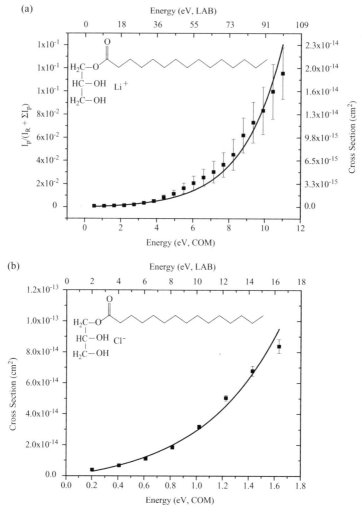

FIGURE 4.13 (*a*) Plot of the ratio of abundances of the dissociated lithium cation to the total product ions [$I_P/(I_R + \sum I_P)$, left y axis] in an energy range that is approximately 3 times greater than the expected dissociation threshold energy of 3.72 eV predicted by computer modeling. The reaction cross sections for the lithium adduct of monopentadecanoin, illustrated on the right y axis, are calculated from the experimentally obtained ratios according to Equation (4.12). The solid line is the best fit of the experimental data using Equation (4.21). The bond dissociation energy for CID of the lithium cation was experimentally determined to be 3.69 ± 0.29 eV (E_{COM}). The theoretical value was calculated to be 3.72 eV. (*b*) Cross sections of the chloride adduct of monopentadecanoin. The bond dissociation energy was experimentally determined to be 1.65 ± 0.05 eV (E_{COM}) for CID of the chloride anion. The theoretical value was calculated to be 1.56 eV.

the mass of the stationary target collision gas and M is the mass of the analyte ion.

E_{LAB} (eV)	m (amu)	M (Thompson)	E_{COM} (eV)
10.63	Nitrogen (N_2)	128.63	
12.53	Argon (Ar)	633.65	
9.66	Helium (He)	155.36	
18.52	Argon (Ar)	210.03	
15.63	Neon (Ne)	330.62	
12.55	Xenon (Xe) 131.293	256.83	
22.06	Nitrogen (N_2)	402.36	
4.55	Helium (He)	680.23	
14.53	Xenon (Xe) 131.293	596.63	

4.7.3 Problem 3

The following product ion dissociation data was collected at a constant temperature of 25°C and collision energy of $E_{LAB} = 15$ eV. Using a reaction path length of 3.2 cm, calculate the collision cross section of the product using the relationship $\sigma_P =$ slope/l (*Hint*: consider constructing a graph similar to Fig. 4.14).

Product Ion Intensity (I_P)	Total Incident Ion Intensity ($I_R + \Sigma I_P$)	Gas Cell Pressure (μ bar)
4.06	10055	0.05
6.47	9863	0.10
8.41	9755	0.14
10.98	8750	0.21
14.11	9025	0.27
16.65	8875	0.32
17.88	8456	0.36
19.72	8166	0.41
19.84	7569	0.45

4.7.4 Problem 4

Figure 4.15 illustrates the structure of the lithium (Li^+) adduct of 1-stearin,2-palmitin diacylglycerol. Figure 4.16 is the product ion spectrum of the Li^+ adduct of 1-stearin,2-palmitin diacylglycerol at 3.10 eV (E_{COM}). Although the spectrum shows a considerable number of lower abundance products formed during CID, we are interested in calculating an experimental value for the noncovalent BDE of the lithium cation adduct. The theoretical value for the BDE of the lithium cation adduct has been determined by modeling to be 3.98 eV (E_{COM}). Calculate the experimental bond

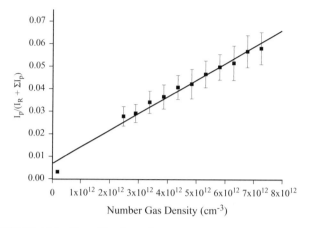

FIGURE 4.14   Plot of ion intensity ratios versus number gas density.

dissociation energy (BDE) by plotting the values in Table 4.3, fitting the curve to the form of Equation (4.21), extracting the values of σ_0 and α, and finally substituting into Equation (4.23).

$$\sigma_p(E) = \sigma_0 \exp\left(\frac{E_{COM}}{\alpha}\right) + y_0 \tag{4.21}$$

$$\text{BDE (eV)} = 0.78051 - 1.7180 \times 10^{13}\,\sigma_0 + 1.4627\alpha - 0.16327\alpha^2 \tag{4.23}$$

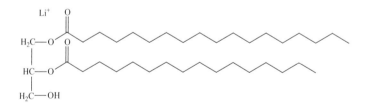

FIGURE 4.15   Structure of the lithium (Li$^+$) adduct of 1-stearin,2-palmitin.

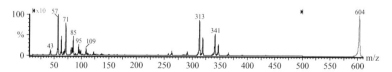

FIGURE 4.16   Product ion spectrum of the Li$^+$ adduct of 1-stearin,2-palmitin at 3.10 eV (E_{COM}).

**TABLE 4.3   Experimental Collision Energies
and Cross Sections**

Collision Energy (eV, COM)	Cross Section (cm^2)
0.31	1.71×10^{-16}
0.62	2.57×10^{-16}
0.93	2.56×10^{-16}
1.24	2.27×10^{-16}
1.55	2.45×10^{-16}
1.86	2.41×10^{-16}
2.17	3.42×10^{-16}
2.48	4.59×10^{-16}
2.79	5.51×10^{-16}
3.10	8.74×10^{-16}
3.41	8.99×10^{-16}
3.72	1.08×10^{-15}
4.03	1.27×10^{-15}
4.34	1.34×10^{-15}
4.66	1.59×10^{-15}
4.97	2.08×10^{-15}
5.28	2.31×10^{-15}
5.59	2.77×10^{-15}
5.90	2.76×10^{-15}
6.21	2.86×10^{-15}

4.7.5   Problem 5

Figure 4.17 shows the chloride (Cl^-) adduct of 1-stearin,2-palmitin diacylglycerol. The theoretical value for the BDE of the lithium cation adduct has been determined by modeling to be 1.37 eV (E_{COM}). Calculate the experimental bond dissociation energy (BDE) by plotting the values in Table 4.4, fitting the curve to the form of Equation (4.21), extracting the values of σ_0 and α, and finally substituting into Equation (4.23).

$$\sigma_p(E) = \sigma_0 \exp\left(\frac{E_{COM}}{\alpha}\right) + y_0 \tag{4.21}$$

$$BDE(eV) = 0.78051 - 1.7180 \times 10^{13}\sigma_0 + 1.4627\alpha - 0.16327\alpha^2 \tag{4.23}$$

FIGURE 4.17   Chloride adduct (Cl^-) adduct of 1-stearin,2-palmitin.

TABLE 4.4 Experimental Collision Energies and Cross Sections

Collision Energy (eV, COM)	Cross Section (cm^2)
0.119	8.87×10^{-15}
0.238	9.66×10^{-15}
0.357	1.15×10^{-14}
0.476	1.50×10^{-14}
0.595	1.97×10^{-14}
0.714	2.42×10^{-14}
0.833	3.07×10^{-14}
0.952	3.81×10^{-14}
1.071	4.58×10^{-14}

4.8 BDE DETERMINATION OF Li$^+$ LIPID DIMER ADDUCTS

We shall now look at an alternative method for calculating BDE values of noncovalent complexes in the gas phase using mass spectrometry. For the CID tandem mass spectrometric determination of BDE values for dimers, the primary method used is to equate the point of 50% dissociation, measured as ML/(ML + M$_2$L), to the center-of-mass collision energy.[13,36] The monoacylglycerol lipid, the 1,2-diacylglycerol, and the 1,3-diacylglycerols, have been observed to form lithium-bound noncovalent dimers of the form [M$_2$Li]$^+$. Using the high-molecular-weight growth model, the energy for the dissociation of the monopentadecanoin dimer the BDE was measured to be 1.43 ± 0.04 eV. Experiments for the mass spectrometric determination of the BDE for the lithium-bound monopentadecanoin dimer is shown in Figure 4.18b where the study was performed in the range of the threshold energy for dissociation. Experiments were also performed to determine the 50% dissociation point of the dimer. A plot of the ratio of the product ion abundance obtained from the dissociation of the monoacylglycerol dimer versus the total ion abundance is illustrated in Figure 4.18a. The total ion abundance was determined as MLi$^+$/(MLi$^+$ + M$_2$Li$^+$) to approximate the precursor ion abundance. The BDE for the monopentadecanoin dimer was obtained by fitting a fourth-order polynomial to the MLi$^+$/(MLi$^+$ + M$_2$Li$^+$) ratio data versus collision energy. The obtained fourth-order polynomial was then set to equal 0.50 and solved for the respective roots (always an obvious choice as to the correct root). The fitting of the fourth-order polynomial for 50% dissociation of the monopentadecanoin dimer in Figure 4.18a gave an E_{COM} of 1.31 ± 0.32 eV (the exponential growth curve model predicted a BDE = 1.43 ± 0.04 eV). This is within an acceptable agreement of experimentally determined BDE values.

Experiments were also performed for the dissociation of the 1,3-dipentadecanoin lithium-bound dimer. The results are illustrated in Figure 4.19b, which is a graph that plots the experimentally determined cross sections versus collision energy. Previously, the exponential growth curve model calculates a dissociation energy of 1.43 ± 0.01 eV for the 1,3-dipentadecanoin lithium-bound dimer. A plot of the

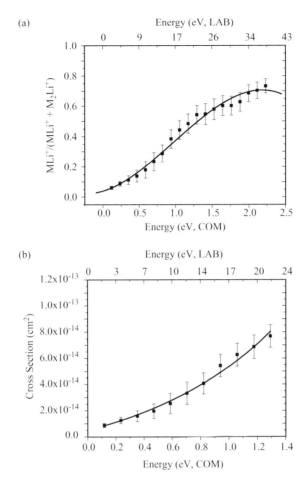

FIGURE 4.18 Cross section versus collision energy for the lithium-bound monopentade-canoin dimer. (*a*) The BDE extracted from the ratio of MLi^+ to $(MLi^+ + M_2Li^+)$ at the point of 50% dissociation gives a value of 1.31 ± 0.32 eV E_{COM}. (*b*) The BDE of 1.43 ± 0.04 eV (E_{COM}) for dimer dissociation by CID was obtained from the exponential growth curve model.

ratio of the product ion from the dissociation of one of the 1,3-diacylglycerol dimers versus the parent ion intensity, $MLi^+/(MLi^+ + M_2Li^+)$ is illustrated in Figure 4.19*a*. For 50% dissociation, an E_{COM} value of 0.90 ± 0.09 eV was calculated from a fourth-order polynomial. The exponential growth curve model predicted a BDE = 1.43 ± 0.01 eV.

The results of an experiment of a 1-stearin,2-palmitin diacylglycerol lithium-bound dimer dissociation study is illustrated in Figure 4.20. The exponential growth curve model shown in Figure 4.20*b* determined an estimated BDE value of 1.38 ± 0.01 eV. The plotted results of the 1-stearin,2-palmitin diacylglycerol lithium-bound dimer is shown in Figure 4.20a. The fitting of the fourth-order polynomial and subsequent root determination gave a 50% dissociation E_{COM} value of 1.16 ± 0.07 eV. This compares quite well with the exponential growth curve model predicted BDE of 1.38 ± 0.01 eV. The experimentally obtained energy values corresponding to 50% dissociation agree well with the growth curve model predicted BDEs. A comparison between the exponential growth curve model determined BDE values and the 50% dissociation method is presented in Table 4.5.

Table 4.5 is a compilation of the results obtained from the newly developed single exponential cross section versus energy method as compared to the theoretical computer modeling method and the "50% dissociation of dimers" method.

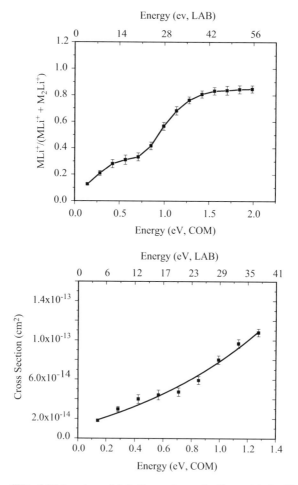

FIGURE 4.19 CID of lithium-bound 1,3-dipentadecanoin dimer. (*a*) At 50% dissociation the BDE was measured to be 0.90 ± 0.09 eV. (*b*) The growth curve model yielded a dissociation energy of 1.43 ± 0.01 eV.

The cationic lithium adducts, the anionic chloride adducts, and the dimers are relatively large systems, spanning molecular weights from 323 to 603 Da, that contain substantial numbers of vibrational degrees of freedom. This is reflected in the magnitude of observed coefficients of variation (CV) for the studied systems such as 7.6% for the lithium adduct of monopentadecanoin, 10.5% for the lithium adduct of dipentadecanoin diacylglycerol, and 4.8% for the lithium adduct of 1-stearin,2-palmitin diacylglycerol. Overall, excellent agreement is observed between the theoretical BDEs obtained from computational molecular modeling, and those obtained by the multivariate growth curve-fitting method using one second-order polynomial [Equation (4.23)] to evaluate all three types of complexes. The relative uncertainty of the experimentally determined bond dissociation energies using

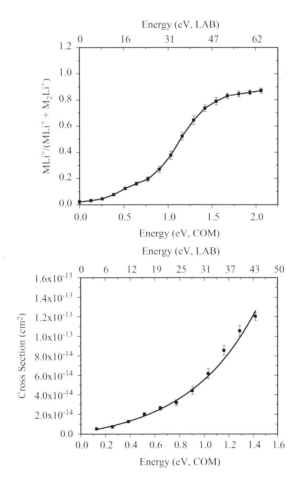

FIGURE 4.20 CID of the lithium-bound 1-stearin,2-palmitin diacylglycerol dimer. (*a*) 50% dissociation energy value of 1.16 ± 0.07 eV. (*b*) The growth curve model yielded a dissociation energy BDE value of 1.38 ± 0.01 eV.

TABLE 4.5 Comparison between the Exponential Growth Curve Model Determined BDE Values and the 50% Dissociation Method

Adduct	Theoretical BDE (eV)	Growth Curve Model BDE (eV)	50% Dissociation Method (eV)
Li^+ monopentadecanoin	3.72	3.69 ± 0.29	—
Cl^- monopentadecanoin	1.56	1.65 ± 0.05	—
Li^+ monopentadecanoin dimer	NA[a]	1.43 ± 0.04	1.31 ± 0.32
Li^+ 1,3-dipentadecanoin	3.52	3.78 ± 0.39	—
Cl^- 1,3-dipentadecanoin	1.48	1.63 ± 0.23	—
Li^+ 1,3-dipentadecanoin dimer	NA[a]	1.43 ± 0.01	0.90 ± 0.09
Li^+ 1-stearin-2-palmitin	3.98	3.59 ± 0.18	—
Cl^- 1-stearin-2-palmitin	1.37	1.26 ± 0.03	—
Li^+ 1-stearin-2-palmitin dimer	NA[a]	1.38 ± 0.01	1.16 ± 0.07

[a] NA = not analyzed.

this method is approximated to be ±7%, while absolute uncertainties are estimated to range between 10 and 20%. Rodgers and Armentrout[10] reported relative uncertainties of ±5% and cross-section absolute uncertainties of ±20% for their guided ion beam mass spectrometer. The variation in their cross-section measurements was attributed to errors in estimating interaction region lengths and errors in gas cell pressure measurements. In the method reported here, gas cell pressure uncertainty still exists, but the calculation of an effective reaction path length is believed to reduce the magnitude of the path length's error contribution to uncertainty in BDE estimations.

4.9 COVALENT APPARENT THRESHOLD ENERGIES OF Li^+ ADDUCTED ACYLGLYCEROLS

The use of electrospray[23,25] tandem mass spectrometry (ES-MS/MS) is an established and efficient analytical tool for compound identification and structural elucidation.[37–39] A wide range of mass spectrometers have been used to study thermochemistry in the gas phase employing collision-induced dissociation (CID) tandem mass spectrometry.[13] Examples include noncovalent receptor–ligand complexes by electrospray triple quadrupole,[40–42] non-covalent bond dissociation energies by guided ion beam tandem mass spectrometry,[3–11] critical energies for ion dissociation by quadrupole ion trap[43] or flowing afterglow triple quadrupole,[15,16] reaction enthalpies and ion–ligand bond energies by electrospray triple quadrupole,[18,19] and binding energies from gas-phase ion-equilibrium[21] studies by electrospray–ion reaction chamber–triple quadrupole.[20,22]

The precursor ion of interest is isolated by the first quadrupole acting as a mass filter, subjected to collision-induced dissociation in the RF-only second hexapole, and finally the product ions are scanned in the third quadrupole for structural elucidation by quadrupole–hexapole–quadrupole (QhQ) tandem mass spectrometry. The products from the collision–induced fragmentation are formed under multiple collision conditions when collision energy and collision cell gas pressure are chosen to achieve 20–70% attenuation of the precursor ion abundance. Single-collision conditions ($\sim$10% attenuation of precursor beam) are employed in the quadrupole–hexapole–quadrupole mass spectrometer in energy-resolved mass spectrometry studies[44–51] where the specific kinetics of the fragmentation pathways are of interest, and a series of spectra are obtained at increasing activation energies controlled by the energy offset of the central hexapole. A wealth of information is being collected that can be used to qualitatively describe the unimolecular fragmentation pathways through the use of breakdown graphs during energy-resolved mass spectrometry studies, where the percentage of the total abundance for each product ion is plotted versus collision energy. The comparative abundances of the product ions at specific collision energies is one quantitative aspect of the breakdown graphs. The observance of a higher relative product ion abundance, expressed as the percentage of total ions measured, equates to a more favored reaction pathway at that energy as compared to other product ions under the same condition. The calculation and graphing of branching ratios[52] and product cross sections are obtained through the acquisition of spectra at incrementally increasing energy offset values. The activation energies required for dissociation according to the RRKM theory[30] of unimolecular reactions are directly related to the product cross sections of unimolecular reactions, which occur during fragmentation studies employing CID.

In the beginning of the chapter we saw that taking advantage of the thin target limit relationship to collision cross sections[4] resulted in the development of a predictive model for bond dissociation energies of alkali metal or halide ion adducts,[53] where only the simple dissociation of the alkali metal cation or halide anion was considered. In the remaining part of the chapter we will explore the extension of this model to applying it to the description of *covalent* apparent threshold energies (ΔE_0) implicated in unimolecular fragmentations of lithium adducts of acylglycerols.

4.9.1 Apparent Threshold Energy Predictive Model

Most experiments performed using low-energy CID have collision energy values ranging from 0 eV up to 200 eV (E_{LAB}). A typical experiment for energy-resolved experiments consists of acquiring product ion spectra at 0 eV (E_{LAB}), as set by the energy offset of the central hexapole, followed by increasing offset values in 1 eV (E_{LAB}) increments. The laboratory (LAB) frame of reference to the center-of-mass (COM) frame of reference is converted from the supplied energy for the collision according to $E_{COM} = E_{LAB}[m/(m + M)]$, where E_{LAB} is the measured offset for the central hexapole, m is the mass of the stationary target Ar atom, and M is the mass of the analyte. Using the thin target limit relationship[4]:

$$\frac{I_P}{I_0} = \frac{I_P}{I_R + \Sigma I_P} = \sigma_P n \ell \qquad (4.14)$$

where I_P is the transmitted product ion abundance, I_R is the transmitted reactant ion abundance, σ_P is the product cross section (cm^2), n is the number gas density (cm^{-3}), I_0 is the incident (precursor) ion abundance, and ℓ is the collision cell path length (cm); individual product cross sections can be obtained for each product ion at a specified collision energy. The magnitude of the change of the $I_P/I_0 = I_P/(I_R + \Sigma I_P)$ ratio using the thin target limit relationship can be used to predict the subsequent decreasing cross sections. To calibrate an effective path length for the experimental unimolecular dissociations used for cross-section calculations only (not for direct energy threshold calculation) (see Ham and Cole[53] for a more detailed description of the effective path length calibration), the thin target limit relationship is used. In the center-of-mass frame of reference the product cross sections are plotted versus the collision energy, and the following exponential growth equation:

$$\sigma_P(E) = \sigma_0 \exp\left(\frac{E_{COM}}{\alpha}\right) + y_0 \tag{4.21}$$

is fit to the data in the threshold activation energy region for product ion formation [between approximately 0.06–4 eV (E_{COM})]. While Equation (4.21) is similar to the Arrhenius exponential equation $\sigma_P(E) = \sigma_0 \exp(-a/E)$, which can be used to directly calculate the dissociation reaction activation barrier where σ_0 is a preexponential factor and a is the activation energy, Equation (4.21) does not directly predict the dissociation reaction activation barrier. Equation (4.21) is used to describe the behavior, or form, of the unimolecular dissociation reaction in the threshold region where the parameters of the fitted curve are then used in the following general multivariate growth curve model (GCM–polynomial regression)[31] developed for precursor ions having m/z values in the 323 to 603 range[53]:

$$BDE\,(eV) = 0.78051 - 1.7180 \times 10^{13}\sigma_0 + 1.4627\alpha - 0.16327\alpha^2 \tag{4.23}$$

to calculate the apparent bond dissociation energy from unimolecular fragmentation for that particular product ion. The relationship between Equations (4.21) and (4.23) is empirical, not physical, based upon a general multivariate growth curve model that has previously been demonstrated to be statistically significant.[53] It should be noted that Equation (4.23) was derived using BDE values for noncovalent dissociations of alkali metal (Li^+) adducts of acylglycerol lipids in identical settings used for the determination of apparent threshold energies.

Equation (4.21) indicates that cross sections are measured below the reaction endoergicity of the unimolecular decomposition reaction being measured (in other words, an activation energy is measured, even though quite small, at all offset energy values). The appearance of decomposition products at an energy lower than the threshold energy for dissociation can be the result of a combination of factors that contribute to error in the direct measurement of a threshold energy value. Briefly, as discussed in Ham and Cole,[53] uncertainty in mass spectrometrically estimating dissociation energies has been attributed to: (1) initial internal energy content, which may lead to metastable decompositions (resulting in underestimation of BDEs), (2) non-neglible translational energy at 0 eV (E_{LAB}) collision energy (underestimation of BDEs), (3) collisions outside of the reaction cell (e.g., in Q_1) (underestimation of BDEs), (4) instrumental bias against observation of "slow kinetics" reactions (overestimation of BDEs), (5) deposition of energy in competitive vibrational excitation modes (overestimation of BDEs), and target gas thermal motion producing energy broadening (underestimation of BDEs). Secondly, the model is stating that statistically, even though very slight, there is the possibility of decomposition at all input energy values. In practice, the activation energy required for decomposition is observed at input energy values at or above the threshold region.

4.9.2 Apparent Threshold Energies for Lithiated Monopentadecanoin

The lithium adduct of monopentadecanoin was the first acylglycerol to which the derived effective reaction path length approach for the measurement of apparent threshold energies was applied. The product ion spectrum of the collisionally activated monopentadecanoin lithium adduct is illustrated in Figure 4.21a. The spectrum was acquired at 40 eV (E_{LAB}) and 3×10^{-4} mbar gas cell pressure (multiple collision conditions), showing the major fragments produced during unimolecular dissociation. Figure 4.21b lists the four major product ions observed and their respective assignments. The most abundant product ion in the spectrum is the m/z 99 product ion that is the result of the neutral loss of the C15 fatty acyl chain as a ketene from the precursor ion. The next are the m/z 81 and m/z 63 product ions that are assigned as consecutive decompositions of m/z 99 produced from one water loss and two water losses, respectively. The m/z 57 product ion is assigned as LiOH loss from the m/z 81 ion and is depicted in Figure 4.21b along with a potential minor contribution from the charge-remote fragmentation pathway that produces the hydrocarbon series $C_nH_{2n+1}^+$ ($n = 4$). The "percent total ion abundance" over a wide range of collision energies of the four product ions follows the trend: m/z 99 > m/z 81 > m/z 63 > m/z 57 (see Table 4.7), as is visually apparent in the breakdown graph presented in Figure 4.23. The CID spectra were acquired under single-collision conditions for these energy-resolved determinations; therefore, the percent total ion abundances for the product ion fragments are very low. To further contrast the precursor versus product ion abundances during these experiments the precursor percent total ion abundance was also plotted (but note the discontinuity in the y axis).

From the theory of unimolecular ion decompositions[30] it is known that the bond with the lowest critical energy for dissociation is not always the one to rupture at each respective internal energy state. Due to simultaneous formation of new bonds during the fragmentation process, the kinetic considerations where the energy barrier required to cleave a bond for a competing dissociation may be lowered, thus resulting in a more abundant product ion at a certain internal energy state. Because the latter benefit from stabilization of the transition state owing to new bond formation, fragment ions produced by simple cleavage from a "loose complex" are often observed at higher threshold energies than those formed via rearrangements involving "tight

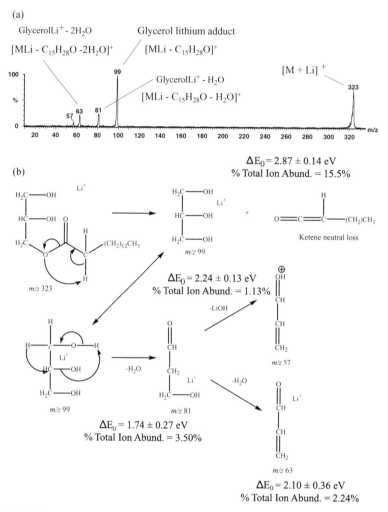

(a)

GlycerolLi$^+$ - 2H$_2$O

[MLi - C$_{15}$H$_{28}$O -2H$_2$O]$^+$

Glycerol lithium adduct

[MLi - C$_{15}$H$_{28}$O]$^+$

GlycerolLi$^+$ - H$_2$O

[MLi - C$_{15}$H$_{28}$O - H$_2$O]$^+$

[M + Li]$^+$

ΔE_0= 2.87 ± 0.14 eV
% Total Ion Abund. = 15.5%

(b)

Ketene neutral loss

m/z 323

m/z 99

ΔE_0 = 2.24 ± 0.13 eV
% Total Ion Abund. = 1.13%

-LiOH

m/z 57

m/z 99

-H$_2$O

m/z 81

-H$_2$O

ΔE_0 = 1.74 ± 0.27 eV
% Total Ion Abund. = 3.50%

m/z 63

ΔE_0 = 2.10 ± 0.36 eV
% Total Ion Abund. = 2.24%

FIGURE 4.21 (a) Product ion spectrum of the collision-activated monopentadecanoin lithium adduct [acquired at 40 eV (Lab) under multiple collision conditions]. Major fragments produced: (1) m/z 57 formed by loss of LiOH from m/z 81 and from part of the C$_n$H$_{2n+1}^+$ hydrocarbon series from the fatty acid acyl chain, (2) m/z 81 produced by water loss from m/z 99, (3) m/z 63 produced by two water losses from m/z 99, and (4) m/z 99 from the neutral loss of C15 fatty acyl chain as ketene, producing lithiated glycerol. (b) Fragmentation pathways, structures, apparent threshold energies, and percent total ion abundance for the four major product ions formed from CID of lithiated monopentadecanoin. The m/z 81 and m/z 63 ions are produced from one water loss, then two water losses, respectively. The m/z 81 ion has the lowest threshold energy at 1.74 ± 0.27 eV (E_{COM}). The m/z 99 ion is the result of the neutral loss of C15 fatty acyl chain as ketene from the precursor ion and is the most abundant ion in the spectrum, at an apparent threshold energy of 2.87 ± 0.14 eV (E_{COM}). The m/z 57 ion is formed primarily by LiOH loss from the m/z 81 ion.

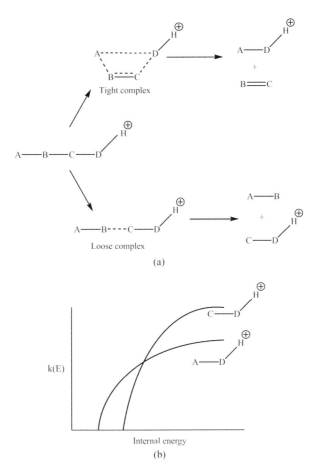

FIGURE 4.22 (*a*) Unimolecular fragmentation pathways through a tight complex and through a loose complex. (*b*) Dependence of $k(E)$ upon tight or loose complex pathways.

complexes." Disfavored in cases where substantial amounts of excess energy are present, decomposition reactions that proceed through tight complexes have unfavorable entropic factors, and insufficient time is available to achieve a proper conformation of the tight complex. This is known as kinetic versus thermodynamic effects. The probability that a unimolecular decomposition will take place is expressed by the rate constant k. To illustrate the effects of kinetic versus thermodynamic, we will consider the two processes taking place in Figure 4.22. In Figure 4.22*a* two decomposition pathways are illustrated where the top pathway represents an intermediate that is formed through a tight complex. In the tight complex intermediate pathway there are two new bonds being formed that will effectively offset the energy of the bond breakage required to form the products. The bottom fragmentation pathway passes through a loose complex where the middle bond breakage is the only bond change taking place without the formation of any new bonds. Thus the energy of the bond breakage is not offset by the formation of any new bonds. In summing up the energy involved in these unimolecular fragmentation pathways, the top pathway through the tight complex has a more favorable enthalpy. However, because the complex must align itself into the tight complex formation, the fragmentation pathway has less favorable entropy. The effect of these two pathways on the probability rate constants, $k(E)$, with respect to internal energy is illustrated in Figure 4.22b. At lower internal energy the $k(E)$ for the tight complex fragmentation pathway is encountered first and is the dominant pathway. As the internal energy is increased, the $k(E)$ for the loose complex becomes the dominant pathway. The accumulation of energy to the point of activation in the bond breakage of the loose complex is the primary requirement for this fragmentation pathway to be observed while for the tight complex two conditions must be met, the required activation energy and the orientation of the bonds in space. At higher energies, the loose complex will win the competition for the unimolecular fragmentation pathway favored.

The predictive model for calculating apparent threshold energies for fragmentation pathways from energy-resolved mass spectral data gives a quantitative aspect to describe the product ions formed during CID experiments. For the various decompositions illustrated in Figure 4.21b at 7 eV (E_{COM}) collision energy where product ion abundances are near maximal, what can be deduced from the values of ΔE_0 are that the calculated percent total ion abundances are not inversely related to the apparent threshold energies (ΔE_0). The product ions are produced through secondary or competitive processes is being suggested from this observation. It can be seen that the production of the m/z 81 ion, $\Delta E_0 = 1.74 \pm 0.27$ eV (E_{COM}), requires the lowest energy compared to the other three product ion pathways. This is followed by the m/z 63 ion at $\Delta E_0 = 2.10 \pm 0.36$ eV (E_{COM}), the m/z 57 ion at $\Delta E_0 = 2.24 \pm 0.13$ eV (E_{COM}), and finally the m/z 99 ion at $\Delta E_0 = 2.87 \pm 0.14$ eV (E_{COM}). The m/z 99 product ion has a much greater rate of formation than the other three product ions at elevated collision energies as shown in the percent total ion abundance of 15.5%, suggesting that the m/z 99 product ion is the most entropically favored fragmentation pathway. Figure 4.23 is an energy-resolved breakdown graph for the dissociation of the lithiated monopentadecanoin diacylglycerol under single-collision conditions that graphically illustrates this.

4.9.3 Apparent Threshold Energies for Lithiated 1-Stearin,2-Palmitin

Next to be studied using energy-resolved mass spectra was the lithium adduct of 1-stearin,2-palmitin. Figure 4.24a, at a collision energy of 50 eV (E_{LAB}) and gas cell pressure of 3×10^{-4} mbar (multiple collision conditions) illustrates the product ion spectrum of the collision-induced dissociation of 1-stearin,2-palmitin. From approximately m/z 43 to m/z 123 the spectrum shows a number of lower molecular weight fragment ions and higher molecular weight product ions ranging from m/z 257 to m/z 365. Assigned in Figure 4.24b (the protonated C18 fatty acid substituent at m/z 285 was very minor and is not listed) is the latter higher molecular weight range that contains nine product ions. The proposed mechanisms each constitute a single

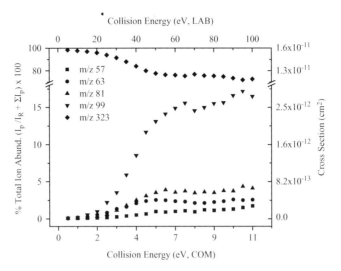

FIGURE 4.23 Single-collision conditions energy-resolved breakdown graph for the dissociation of the lithiated monopentadecanoin. Percent total ion abundances follow the trend of m/z 99 > m/z 81 > m/z 63 > m/z 57. Due to collection under single-collision conditions the percent total ion abundances are very low. The precursor percent total ion abundance is included in the plot to illustrate the small percentage losses during these experiments.

dissociation involving one H transfer in concert with cleavage to produce the array of product ions listed in Figure 4.24b, and all nine of these higher mass product ions appear to have plausible fragmentation pathways that do not arise from consecutive dissociations.

The product ions from 1-stearin,2-palmitin dissociations generally appear in percent total ion abundances that are roughly inversely related to the apparent threshold energies. This is in contrast to the product ions of the lithium adduct of monopentadecanoin presented above. Table 4.6 is a listing of the derived apparent threshold energies versus percent total ion abundance, which shows the inverse relationship. For this set of fragmentation pathways a lower threshold energy equates to a higher percent total ion abundance, except for anomalies with the two lowest abundance ions (m/z 365 and m/z 257). The breakdown graph of the higher molecular weight product ions is illustrated in Figure 4.25a, which shows that each fragmentation product increases in abundance in a roughly similar pattern with increasing collision energy, that is, little curve crossing occurs. The curves in Figure 4.25a do not cross, indicating that the relative importance of entropic versus enthalpic factors, which govern the rates of decomposition for each process, does not change substantially as higher internal energy uptake is achieved. From the figure it can be seen that the m/z 313 product ion has the greatest abundance over a wide range of energies, followed by the m/z 341 ion. The processes leading to formation of these two product ions are related in that each has lost a neutral lithium fatty acetate molecule. The loss of the 1-position substituent is the favored pathway (giving rise to the m/z 313 product ion), as has been observed previously,[54] a conclusion that the current studies also support.

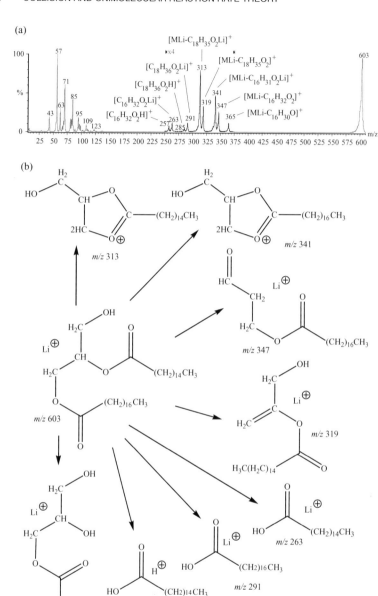

FIGURE 4.24 (*a*) Product ion spectrum of the collision-induced dissociation (multiple collisions) of lithiated 1-stearin,2-palmitin, at collision energy of 50 eV (Lab). From approximately *m/z* 43 to *m/z* 123, primarily representing the hydrocarbon series $C_nH_{2n+1}{}^+$ from σ-bond fatty acyl chain cleavage, are the lower molecular weight fragment ions. Ranging from *m/z* 257 to *m/z* 365 are the higher molecular weight product ions produced from single-cleavage reactions of fatty acyl loss in several forms (fatty acyl chain as ketene, lithium fatty acetate, and fatty

TABLE 4.6 Percent Total Ion Abundances, Apparent Threshold Energies, and Cross Sections of High-Molecular-Weight Product Ions for Lithium Adduct of 1-Stearin, 2-palmitin

Product Ion (m/z)	% Total Ion Abundance[a] (%)	Apparent Threshold Energy (ΔE_0, eV)	Cross Section (cm^2)
313	4.74	1.35 ± 0.13	7.74×10^{-13}
341	3.23	1.60 ± 0.13	5.28×10^{-13}
319	2.38	1.66 ± 0.15	3.90×10^{-13}
347	1.77	1.79 ± 0.16	2.90×10^{-13}
263	1.01	1.97 ± 0.26	1.65×10^{-13}
291	0.99	2.26 ± 0.01	1.62×10^{-13}
365	0.82	2.02 ± 0.11	1.34×10^{-13}
257	0.78	1.86 ± 0.26	3.55×10^{-14}

[a] Listed in descending order.

The m/z 313 product ion has the lowest apparent threshold energy at 1.35 ± 0.13 eV (E_{COM}), followed by the m/z 341 product ion at 1.60 ± 0.13 eV (E_{COM}).

The pair of product ions with the highest abundances, and lowest apparent threshold energies, arise from the neutral loss of the lithium fatty acetates at m/z 313 and 341 for the losses involving the fatty acid substituents from this diacylglycerol. With somewhat higher apparent threshold energies, the next most abundant pair are product ions formed by the neutral loss of the fatty acid substituent (acid form) at m/z 347 and 319. The next highest abundance and even greater apparent threshold energy is the neutral loss of the fatty acyl chain as a ketene at m/z 365, suggesting the following competitive ranking for fragmentation pathways: loss of lithium fatty acetate > loss of fatty acid > loss of fatty acyl chain as ketene.

Illustrated in Figure 4.25b is the breakdown graph of the lower molecular weight product ions, and the corresponding apparent threshold energies are listed in Table 4.7. Low-molecular-weight product ions in the m/z 43 to 127 region are assignable as alkyl chain fragmentations through charge-remote σ-bond cleavages of the fatty acid hydrocarbon chain producing the series $C_nH_{2n+1}^+$. Figure 4.24a illustrates that the hydrocarbon series from σ-bond cleavages of the saturated fatty acyl chain consists of m/z 43, m/z 57, m/z 71, m/z 85, and m/z 99. The m/z 57 product ion has the lowest apparent threshold energy at 1.35 ± 0.26 eV (E_{COM}) and the highest percent total ion abundance at 16.0%. This is analogous to the fragmentation

acid). (b) CID of 1-stearin,2-palmitin, ranging from m/z 257 to m/z 365, fragmentation pathways, and structures for the product ions. Paired products include the neutral loss of C18 and C16 fatty acid at m/z 319 and m/z 347, and neutral loss of C18 and C16 lithium fatty acetate at m/z 313 and m/z 341; m/z 365 is formed by the neutral loss of C16 fatty acyl chain as ketene. Fragmentation ranking of loss of lithium fatty acetate > loss of fatty acid > loss of fatty acyl ketene.

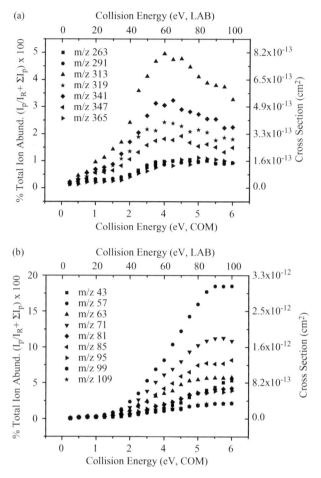

FIGURE 4.25 Single-collision condition energy-resolved breakdown graph of: (*a*) 1-stearin,2-palmitin higher molecular weight product ions from CID. The fragmentation pathways primarily arise from losses of fatty acid substituents in the form of lithium fatty acetate, fatty acyl chain as ketene, and fatty acid. In a similar pattern for increasing collision energy all increase in product ion intensity. (*b*) 1-Stearin,2-palmitin lower molecular weight product ions from CID. Peak series *m/z* 43, *m/z* 57, *m/z* 71, *m/z* 85, and *m/z* 99 result from σ-bond cleavage of the fatty acid hydrocarbon chains producing the series $C_nH_{2n+1}^+$ through alkyl chain fragmentation.

of monopentadecanoin in Figure 4.21*b* where the *m/z* 57 product ion may also be formed through a multistep process requiring first losses of both fatty acyl chains to form *m/z* 99, then loss of water, followed finally by loss of LiOH. Two processes may be responsible for formation of the *m/z* 57 product ion. The *m/z* 81 and *m/z* 63 product ions from the losses of one, and then two waters, respectively, from the

TABLE 4.7   Percent Total Ion Abundance and Apparent Threshold Energy for Lithium Adduct of Monopentadecanoin, 1-Stearin,2-Palmitin, and 1,3-Dipentadecanoin

Product Ion (m/z)	Monopentadecanoin		1-Stearin,2-Palmitin		1,3-Dipentadecanoin	
	Abundance (%)	ΔE_0 (eV)	Abundance (%)	ΔE_0 (eV)	Abundance (%)	ΔE_0 (eV)
43	—	—	3.54	1.78 ± 0.03	2.36	1.87 ± 0.30
57	1.13	2.24 ± 0.13	16.0	1.35 ± 0.26	8.16	1.69 ± 0.10
63	2.24	2.10 ± 0.36	5.43	1.71 ± 0.03	3.30	1.68 ± 0.19
71	—	—	10.2	1.55 ± 0.09	5.08	1.53 ± 0.07
81	3.50	1.74 ± 0.27	3.54	1.87 ± 0.18	5.07	1.51 ± 0.26
85	—	—	7.46	1.70 ± 0.05	3.34	1.70 ± 0.15
95	—	—	3.63	1.88 ± 0.05	1.48	1.65 ± 0.35
99	15.5	2.87 ± 0.14	1.84	2.07 ± 0.35	2.16	1.62 ± 0.15

lithiated glycerol backbone, are also included in the low-molecular-weight series are in a manner analogous to the pathways illustrated in Figure 4.21b.

4.9.4   Apparent Threshold Energies for Lithiated 1,3-Dipentadecanoin

Acquired at 60 eV (E_{LAB}) and collision cell pressure of 3×10^{-4} mbar (multiple collision conditions), Figure 4.26a is a product ion spectrum for the collision-induced dissociation of the lithium adduct of 1,3-dipentadecanoin. As is similar to the 1-stearin,2-palmitin CID spectrum, the 1,3-dipentadecanoin produces low-molecular-weight product ions ranging from m/z 29 to m/z 99 by the same types of processes that were discussed above. Five principal product ions are observed in the higher molecular weight range of m/z 249 to m/z 323. This is also similar to the 1,2-diacylglycerol described above where product ions are formed primarily through single cleavages of a fatty acid substituent with concomitant H transfer resulting in the losses of various neutrals, such as loss of the fatty acyl chain as a ketene, loss of lithium fatty acetate, and loss of fatty acid (acid form). Assignments for the product ions are illustrated in Figure 4.26b. As is similar to 1-stearin,2-palmitin, there is an inverse relationship observed between percent total ion abundance and apparent threshold energy, as listed in Table 4.8 . The product ion with the highest abundance and lowest threshold results from the neutral loss of the fatty acyl chain as a ketene at m/z 323 is observed among the competitive decomposition processes. At m/z 299 is observed the next highest abundance, slightly higher threshold energy product ion from the neutral loss of lithium fatty acetate. At m/z 305 is the neutral loss of the fatty acid (acid form), the third in this fatty acid loss series, with the next highest abundance, and even higher reaction threshold, suggesting the following ranking of favored fragmentation pathways: neutral loss of fatty acyl chain as ketene $>$ neutral loss of lithium fatty acetate $>$ neutral loss of fatty acid. Produced from the consecutive neutral losses of H_2O followed by C15:1 α-β unsaturated fatty acid (this order was established by precursor scans of m/z 289[54]), thus resulting in the lowest percent total ion abundance,

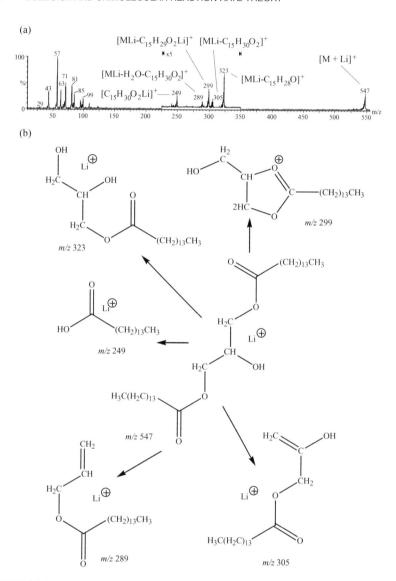

FIGURE 4.26 (*a*) Acquired at 60 eV (E_{LAB}) under multiple-collision conditions product ion spectrum of the collision-induced dissociation of lithiated 1,3-dipentadecanoin. Expressed by the hydrocarbon series $C_nH_{2n+1}{}^+$ produced from σ-bond fatty acyl chain cleavage are the low-molecular-weight product ions ranging from m/z 29 to m/z 99; m/z 249 to m/z 323 produced from single-cleavage reactions of fatty acyl loss in several forms (fatty acyl chain as ketene, lithium fatty acetate, and fatty acid). (*b*) High-molecular-weight range product ions produced from the CID of 1,3-dipentadecanoin, ranging from m/z 249 to m/z 323, fragmentation pathways, and structures. Containing the highest abundance and lowest threshold energy is the neutral loss of fatty acyl chain as ketene at m/z 323, followed by neutral loss of lithium fatty acetate at m/z 299, then neutral loss of the fatty acid at m/z 305. Fragmentation ranking: loss of fatty acyl ketene > loss of lithium fatty acetate > loss of fatty acid. Produced from the consecutive neutral losses of H_2O followed by C15:1 α-β unsaturated fatty acid is the m/z 289 product ion.

TABLE 4.8   Percent Total Ion Abundances, Apparent Threshold Energy, and Cross Sections of High-Molecular-Weight Product Ions for Lithium Adduct of 1,3-Dipentadecanoin

Product Ion (m/z)	% Total Ion Abundance[a] (%)	Apparent Threshold Energy (ΔE_0, eV)	Cross Section (cm^2)
323	3.13	1.66 ± 0.24	5.11×10^{-13}
299	2.06	1.79 ± 0.21	3.37×10^{-13}
305	1.25	1.97 ± 0.21	2.05×10^{-13}
249	0.91	2.10 ± 0.44	1.49×10^{-13}
289	0.82	2.30 ± 0.33	1.34×10^{-13}

[a] Listed in descending order.

and the highest apparent threshold energy up to this point is the m/z 289 product ion. The breakdown plot for formation of the higher molecular weight product ions is illustrated in Figure 4.27a graphically demonstrating the observed percent total ion abundances versus collision energy.

The low-molecular-weight product ions range from m/z 43 to m/z 99. These are produced primarily from a mixture of the $C_nH_{2n+1}^+$ hydrocarbon series (m/z 43, m/z 57, m/z 71, m/z 85, and m/z 99 formed in part by charge-remote fragmentations), and product ions arising from the glycerol backbone (m/z 63, m/z 81, and m/z 99). Table 4.7 shows the percent total ion abundance and apparent threshold energies for 1,3-dipentadecanoin, as compared with monopentadecanoin, and 1-stearin,2-palmitin. Breakdown curves for the lower molecular weight products resulting from CID of 1,3-dipentadecanoin are shown in Figure 4.27b. Similar to the breakdown graph for 1-stearin,2-palmitin in Figure 4.25b, the m/z 57 product ion is the favored pathway overall for reasons discussed above. The other product ions have similarly shaped breakdown curves, except for the m/z 43 product ion. The m/z 43 product ion is observed to rise more rapidly than the other product ions as the collision energy is increased. The m/z 43 product ion is produced through two processes: charge-remote σ-bond cleavage of the fatty acyl hydrocarbon chain ($[C_3H_7]^+$), or at least two cleavages of the glycerol backbone ($[C_2H_3O]^+$), with the latter route predominating at higher internal energies.

4.10 COMPUTATIONAL REACTION ENTHALPIES AND PREDICTED APPARENT THRESHOLD ENERGIES

To more completely describe reactivity via gas-phase unimolecular decomposition pathways, the derived effective path length method is intended to furnish quantitative information. The additional insight used to resolve mechanistic ambiguities in cases where more than one reaction pathway may lead to production of an ion of a given m/z value will be illustrated in this section. Computations were performed to

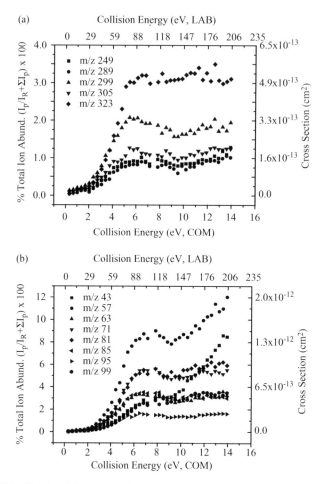

FIGURE 4.27   Single-collision conditions energy-resolved breakdown graphs for: (*a*) high-molecular-weight product ions produced from CID of 1,3-dipentadecanoin, and (*b*) low-molecular-weight product ions produced from CID of 1,3-dipentadecanoin. As the collision energy is increased, *m/z* 43 product ion rapidly increases, produced from the σ-bond cleavage of the fatty acyl hydrocarbon chain and the glycerol backbone $[C_2H_3O]^+$, appearing to be a favored pathway at increasing collision energies; *m/z* 57 product ion is the favored fragmentation pathway due to joint production by glycerol backbone water loss, and fatty acyl chain hydrocarbon cleavage.

calculate surface potential energies employing Gaussian 98[32] to create a benchmark to evaluate the derived effective reaction path length approach for predicting apparent threshold energies of covalent dissociations arising from CID of the acylglycerols. All of the monopentadecanoin CID products listed in Figrue 4.21*b* had final ground-state surface potential energies calculated along with those of the respective neutral molecules formed concomitantly. The computational reaction enthalpy ($\Delta H^{\circ}_{r \times n}$) was

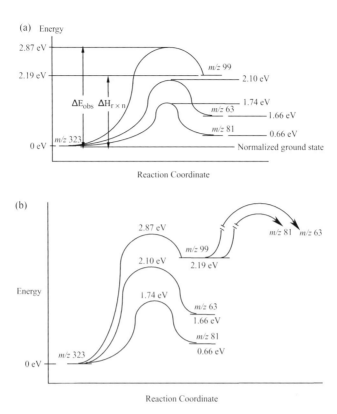

FIGURE 4.28 Energy diagrams for: (*a*) production of the *m/z* 63, 81, and 99 product ions from the monopentadecanoin lithium adduct at low collision energies near threshold regions. The production of the *m/z* 63 and 81 product ions have pathways that do not pass through the *m/z* 99 product ion. (*b*) Production of the *m/z* 63, 81, and 99 product ions from the monopentadecanoin lithium adduct including higher energy pathways of *m/z* 63 and 81, which pass through *m/z* 99.

determined to be 2.19 eV for the pathway producing the *m/z* 99 lithiated glycerol product and the neutral fatty acyl chain as a ketene [the magnitude of uncertainty in potential energy calculations at the B3LYP 6-31G* level has been estimated to be 3.9 kcal·mol^{-1}, or 0.17 eV[33] (mean absolute deviation)]. Illustrated in Figure 4.28*a* the apparent threshold energy (ΔE_0) was experimentally determined to be 2.87 ± 0.14 eV. The ground-state surface potential energy of the precursor *m/z* 323 lithium adduct of monopentadecanoin has been set to a value of 0 eV. The product ions *m/z* 81, 63, and 57 (*m/z* 57, not shown) have lower ΔE_0 values than the *m/z* 99 product ion (Fig. 4.28*b*). An inverse relationship between the percent total ion abundance and the apparent threshold energy ΔE_0 is observed for the product ions *m/z* 81, 63, and 57. These three product ions, however, were proposed to be formed by consecutive decompositions after first step loss of the fatty acyl chain as a ketene, yielding the

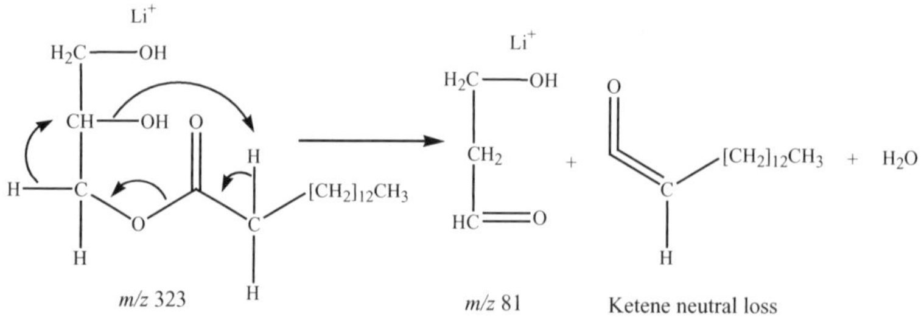

FIGURE 4.29 Direct production of the m/z 81 product ion from the m/z 323 precursor ion proposed pathway.

m/z 99 ion, which initially requires a ΔE_0 of 2.87 ± 0.14 eV. The m/z 99 production pathway involves a single hydrogen transfer from the fatty acyl chain to the glycerol oxygen. If a consecutive loss is assumed for the production of the m/z 81 ion:

$$m/z\,323 \xrightarrow[-\text{Ketene}]{\text{CID}} m/z\,99 \rightarrow m/z\,81 + H_2O \tag{4.25}$$

a second hydrogen transfer is involved, suggesting an additional energy barrier. The initial step requires a (predicted) activation energy of 2.87 ± 0.14 eV, but the m/z 81 formation may occur from m/z 99 after initial ketene loss. Production of the m/z 81 ion from the m/z 323 precursor was calculated to have a ΔE_0 of 1.74 ± 0.27 eV (by the derived effective reaction path length approach) with a ΔH_0 of 0.66 eV (by the Gaussian 98 method). The m/z 81 product ion may be produced through an alternative pathway to that shown in Equation (4.25) with a lower activation energy. A proposed mechanism for the formation of the m/z 81 product ion from the m/z 323 precursor ion is illustrated in Figure 4.29. The product ions at m/z 99, 81, 63, and 57 that are also observed in the CID of the 1,2- and the 1,3-diacylglycerols can give additional information concerning the fragmentation pathways involved in order to elucidate the mechanism of m/z 81 production from monopentadecanoin. The experimentally derived ΔE_0 for the m/z 99, 81, 63, and 57 product ions of the 1,2- and the 1,3-diacylglycerols are all lower than 2.2 eV (most are much lower, i.e., around 1.6−1.7 eV) as can be seen in Table 4.7.

Figure 4.30 illustrates ratio plots of m/z 81, 63, and 57 with reference to m/z 99 for the three acylglycerols. The mechanisms for formation of the m/z 81, 63, and 57 product ions are different for monopentadecanoin (Fig. 4.30a) as compared to the 1,2- (Fig. 4.30b) and the 1,3-diacylglycerols (Fig. 4.30c). The m/z 81, 63, and 57 product ions exhibit immediate increases in abundance relative to m/z 99 as the internal energy uptake of the system increases from 0 eV for the 1,2- and the 1,3-diacylglycerols. If the three lower mass product ions were formed via consecutive decompositions of m/z 99, this would be expected. For monopentadecanoin, the m/z 81, 63, and 57 product ions decrease in production relative to m/z 99 as the internal energy uptake

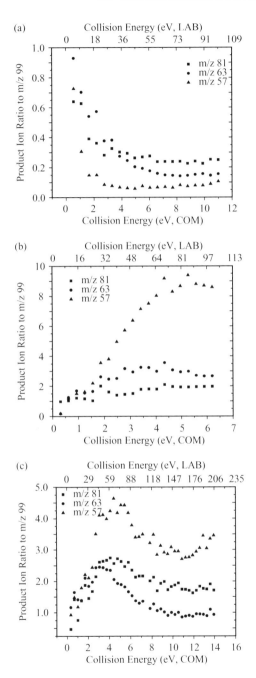

FIGURE 4.30 Ratio plots obtained under single-collision conditions of *m/z* 57, 63, and 81 to *m/z* 99 for: (*a*) monopentadecanoin lithium adduct, (*b*) 1-stearin,2-palmitin lithium adduct, and (*c*) 1,3-dipentadecanoin lithium adduct.

of the system increases. For the monopentadecanoin, this suggests that the m/z 81, 63, and 57 product ions have pathways of formation that are competitive with the m/z 99 product ion at lower internal energies (i.e., they are not formed by consecutive decompositions after initial m/z 99 formation). The routes to these products may have lower activation energies than the m/z 99 product ion as is reflected in the lower ΔE_0 values predicted by the derived effective reaction path length approach. Figure 4.28b is an energy diagram describing the three pathways for the production of the m/z 99, 81, and 63 ions from the monoacylglycerol m/z 323 precursor. The energy diagram depicts that the low-energy pathways to m/z 81 and 63 do not require initial production of m/z 99. As energy input is increased, the pathways to m/z 81, 63, and 57 become less favorable relative to m/z 99 (Fig. 4.28a). The calculated apparent threshold energies that are listed in Figure 4.28b represent the weighted average of all pathways that result in these product ions. This weighted average most likely corresponds to a single or highly predominant pathway leading to formation of a given m/z ion near the threshold region. At low energies, m/z 81 and 63 production appear to pass through channels that circumvent m/z 99, and their apparent threshold energies are lower as illustrated in Figure 4.28a. At higher collision energies the predominant pathway for the production of the m/z 81 and m/z 63 product ions appears to occur by consecutive decompositions of the m/z 99 product ion as is illustrated in Figure 4.28b. The breaks in the higher energy pathways representing consecutive decompositions of m/z 99 illustrate that the heights of the second barriers leading to production of m/z 81 and 63 are unknown, and the current methodology cannot predict these.

The derived effective path length approach predicts that the apparent threshold energy for the first neutral loss of a fatty acid substituent is lower for the 1,2-dipentadecanoin (loss of lithium fatty acetate forming m/z 313, ΔE_0 of 1.35 ± 0.13 eV, Table 4.6, Figure 4.24b) than for the 1,3-dipentadecanoin (loss of fatty acyl chain as ketene forming m/z 323, ΔE_0 of 1.66 ± 0.24 eV; see Table 4.8 and Fig. 4.26b). This appears to make the former more reactive than the latter. The monoacylglycerol is the least reactive of the three, as dominant ketene loss from the monopentadecanoin has a much higher ΔE_0 (2.87 ± 0.14 eV). The predicted apparent threshold energy ranking for reactivity in the form of a first neutral loss of the fatty acid substituents (i.e., 1,2-dipentadecanoin > 1,3-dipentadecanoin > monopentadecanoin) is another example of the value of this method. This demonstrates that the method offers a comparative quantitative description of the tendency toward fragmentation. By considering that additional steric crowding on the glycerol backbone that increases the propensity for reaction, the higher reactivity of the 1,2-diacyl relative to the 1,3-diacylglycerol may be rationalized.

4.11   CONCLUSIONS

The derived effective reaction path length approach for predicting noncovalent BDEs, and covalent apparent threshold energies, using the quadrupole–hexapole–quadrupole mass spectrometer has demonstrated an effective use in fragmentation pathway studies as an enhancement to traditional energy-resolved mass spectrometry studies.

A more extensive quantitative description of fragmentation pathways is possible by combining the calculated apparent threshold energies with the percent total ion abundance breakdown graphs. A unique interpretation of the product ion results was obtained with the use of the derived effective reaction path length approach to calculate apparent threshold energies for the unimolecular fragmentation reactions of lithiated acylglycerols by quadrupole–hexapole–quadrupole mass spectrometry. The derived effective path length approach was shown to be useful in ranking the reactivity of the acyl-substituted glycerols, with the 1,2-diacyl being more reactive than the 1,3-diacyl, which was more reactive than the monoacylglycerol. A cleavage mechanisms with single hydrogen transfers that are competitive on the basis of threshold energy was obtained through the observance of an inverse relationship between percent total ion abundance versus apparent threshold energies for the dissociation of the fatty acid substituents of the 1,2- and the 1,3-diacylglycerols. Rankings of the various decompositions in the form of the first neutral loss of a fatty acyl substituent was also obtained by the calculated apparent threshold energies. For 1-stearin,2-palmitin, the following favored fragmentation pathway was observed: loss of lithium fatty acetate > loss of fatty acid > loss of fatty acyl chain as ketene. For the 1,3-dipentadecanoin, the following favored fragmentation pathway was observed: loss of fatty acyl chain as ketene > loss of lithium fatty acetate > loss of fatty acid. Low-molecular-weight fragment ion production (m/z 81, 63, 57) exhibited low apparent threshold energies for monopentadecanoin fragmentation, rationalized by considering that multiple routes to their production exist. In general, the derived effective reaction path length approach for apparent threshold energy determinations is relatively fast, straightforward, and easy to use and has demonstrated itself to be applicable to both noncovalent, and covalent bond dissociation studies, making it versatile and general in gas-phase thermochemistry applications.

REFERENCES

1. Noggle, J. H. *Physical Chemistry*, 3rd ed. New York: HarperCollins, 1996, p. 28.
2. Nesatyy, V. J. *J. Mass Spectrom* 2001, *36*, 950–959.
3. Armentrout, P. B., and Beauchamp, J. L. *J. Chem. Phys.* 1981, *74*, 2819–2826.
4. Ervin, K. M., and Armentrout, P. B. *J. Chem. Phys.* 1985, *83*, 166–189.
5. Weber, M. E., Elkind, J. L., and Armentrout, P. B. *J. Chem. Phys.* 1986, *84*, 1521–1529.
6. Aristov, N., and Armentrout, P. B. *J. Chem. Phys.* 1986, *90*, 5135–5140.
7. Hales, D. A., Lian, L., and Armentrout, P. B. *Int. J. Mass Spectrom. Ion Processes* 1990, *102*, 269–301.
8. Schultz, R. H., Crellin, K. C., and Armentrout, P. B. *J. Am. Chem. Soc.* 1991, *113*, 8590–8601.
9. Rodgers, M. T., Ervin, K. M., and Armentrout, P. B. *J. Chem. Phys.* 1997, *106*, 4499–4508.
10. Rodgers, M. T., and Armentrout, P. B. *J. Chem. Phys.* 1998, *109*, 1787–1800.
11. Rodgers, M. T., and Armentrout, P. B. *Mass Spec. Rev.* 2000, *19*, 215–247.
12. Pramanik, B. N., Bartner, P. L., Mirza, U. A., Liu, Y. H., and Ganguly, A. K. *J. Mass Spectrom.* 1998, *33*, 911–920.
13. Shukla, A. K., and Futrell, J. H. *J. Mass Spectrom.* 2000, *35*, 1069–1090.

14. Daniel, J. M., Friess, S. D., Rajagopala, S., Wendt, S., and Zenobi, R. *Int. J. Mass Spectrom.* 2002, *216*, 1–27.

15. Graul, S. T., and Squires, R. R. *J. Am. Chem. Soc.* 1990, *112*, 2517–2529.

16. Sunderlin, L. S., Wang, D., and Squires, R. R. *J. Am. Chem. Soc.* 1993, *115*, 12060–12070.

17. Dougherty, R. C. *J. Am. Soc. Mass Spectrom.* 1997, *8*, 510–518.

18. Anderson, S. G., Blades, A. T., Klassen, J., and Kebarle, P. *Int. J. Mass Spectrom. Ion Processes* 1995, *141*, 217–228.

19. Klassen, J. S., Anderson, S. G., Blades, A. T., and Kebarle, P. *J. Phys. Chem.* 1996, *100*, 14218–14227.

20. Nielsen, S. B., Masella, M., and Kebarle, P. *J. Phys. Chem. A* 1999, *103*, 9891–9898.

21. Kebarle, P. *Int. J. Mass Spectrom.* 2000, *200*, 313-330.

22. Peschke, M., Blades, A. T., and Kebarle, P. *J. Am. Chem. Soc.* 2000, *122*, 10440–10449.

23. Whitehouse, C. M., Dreyer, R. N., Yamashita, M., and Fenn, J. B. *Anal. Chem.* 1985, *57*, 675–679.

24. Fenn, J. B. *J. Am. Soc. Mass Spectrom.* 1993, *4*, 524–535.

25. Cole, R. B. *J. Mass Spectrom.* 2000, *35*, 763–772.

26. Cech, N. B., and Enke, C. G. *Mass Spec. Rev.* 2001, *20*, 362–387.

27. Gidden, J., Wyttenbach, T., Batka, J. J., Weis, P., Jackson, A. T., Scrivens, J. H., and Bowers, M. T. *J. Am. Soc. Mass Spectrom.* 1999, *10*, 883–895.

28. Chen, Y. L., Campbell, J. M., Collings, B. A., Konermann, L., and Douglas, D. J. *Rapid Commun. Mass Spectrom.* 1998, *12*, 1003–1010.

29. Taft, R. W., Anvia, F., Gal, J. F., Walsh, S., Capon, M., Holmes, M. C., Hosn, K., Oloumi, G., Vasanwala, R., and Yazdani, S. *Pure Appl. Chem.* 1990, *62*, 17–23.

30. McLafferty, F. W., and Turecek, F. *Interpretation of Mass Spectra*, 4th ed., Sausalito, CA: University Science Books, 1993, p. 126.

31. Srivastava, M. S. *Methods of Multivariate Statistics*. Hoboken, NJ.: Wiley, 2002, p. 365.

32. Frisch, M. J., Trucks, G. W., Schlegel, H. B., Scuseria, G. E., Robb, M. A., Cheeseman, J. R., Zakrzewski, V. G., Montgomery, J. A., Jr., Stratmann, R. E., Burant, J. C., Dapprich, S., Millam, J. M., Daniels, A. D., Kudin, K. N., Strain, M. C., Farkas, O., Tomasi, J., Barone, V., Cossi, M., Cammi, R., Mennucci, B., Pomelli, C., Adamo, C., Clifford, S., Ochterski, J., Petersson, G. A., Ayala, P. Y., Cui, Q., Morokuma, K., Malick, D. K., Rabuck, A. D., Raghavachari, K., Foresman, J. B., Cioslowski, J., Ortiz, J. V., Stefanov, B. B., Liu, G., Liashenko, A., Piskorz, P., Komaromi, I., Gomperts, R., Martin, R. L., Fox, D. J., Keith, T., Al-Laham, M. A., Peng, C. Y., Nanayakkara, A., Gonzalez, C., Challacombe, M., Gill, P. M. W., Johnson, B., Chen, W., Wong, M. W., Andres, J. L., Gonzalez, C., Head-Gordon, M., Replogle, E. S., and Pople, J. A. Gaussian 98, Revision A.3. Pittsburgh: Gaussian, Inc., 1998.

33. Foresman, J. B. *Exploring Chemistry with Electronic Structure Methods*, 2rd ed. Pittsburgh: Gaussian, Inc., 1996, p. 157.

34. Chupka, W. A. *J. Chem. Phys.* 1959, *30*, 458–465.

35. Armentrout, P. B. *Top Curr. Chem.* 2003, *225*, 233–262.

36. Cai, Y., and Cole, R. B. *Anal. Chem.* 2002, *74*, 985–991.

37. Cao, P., and Stults, J. T. *Rapid Commun. Mass Spectrom.* 2000, *14*, 1600–1606.

38. Ho, Y. P., Huang, P. C., and Deng, K. H. *Rapid Commun. Mass Spectrom.* 2003, *17*, 114–121.

39. Kocher, T., Allmaier, G., and Wilm, M. *J. Mass Spectrom.* 2003, *38*, 131–137.

40. Katta, V., and Chait, B. T. *J. Am. Chem. Soc.* 1991, *113*, 8534–8535.

41. Ganem, B., Li, Y. T., and Henion, J. D. *J. Am. Chem. Soc.* 1991, *113*, 6294–6296.

42. Ganem, B., Li, Y. T., and Henion, J. D. *J. Am. Chem. Soc.* 1991, *113*, 7818–7819.

43. Colorado, A., and Brodbelt, J., *J. Am. Soc. Mass Spectrom.* 1996, *7*, 1116–1125.

44. Reid, C. J. *Organic Mass Spectrom.* 1991, *26*, 402–409.

45. Martinez, R. I., and Ganguli, B. *J. Am. Soc. Mass Spectrom.* 1992, *3*, 427–444.

46. Rogalewicz, F., Hoppilliard, Y., and Ohanessian, G. *Int. J. Mass Spectrom.* 2000, *195/196*, 565–590.

47. Vazquez, S., Truscott, R. J. W., O'Hair, R. A. J., Weimann, A., and Sheil, M. M. *J. Am. Soc. Mass Spectrom.* 2001, *12*, 786–794.

48. Butcher, C. P. G., Dyson, P. J., Johnson, B. F. G., Langridge-Smith, P. R. R., McIndoe, J. S., and Whyte, C. *Rapid Commun. Mass Spectrom.* 2002, *16*, 1595–1598.

49. Harrison, A. G. *Rapid Commun. Mass Spectrom.* 1999, *13*, 1663–1670.

50. Harrison, A. G., Csizmadia, I. G., Tang, T. H., and Tu, Y. P. *J. Mass Spectrom.* 2000, *35*, 683–688.

51. Chass, G. A., Marai, C. N. J., Harrison, A. G., and Csizmadia, I. G. *J. Phys. Chem. A* 2002, *106*, 9695–9704.

52. Martinez, R. I. *J. Res. Natl. Inst. Std. Technol. (U.S.)* 1989, *94*, 281–304.

53. Ham, B. M., and Cole, R. B. *Anal. Chem.* 2005, *77*, 4148–4159.

54. Ham, B. M., Jacob, J. T., Keese, M. M., and Cole, R. B. *J. Mass Spectrom.* 2004, *39*, 1321–1336.

CHAPTER 5

THE MASS SPECTRUM: ODD ELECTRON MOLECULAR ION VERSUS EVEN ELECTRON PRECURSOR ION MASS SPECTRA

5.1 ELECTRON IONIZATION ODD ELECTRON PROCESSES

Electron impact or electron ionization (EI) (see Section 2.2) typically produces a charged species through the ejection of a single electron. The ejection of a single electron results in a molecular ion that is a positively charged radical, possesses an odd number of electrons, and is designated as an odd electron (OE) molecular ion $M^{+\cdot}$. The EI process is an energetic process that often results in significant fragmentation of the molecular ion producing a large number of product ions. The product ions produced in EI are comprised of both OE ions and even electron (EE) ions. Even electron ions contain an even number of valence shell electrons, and in positive ion mode are designated as $[M + nX]^{n+}$, for example, the protonated form of an even electron ion is designated as $[M + H]^+$, the sodium adduct form as $[M + Na]^+$, and the ammonium adduct as $[M + NH_4]^+$. In negative ion mode even electron ions are designated as $[M + nX]^{n-}$, for example, chloride adducts $[M + Cl]^-$ and for deprotonated species as $[M - nH]^{n-}$. Even electron ions are often derived by the softer ionization techniques such as electrospray ionization (ESI), fast atom bombardment (FAB), some types of chemical ionization (CI), and matrix-assisted laser desorption ionization (MALDI). A simplified comparison between the production of odd electron molecular ions and even electron ions is

Even Electron Mass Spectrometry with Biomolecule Applications   By Bryan M. Ham
Copyright © 2008 John Wiley & Sons, Inc.

as follows:

$$\underset{\text{Neutral pentanol}}{\text{\Large pentanol structure } \ddot{\text{O}}-\text{H}} \quad \xrightarrow[-e^-]{\text{EI}} \quad \underset{\text{OE, M}^{+\cdot}}{\text{\Large pentanol structure } \overset{+\cdot}{\text{O}}-\text{H}} \tag{5.1}$$

$$\underset{\text{Neutral pentanol}}{\text{\Large pentanol structure } \ddot{\text{O}}-\text{H}} \quad \xrightarrow[\text{H}^+]{\text{ESI}} \quad \underset{\text{EE, [M+H]}^+}{\text{\Large pentanol structure } \overset{\text{H}^+}{\ddot{\text{O}}}-\text{H}} \tag{5.2}$$

where Equation (5.1) represents electron ionization (removal of one electron) of pentanol, producing an odd electron molecular ion $M^{+\cdot}$, and Equation (5.2) represents the protonation of pentanol, producing an even electron ion $[M + H]^+$. The classification of ease of ionization by removal of an electron generally follows the trend of nonbonding $> \pi > \sigma$. An electron associated with a sigma bond (σ) requires the largest amount of energy to remove an electron and ionize the species. This is followed by pie-bond (π) electrons associated with double and triple bonds, and finally nonbonding electrons that generally require the least amount of energy for removal. The most common use of electron ionization is the coupling of a single quadrupole mass analyzer to gas chromatography instrumentation, which still finds wide use for the analysis of volatile compounds. Due to the reproducibility of the molecular ion spectra obtained from the standardized 70-eV electron ionization beam used in EI sources, the generated molecular ion spectra are often searched against standard libraries for structural identification of the analyte of interest. However, it is useful to know the basics of some of the most common types of fragmentation pathways that occur using EI as the ionization technique (a comprehensive reference for odd electron spectral interpretation is given by McLafferty and Turecek[1]). With EI fragmentation the dissociation is unimolecular where the activation of the fragmentation process has not involved collision-induced dissociation with a stationary target gas or chemical ionization of the ion–molecule type of reaction. There are five basic types of bond cleavage that takes place during the fragmentation of a molecular ion when using EI: sigma cleavage (σ), inductive cleavage (i), alpha cleavage (α), retro-Diels Alder, and hydrogen rearrangement (rH). When decomposition takes place with an odd electron molecular ion involving the breaking of a single bond, an even electron ion is always produced along with a neutral radical. In breaking two bonds an odd electron product ion is produced along with an even electron neutral. The first type of cleavage mentioned, sigma cleavage (σ), is the predominant type of bond breakage that is observed in EI spectra when saturated nonaromatic, aliphatic hydrocarbon species comprise the OE molecular ions. Sigma cleavage fragmentation of an OE molecular ion producing an EE product ion and radical OE neutral is illustrated in Equation (5.3) for the EI ionization of isopentane.

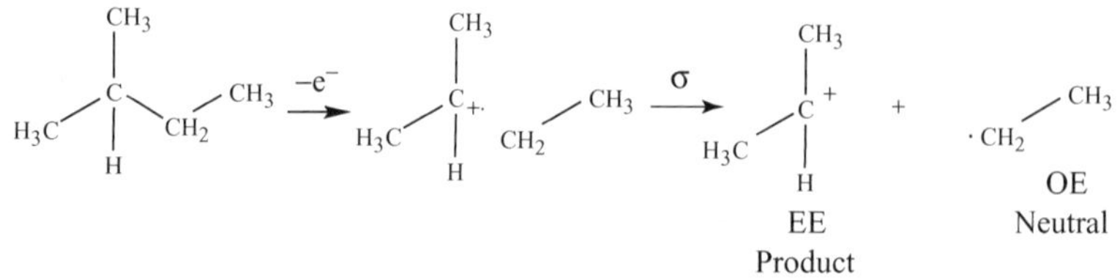

(5.3)

In the mechanistic equations used, such as that found in Equations (5.4)–(5.8), a single fishhook arrow represents the movement of a single electron and represents a homolytic cleavage of the bond.

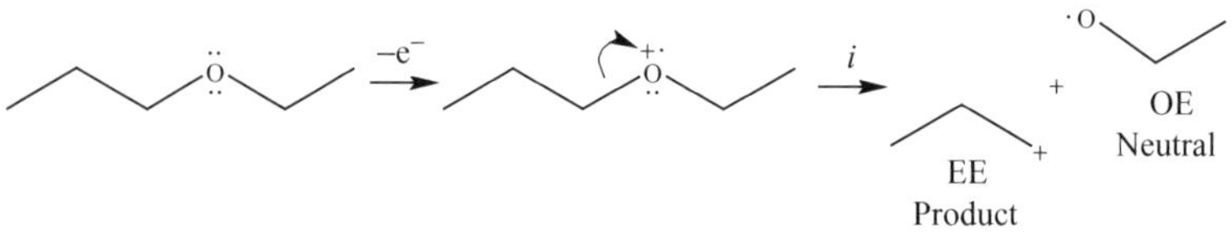

(5.4)

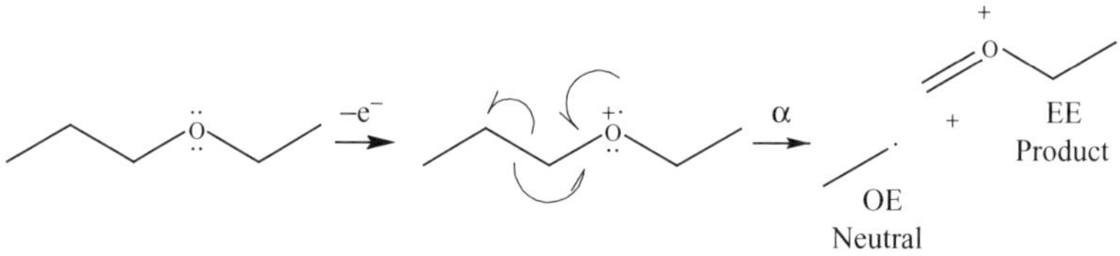

(5.5)

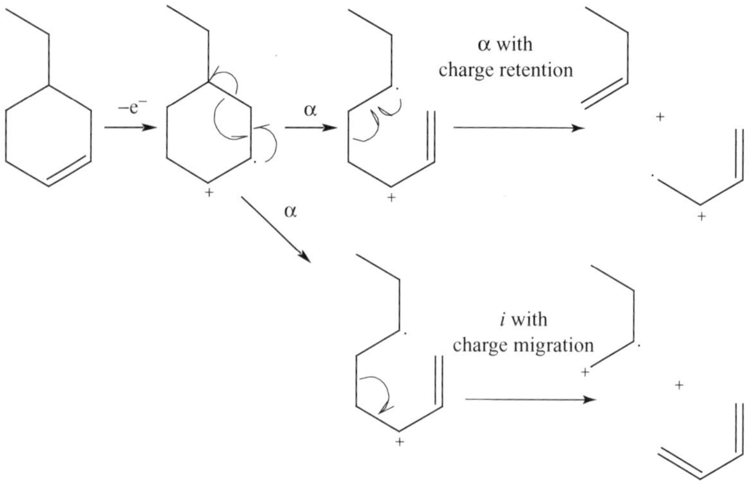

(5.6)

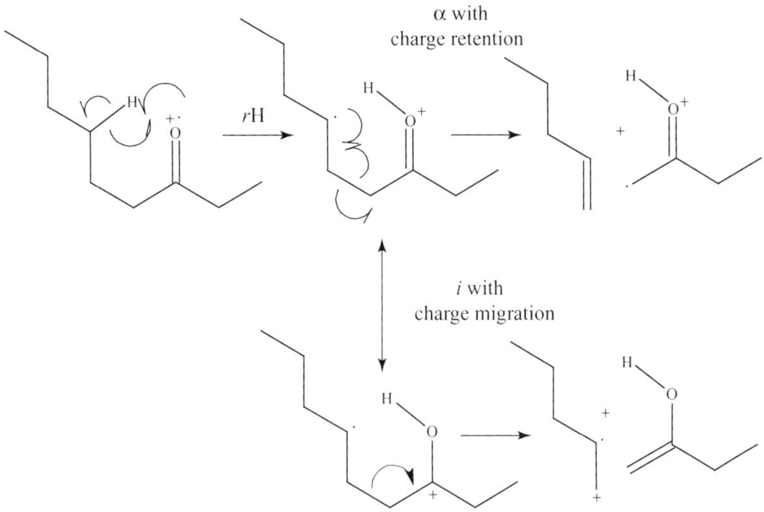

(5.7)

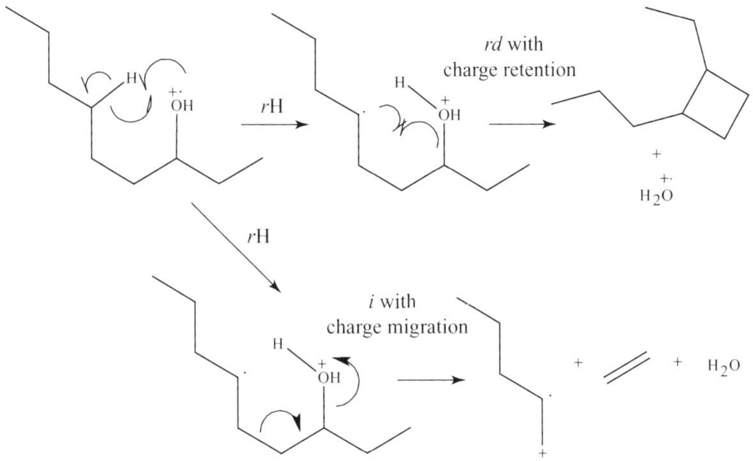

(5.8)

An example of a single electron homolytic cleavage illustrated by a fishhook arrow can be found in the α-bond cleavage in Equation (5.5). When an electron pair is moved, such as in Equation (5.4) for the inductive cleavage, a full arrowhead is used. This type of cleavage represents a heterolytic cleavage that moves the charge site. In combinations involving these processes, when two bonds are broken, an OE product ion is produced along with the corresponding EE neutral fragment. Inductive cleavage (i), also known as charge-site-initiated cleavage, is a unimolecular dissociation that involves the migration of a pair of electrons to the charged site of the molecular ion. The driving force for the movement of the electrons is greater as the electronegativity, or electron-withdrawing tendency of the charged atom in the molecule is increased (i.e., Cl > O > N > C). This type of dissociation also produces an EE product ion and a neutral radical as is also observed for σ cleavages and is illustrated in

Equation (5.4) for the inductive cleavage of propylethyl ether. The EI α cleavage fragmentation pathway involves the movement of one electron to pair with the lone electron of the radical OE species forming a double bond, and a subsequent movement of one electron from a bond that is adjacent (α) to the double bond formed. The driving force behind the electron movement is the pair completion of the lone radical electron, and the greater the electron donation tendency of the radical atom in the molecule (i.e., N > O > C > Cl). Alpha cleavages also produce an EE product ion and a neutral radical. Equation (5.5) illustrates an α cleavage for propylethyl ether. In cyclic structures that contain an unsaturation moiety, the π electrons often allow for the radical site and initial charge with EI ionization. In the ensuing fragmentation pathways, an α cleavage takes place shifting the remaining π electron to the α position, creating a new double bond. This process is illustrated in the first step of the retro–Diels Alder fragmentation process of Equation (5.6). A second α cleavage can take place resulting in a product ion where the charge has remained in its initial location, thus describing an α cleavage with charge retention [Equation (5.6) bottom]. An alternative process can also take place following the initial α cleavage where the charge migrates within the structure and an inductive cleavage is the fragmentation pathway. This is also illustrated in the latter part of Equation (5.6). Finally, a fifth type of fragmentation pathway often associated with the decomposition of an OE molecular ion produced by EI is gamma-hydrogen (γ-H) rearrangement accompanied by bond dissociation (known as the McLafferty rearrangement). The two types of bond dissociation that occur are β cleavage, when an unsaturated charge site is involved, and adjacent cleavage, when there is no unsaturation involved in the decomposition pathway. As was also the case with the retro–Diels Alder decomposition pathway, if the second step in the γ-H rearrangement is α cleavage, charge retention will take place; if the second step is an inductive cleavage, charge migration will take place. The γ-H rearrangement decomposition pathways are illustrated in the top half of Equation (5.7) for the singly unsaturated molecular ion of the aliphatic ketone nonanone resulting in the γ-H rearrangement and β cleavage. The adjacent cleavage, which takes place for a saturated molecular ion following γ-H rearrangement, is illustrated in Equation (5.8) for 3-nonanol. This process also involves two mechanisms describing a product ion with charge retention and a product ion formed through charge migration. This completes our basic introduction to odd electron mass spectral processes observed using electron ionization as the source. We will next look at a specific example of spectral interpretation of an EI-generated mass spectrum.

5.2 OLEAMIDE FRAGMENTATION PATHWAYS—ODD ELECTRON M$^{+\cdot}$ BY GAS CHROMATOGRAPHY/ELECTRON IONIZATION–MASS SPECTROMETRY (GC/EI-MS)

Fatty acid primary amides (FAPA) such as cis-9-octadecenamide (oleamide),[2] and the related N-acyl ethanolamines (NAEs)[3–7] are a special class of lipids that act as messengers or signaling molecules. Oleamide was first reported observed in a biological sample in a study by Arafat et al.[8] where oleamide was isolated from

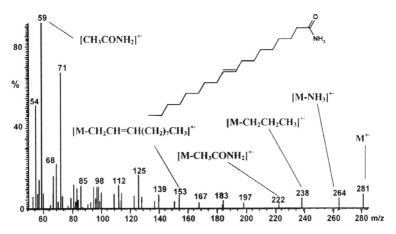

FIGURE 5.1 EI mass spectrum of cis-9-octadecenamide (oleamide).

plasma. Six years later it was reported that oleamide was observed in cerebrospinal fluid of mammals experiencing sleep deprivation by Cravatt et al.[9] Following this work, others have also reported the apparent function of oleamide as an endogenous sleep-inducing lipid.[10, 11] Figure 5.1 is an EI spectrum of oleamide obtained by gas chromatography electron ionization single quadrupole mass spectrometry (GC/EI-MS) with the molecular ion M$^{+\bullet}$ at m/z 281. The spectrum contains seven fragment ions that are illustrative of the diagnostic product ions observed in the EI fragmentation spectra of oleamide and the mechanistic pathways described previously for EI [Equations (5.3) to (5.9)].

$$(5.9)$$

The first predominant product ion observed in Figure 5.1, the m/z 264 product ion, is derived from ketene formation through neutral loss of ammonia NH_3 (-17 amu)

$[M - NH_3]^{+\cdot}$. This mechanism is illustrated in Equation (5.9) where the first step in the fragmentation pathway is a β-hydrogen shift to the amine group (α to the carbonyl and β to the amine) followed by inductive cleavage of ammonia. The m/z 238 product ion is formed through σ cleavage, resulting in an aliphatic hydrocarbon loss $[M\text{-}CH_2CH_2CH_3]^+$.

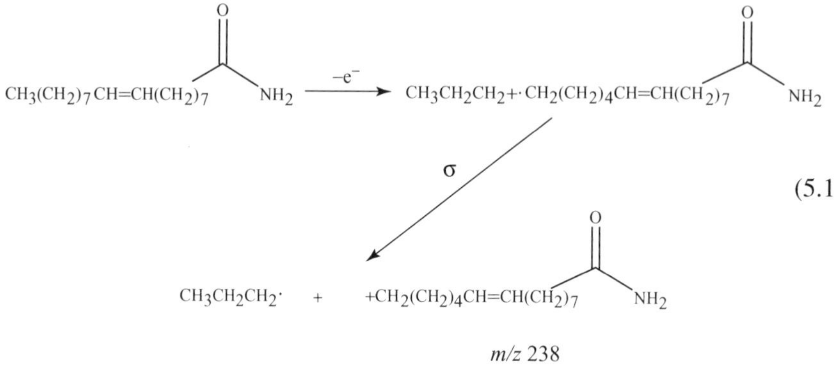

$$(5.10)$$

m/z 238

The m/z 59 and the m/z 222 product ions are formed through γ-hydrogen migration with subsequent inductive cleavages, as illustrated in Equations (5.7) and (5.8) for 3-nonanone and 3-nonanol, respectively. The m/z 59 is a γ-hydrogen rearrangement mechanism accompanied by β-cleavage bond dissociation (McLafferty rearrangement). This is due to the unsaturated charge site being involved (carbonyl oxygen is ionized) with charge retainment producing $[CH_2COHNH_2]^{+\cdot}$. The fragmentation pathway mechanism for the production of the m/z 59 product ion is illustrated in Equation (5.11).

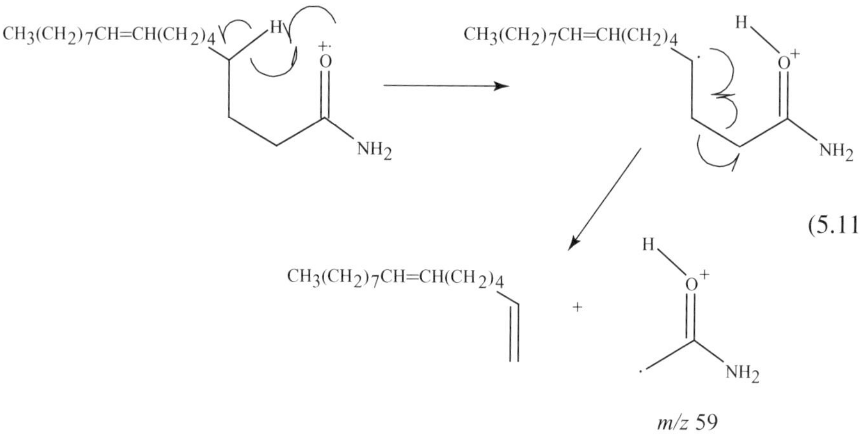

$$(5.11)$$

m/z 59

For the production of the m/z 222 product ion the molecular ion is ionized at the amine nitrogen, which is a saturated moiety. This fragmentation pathway mechanism is γ-hydrogen migration with adjacent inductive cleavage involving charge migration $[M-CH_2CO-NH_3]^{+\cdot}$. The fragmentation pathway mechanism for the production of the m/z 222 product ion is illustrated in Equation (5.12).

$$CH_3(CH_2)_7CH=CH(CH_2)_4 \quad\quad\quad \overset{H}{\underset{NH_2}{}} \quad\longrightarrow\quad CH_3(CH_2)_7CH=CH(CH_2)_4 \quad\quad \overset{H}{\underset{NH_2}{}}$$

$$CH_3(CH_2)_7CH=CH(CH_2)_4 \qquad\qquad + \qquad \overset{}{\diagup\diagup}O \;+\; NH_3 \qquad\qquad (5.12)$$

$$m/z\ 222$$

Finally, the product ion at m/z 154 is derived from a radical-site-initiated α cleavage (allylic cleavage) between the C7 and the C8 carbons. This is not a σ cleavage due to initial ionization of the σ bond but is driven by the initial removal of a π electron from the unsaturation site between C9 and C10. Following the cleavage of the C7 and C8 σ bond, there is a hydrogen migration from the C7 to the C8 carbon. A probable mechanism describing the radical-site-initiated α-cleavage fragmentation pathway for the production of the m/z 154 product ion is illustrated in Equation (5.13):

$$CH_3(CH_2)_7CH=CH(CH_2)_7 \quad NH_2 \qquad \overset{-e^-}{\longrightarrow} \qquad CH_3(CH_2)_7CH\overset{+\cdot}{-}CH-CH_2-CH_2(CH_2)_5 \quad NH_2$$

$$CH(CH_2)_5 \quad NH_2$$

$$+$$

$$CH_3(CH_2)_7\overset{+}{CH}=\overset{\cdot}{CH}CH_3 \qquad\qquad \longleftarrow \qquad CH_3(CH_2)_7\overset{+}{CH}CH=CH_2 \qquad \cdot\overset{H}{\underset{H}{C}}(CH_2)_5 \quad NH_2$$

$$m/z\ 154$$

$$(5.13)$$

5.3   OLEAMIDE FRAGMENTATION PATHWAYS—EVEN ELECTRON [M + H]⁺ BY ELECTROSPRAY IONIZATION/ION TRAP MASS SPECTROMETRY (ESI/IT-MS)

We have now covered the electron ionization (EI) odd electron (OE) mass spectrum of cis-9-octadecenamide (oleamide), including fragmentation pathways that describe the product ions produced. We will now compare the electrospray ionization (ESI) mass spectrum of the even electron (EE) protonated form of cis-9-octadecenamide (oleamide) as the precursor ion at m/z 282.6 $[C_{18}H_{35}NO + H]^+$. The protonated form of cis-9-octadecenamide was subjected to collision-induced dissociation (CID) for structural information using a three-dimensional quadrupole ion trap mass spectrometer (ESI-IT/MS) and the product ion spectrum is illustrated in Figure 5.2. In electrospray ionization mass spectrometry the precursor ions and the product ions are almost exclusively comprised of EE species. Often analytes of interest are neutral; therefore, a cationizing agent (for positive ion mode analysis) is added to promote the gas-phase ionization of the analyte and to also help support the electrospray process. In positive ion mode some typical cationizing agents added to the analyte solution prior to electrospray are volatile organic acids such as acetic acid or formic acid. These volatile organic acids will promote the gas-phase protonation of the analyte, producing the even electron precursor ion $[M + H]^+$. Metals are also often used as cationizing agents such as lithium $[M + Li]^+$ or sodium $[M + Na]^+$. Sometimes it is also advantageous to use ammonium acetate as the solution additive prior to electrospray, which promotes the formation of ammonium adducts $[M + NH_4]^+$. The ESI product ion spectrum of oleamide at m/z 282.6 illustrated in Figure 5.2 was obtained in an acidified (1% acetic acid) 1 : 1 chloroform/methanol solution.

The first major product ion observed in the ESI mass spectrum of Figure 5.2 is the m/z 265.6 peak. This is formed through neutral loss of ammonia (NH₃, −17 amu) from the EE precursor ion of protonated oleamide at m/z 282.6 $[M + H - NH_3]^+$. The protonation of the oleamide lipid species will take place as an association of the positive charge on the proton with the most basic site on the lipid. The most basic

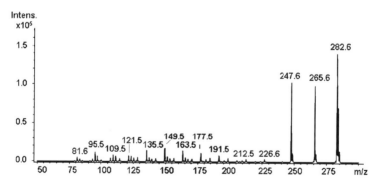

FIGURE 5.2   Electrospray ionization (ESI) product ion mass spectrum of the even electron (EE) protonated form of cis-9-octadecenamide (oleamide) as the precursor ion at m/z 282.6 $[C_{18}H_{35}NO + H]^+$.

site is associated with the amide moiety of the lipid with a closer electrostatic sharing of the positive charge on the proton with the nonbonding electron pair of the amide nitrogen. This association is illustrated in Equation (5.14) where the amide nitrogen is drawn as protonated.

$$ (5.14) $$

m/z 265.6

This same neutral loss of ammonia was also observed in the EI mass spectrum illustrated in Figure 5.1. The mechanism that described the production of the *m/z* 264 product ion in Figure 5.1 involved a two-step process first of hydrogen transfer from the β carbon to the amide nitrogen followed by inductive cleavage to release the neutral NH_3 and subsequently producing an odd electron product ion as illustrated in Equation (5.9). In Equation (5.14) the production of the neutral loss of NH_3 from the protonated form of the precursor ion of oleamide is through a one-step inductive, heterolytic cleavage of the carbon–nitrogen bond producing an even electron product ion. Fragmentation pathway mechanisms involved in the dissociation of even electron precursor ions almost exclusively occur through an inductive, heterolytic cleavage of σ bonds. A second predominant product ion observed in the ESI tandem mass spectrum of Figure 5.2 is the *m/z* 247.6 ion that on first inspection appears to be produced through neutral loss of ammonium hydroxide $[M + H - NH_4OH]^+$ (loss of 35 amu). The product ion spectrum illustrated in Figure 5.2 was obtained using an ion trap mass spectrometer. This type of mass analyzer allows the collection of multiple product ion spectra in the form of MS^n (where $n = 2, 3$, up to 9 maximum). This type of mass analysis involves the isolation of a precursor ion, product ion generation of the precursor, isolation of a first-generation product ion, and subsequent production of second-generation product ions, and so on. Figure 5.3 illustrates the product ion spectrum of the *m/z* 265.6 product ion obtained from the *m/z* 282.6 precursor protonated ion of oleamide. This is an MS^3 product ion spectrum where the production of the *m/z* 282.6 precursor protonated ion of oleamide is the first mass analysis stage designated as MS, the production of the *m/z* 265.6 product ion is the second mass analysis stage designated as MS^2, and the product ion spectrum illustrated in Figure 5.3 is the third mass analysis stage designated as MS^3. The product ions observed in the MS^3 product ion spectrum illustrated in Figure 5.3 demonstrate that the production of the *m/z* 247.6 product ion is in two steps: the

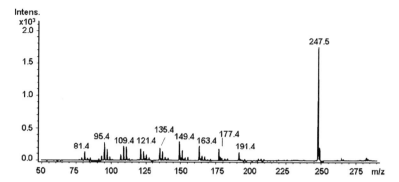

FIGURE 5.3 MS^3 product ion spectrum of the precursor protonated ion of oleamide at m/z 282.6, $[M + H]^+$. In the second mass analysis stage the m/z 265.6 product ion was isolated and subjected to collision-induced dissociation.

first step is neutral loss of ammonia from the m/z 282.6 precursor ion followed by a second step of neutral loss of H_2O, $[M + H - NH_3 - H_2O]^+$. This is effectively loss of ammonium hydroxide (NH_4OH) in what requires a series of steps. The fragmentation pathway mechanism for the production of the m/z 247.6 product ion has similarities to the production of the m/z 265.6 product ion [Equation (5.14)] in its first step of loss of ammonia [Equation (5.15)].

$$(5.15)$$

However, the m/z 265.6 product ion continues to decompose in a series of steps producing the m/z 247.6 product ion.

Also observed in both Figures 5.2 and 5.3 is a series of peaks ranging from m/z 81.6 to m/z 226.6 that is often described as a "picket fence" series primarily from hydrocarbon chain decomposition (four series are included within m/z 81.6 to m/z 226.6). The predominant peaks in the picket fence series are odd m/z product ions that would appear to be derived from the fatty acyl chain hydrocarbon series. In odd electron spectra [e.g., electron impact (EI) spectra], hydrocarbon series are typically represented by the $C_nH_{2n+1}^+$ series.[12] Typically, for a straight-chain saturated aliphatic hydrocarbon series the picket fence product ions will have a maximum around C_4–C_5. Examples include C_3H_7 and C_5H_{11} for the $C_nH_{2n+1}^+$ series and C_3H_5 and C_5H_9 for the $C_nH_{2n-1}^+$ series. Notice that the picket fence series type of product ions produced through aliphatic hydrocarbon chain cleavage are even electron product ions. Aliphatic hydrocarbon chain cleavage in OE spectra are produced by σ-bond ionization followed by carbon–carbon cleavage. Aliphatic hydrocarbon chain cleavage product ions are known to undergo rearrangements that are random, resulting in cyclic formation and unsaturation. In the product ion spectra illustrated in Figures 5.2 and 5.3, there is also observed the production of aliphatic hydrocarbon chain cleavage. Unlike the case in OE/EI mass spectra where aliphatic hydrocarbon chain ionization has taken place (thus resulting in charge-driven fragmentation) the production of the m/z 81.6–226.6 product ions in Figures 5.2 and 5.3 are the result of charge remote fragmentation processes, while the production of the m/z 265.6 and m/z 247.6 product ions are localized charge-driven fragmentation processes. Gross[13] has also observed charge remote fragmentation for fatty acids as reported recently in an account of product ion mass spectra of lipids obtained using high-energy collisions. In the CID product ion spectrum of oleamide (Figs. 5.2 and 5.3), there is also the observance of charge remote fragmentation, however, these spectra were obtained at low energy collisions with ES ion trap mass spectrometry. The EE product ion spectrum of the protonated form of oleamide as the precursor ion is illustrated in Figure 5.4,

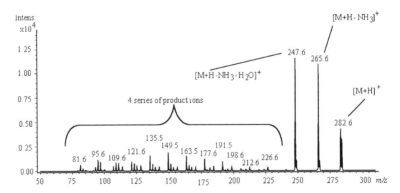

FIGURE 5.4 Product ion spectrum of the EE protonated form of oleamide as the precursor ion listing the production of the m/z 265.6 and 247.6 product ions and the m/z 81.6–226.6 picket fence product ion series.

listing the product ions discussed thus far, including the picket fence series range from m/z 81.6–226.6. The most predominant series as illustrated in Figures 5.2 and 5.3 is comprised of an aliphatic hydrocarbon product ion series of the form $C_nH_{2n-5}^+$. For example, the m/z 93.6 product ion has the product ion formula of $C_7H_9^+$. Using the rings plus double-bond formula results in a value of 3. This indicates that the series is comprised of EE product ions that contain three rings plus double bonds. As with OE/EI aliphatic hydrocarbon chain cleavages, the three rings plus double bonds value for the m/z 93.5 product ion indicates that cyclization, rearrangement, and isomerization is also taking place. The rest of the series includes m/z 107.5, 121.5, 135.5, 149.5, and so forth. A second product ion in series observed in Figures 5.2 and 5.3 includes m/z 81.6, 95.5, 109.5, 123.5, 137.5, and 151.5. This series is represented by $C_nH_{2n-3}^+$ such as $C_6H_9^+$ for the m/z 81.6 product ion. An application of the rings plus double-bond equation results in a value of 2, indicating that this is an EE product ion series that contains 2 rings plus double bonds. In this product ion series from m/z 81.6 to 151.5 there is also cyclization, rearrangement, and isomerization taking place forming the product ions. Also included in the spectra is a series comprised of the product ions m/z 83.6, 97.5, 111.5, 125.5, 139.5, and 153.5. This is an acylium product ion series observed in the form of $C_nH_{2n-3}O^+$ such as $C_6H_9O^+$ for the m/z 97.5 product ion. The fragmentation pathway mechanism for the production of this series of product ions would be derived through a two-step process involving the production of the m/z 265.6 product ion, then neutral loss of C_nH_{2n+2} from an initial acyllium product ion [Equation (5.16)]:

$$(5.16)$$

Lastly, also observed in the m/z 81.6 to m/z 226.6 series are even molecular weight product ions indicating the retainment of the amide moiety. These product ions are produced through single-step neutral losses of the C_nH_{2n} hydrocarbon series from the precursor ion at m/z 282.6, $[M + H - C_nH_{2n}]^+$. Examples include the m/z 226.6 product ion produced through neutral loss of C_4H_8, $[M + H - C_4H_8]^+$, the m/z 184.5 product ion produced through neutral loss of C_7H_{14}, $[M + H - C_7H_{14}]^+$, and the m/z 156.5 product ion produced through neutral loss of C_9H_{18}, $[M + H - C_9H_{18}]^+$. Equation (5.17) illustrates the production of the m/z 226.6 product ion through neutral loss of C_4H_8:

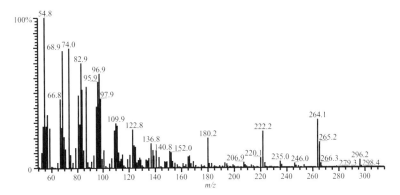

$$(5.17)$$

FIGURE 5.5   Odd electron product ion spectrum of metyl oleate ($C_{19}H_{36}O_2$) at m/z 296 as M$^{+\cdot}$ collected by electron impact ionization.

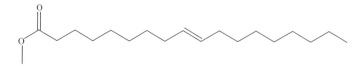

FIGURE 5.6   Structure of methyl oleate.

5.4   PROBLEM: METHYL OLEATE EI MASS SPECTRUM

Figure 5.5 shows the odd electron product ion spectrum of methyl oleate ($C_{19}H_{36}O_2$) at m/z 296 as $M^{+\cdot}$ collected by electron impact ionization. Also included is the structure of methyl oleate along with its molecular formula and mass (Fig. 5.6). Try to use the example we just covered of cis-9-octadecenamide to assign structures to as many of the product ions in the spectrum that were generated by electron impact of methyl oleate. You will need to use Equations (5.3)–(5.8).

REFERENCES

1. McLafferty, F. W., and Turecek, F. *Interpretation of Mass Spectra*, 4th ed. Sausalito, CA: University Science Books, 1993.
2. Di Marzo V. *Biochim. Biophys. Acta* 1998, *1392*, 153–175.
3. Bachur, N. R., Masek, K., Melmon, K. L., and Udenfriend, S. *J. Biol. Chem.* 1965, *240*, 1019–1024.
4. Epps, D. E., Natarajan, V., Schmid, P. C., and Schmid, H. H. *Biochem. Biophys. Acta* 1980, *618*, 420–430.
5. Epps, D. E., Palmer, J. W., Schmid, H. H., and Pfeffer, D. R. *J. Biol. Chem.* 1982, *257*, 1383–1391.
6. Epps, D. E., Schmid, P. C., Natarajan, V., and Schmid, H. H. *Biochem. Biophys. Res. Commun.* 1979, *90*, 628–633.
7. Devane, W. A., Hanus, L., Breuer, A., Pertwee, R. G., and Stevenson, L. A. *Science* 1992, *258*, 1946–1949.
8. Arafat, E. S., Trimble, J. W., Anderson, R. N., Dass, C., and Desiderio, D. M. *Life Sci.* 1989, *45*, 1679–1687.
9. Cravatt, B. F., Prospero-Garcia, O., Siuzdak, G., Gilula, N. B., Henriksen, S. J., Boger, D. L., and Lerner, R. A. *Science* 1995, *268*, 1506–1509.
10. Basile, A. S., Hanus, L., and Mendelson, W. B. *NeuroReport* 1999, *10*, 947–951.
11. Stewart, J. M.; Boudreau, N. M.; Blakely, J. A., and Storey, K. B. *J. Therm. Biol.* 2002, *27*, 309–315.
12. McLafferty, F. W., and Turecek, F. *Interpretation of Mass Spectra*, 4th ed. Sausalito, CA: University Science Books, 1993, p. 226.
13. Gross, M. L. *Int. J. Mass Spectrom.* 2000, *200*, 611–624.

CHAPTER 6

PRODUCT ION SPECTRAL INTERPRETATION

6.1 INTRODUCTION TO PRODUCT ION SPECTRAL INTERPRETATION

The mass spectrometrist is often confronted with a product ion spectrum representing the fragmentation of an odd electron (OE) molecular ion or a charged (anion or cation) even electron (EE) precursor of interest for identification. Within the product ion spectrum there are clues as to the identification of the unknown species. Even electron precursors will fragment generally according to reasonable bond breakage within the structure of the unknown under low-energy collision studies. Low-energy collision is generally the case with tandem mass spectrometry studies using triple quadrupole mass spectrometers, ion trap mass spectrometers, quadrupole time-of-flight mass spectrometers, and FTICR mass spectrometers. In collision-induced dissociation energy is obtained through a hard body collision between the precursor ion that often has been given increased kinetic energy through acceleration and a stationary target gas. The energy obtained by the precursor ion through the collision is distributed across the precursor ion and dissipated through the vibrational modes of the precursor ion with resultant fragmentation when the activation energy is surpassed for a bond cleavage process. Larger molecules have a greater number of vibrational modes in which to spread and dissipate internal energy and tend therefore to fragment less compared to smaller molecules, which posses a lower amount of vibrational degrees of freedom (see Chapter 4 for more detailed discussion). The bond cleavage will generally take place at the most

Even Electron Mass Spectrometry with Biomolecule Applications   By Bryan M. Ham
Copyright © 2008 John Wiley & Sons, Inc.

labile bonds, which in the case of typical biomolecular compounds will be between carbon and oxygen, carbon and nitrogen, and also carbon–carbon bonds. In the identification of an unknown precursor ion by its product ion spectrum, the mass spectrometrist first looks for familiar losses that will give a first indication of the type of biomolecule the unknown ion may be. For example, the loss of 18 amu is associated with water loss (H_2O) and may indicate an oxidized or hydroxyl containing species, the observance of a product ion at m/z 184 almost exclusively represents a protonated phosphatidylcholine headgroup, and repeated losses observed in a product ion spectrum typically represent symmetry such as repeated monomer loss from a polymer. This ability of the mass spectrometrist to recognize structural losses, of course, comes from the experience of working with many product ion spectra. An invaluable resource to the mass spectrometrist for the identification of an unknown through its product ion spectrum is the use of the fragmentation behavior of a known compound. The interpretation of the fragmentation pathways of the known compound can greatly facilitate the description of the fragmentation pathway of an unknown if they are similar. In fact this is often used as proof that the identification of the unknown is based off of similar fragmentation pathway behavior of a known compound. This brings us to a point where a listing of some initial questions concerning the unknown can help to facilitate the interpretation of the product ion spectrum and a reasonable assignment of a structure to the unknown.

1. First question to ask is: From where did the sample originate? Is it a biological extract from cells, tissue, or fluid, a natural compound from a plant extract, an environmental sample from soil, wastewater, or runoff, a protein digestion or gel extract, or a synthesis product? Is any other information about the origin of the sample known? This can greatly facilitate in the initial guess or reasonableness of the particular class of biomolecule that the unknown(s) may be suspected to be such as a sugar or polysaccharide, a protein or peptide, an oligonucleotide, a polymer, a lipid, a hydrocarbon, or a natural product, just as some examples.

2. Of what is the sample matrix comprised? For many mass spectrometric analyses of biomolecules the sample must first be placed into a suitable matrix (current-conducting salt solution for electrospray ionization, suitable crystal or ionic liquid matrix for matrix-assisted laser desorption ionization, etc.). If you have direct control of this the more the better, otherwise the product ion spectrum was collected by another previously, and you must rely on the sample matrix information given. This can help to identify whether the precursor ion, in positive ion mode, is a proton (acid) adduct $[M + H]^+$, a metal adduct such as sodium $[M + Na]^+$, or potassium $[M + K]^+$, or an ammonium adduct $[M + NH_4]^+$, or in the negative ion mode as a chloride adduct $[M + Cl]^-$, or deprotonated $[M - H]^-$. This, of course, is crucial to know as the proposed precursor ion formula must contain the adduct species. For a simple example, protonated water would have a precursor ion at m/z 19 for H_3O^+, sodiated water would have a precursor ion at m/z 41 for H_2ONa^+, and not at m/z 18 for neutral H_2O.

3. What type of tandem mass spectrometric methodology was used to obtain the product ion spectrum? Different ionization techniques and different mass spectrometric approaches tend to give product ion spectra that are different. In general though, structural losses will follow reasonable pathways that can be assigned to structures if a spectrum is to be solved without the aid of standard spectra. Also remember that electron ionization (EI) produces radical OE molecular ions while most other ionization techniques such as ESI and MALDI produce EE precursor ions.

4. Concerning the mass of the precursor ion, is there an exact mass value to help in the assignment of a possible molecular formula? If there is, what is the number of rings plus double bonds calculated from the molecular formula (see Chapter 1)? Also, is the mass of the unknown ion even or odd? The application of the nitrogen rule to determine the number of nitrogen can greatly facilitate structural identification (Chapter 1).

5. Finally, are there product ion spectra of related compounds to that of the suspected unknown that can be used for structural elucidation?

6.2   STRUCTURAL ELUCIDATION OF 1,3-DIPENTADECANOIN

We shall now look at reasonable fragmentation pathways for the symmetric diacylglycerol 1,3-dipentadecanoin. The following will be an example of using a known standard compound to assign postulated product ions according to the precursor ion's structure. This example also emphasizes the indispensable utility of known compound product ion spectra interpretation toward the basic understanding of how biomolecular species dissociate during collision-induced dissociation. Figure 6.1 illustrates a simple, straightforward representation of 1,3-dipentadecanoin that is drawn for the purposes of ease of structural view (i.e., a simple force field optimization of the bond angles has not been applied). The molecular formula for neutral 1,3-dipentadecanoin is $C_{33}H_{64}O_5$ and the monoisotopic exact mass is 540.4754 Da.

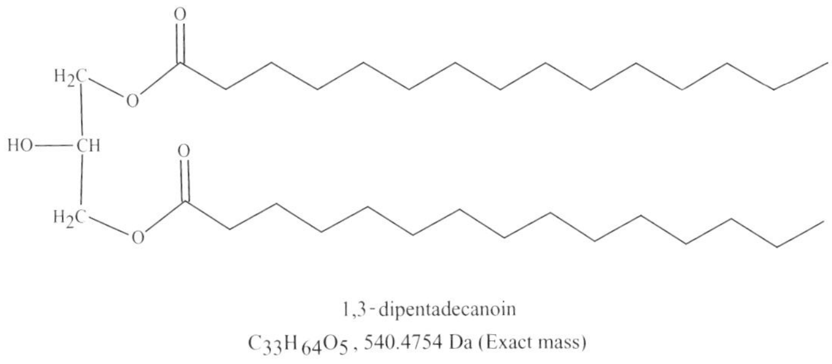

1,3-dipentadecanoin
$C_{33}H_{64}O_5$, 540.4754 Da (Exact mass)

FIGURE 6.1   Structure of neutral 1,3-dipentadecanoin with molecular formula $C_{33}H_{64}O_5$ and exact mass 540.4754 Da.

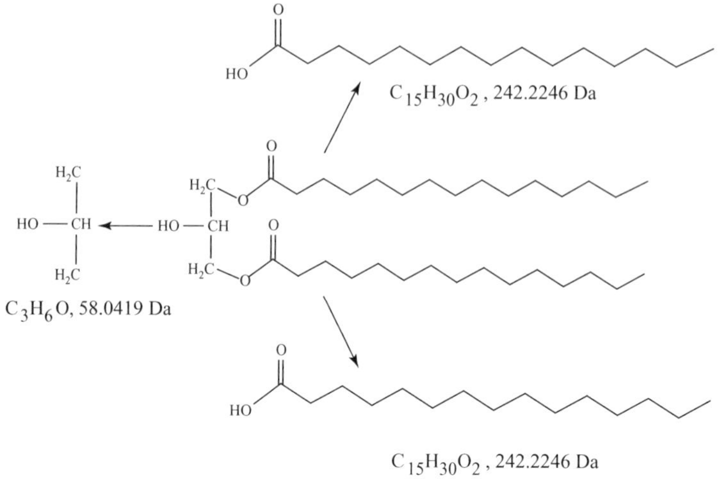

C$_{15}$H$_{30}$O$_2$, 242.2246 Da

C$_3$H$_6$O, 58.0419 Da

C$_{15}$H$_{30}$O$_2$, 242.2246 Da

FIGURE 6.2   Exploded view of the structure of 1,3-dipentadecanoin into three major components: the C15:0 fatty acid substituents and the glycerol backbone.

It is often helpful to break the structure into its main components in a kind of exploded view of the molecule. This is illustrated in Figure 6.2 along with associated molecular formulas and molecular weights. The structure is comprised of three components: the two C15:0 fatty acid substituents with molecular formula C$_{15}$H$_{30}$O$_2$ and exact mass 242.2246 Da and the glycerol backbone with molecular formula C$_3$H$_6$O and exact mass 58.0419 Da. It can be expected that the fragmentation pathways produced in the collision-induced dissociation product ion spectrum will involve primarily these three components of the 1,3-dipentadecanoin structure. Figure 6.3 is the product ion spectrum of the lithium adduct of 1,3-dipentadecanoin at m/z 547, [C$_{33}$H$_{64}$O$_5$ + Li]$^+$, which was obtained on a triple quadrupole mass spectrometer using electrospray as the ionization source (ESI/QhQ MS). Remember that this is now the lithium adduct of 1,3-dipentadecanoin, and we must add lithium to the mass of 1,3-dipentadecanoin to obtain the mass of the precursor ion. That is why we see 1,3-dipentadecanoin at

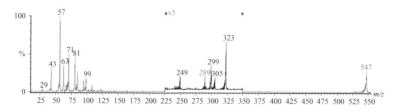

FIGURE 6.3   Product ion spectrum of the lithium adduct of 1,3-dipentadecanoin at m/z 547, [C$_{33}$H$_{64}$O$_5$ + Li]$^+$.

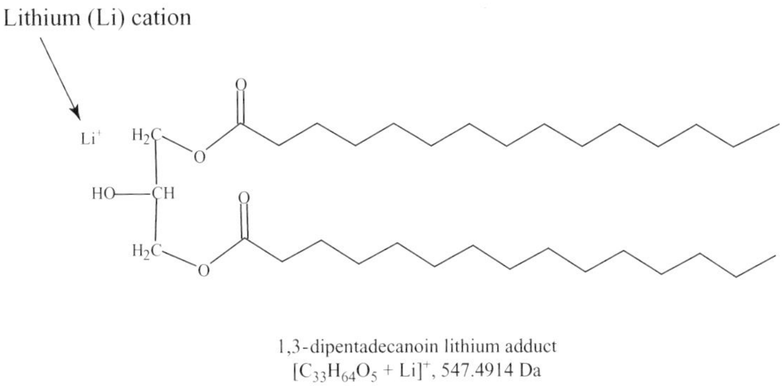

1,3-dipentadecanoin lithium adduct
$[C_{33}H_{64}O_5 + Li]^+$, 547.4914 Da

FIGURE 6.4   Structural formula representation of the lithium adduct of 1,3-dipentadecanoin $[C_{33}H_{64}O_5 + Li]^+$ at 547.4914 Da.

m/z 547 and not at m/z 540, which would be the neutral weight of 1,3-dipentadecanoin. The lithium adduct of 1,3-dipentadecanoin is illustrated in Figure 6.4.

Notice that there are two areas of product ion peaks: approximately m/z 29 to m/z 99 and m/z 249 to m/z 323. The first major product ion peak in Figure 6.3 is the m/z 323 peak. This is 224 amu different from the precursor ion's mass at 547 Thompson. In Figure 6.2 we see that the mass of one of the C15:0 fatty acid substituents is 242 amu, which is close to the neutral loss of 224 amu for the production of the m/z 323 product ion. Therefore, it is resonable to speculate that the fragmentation pathway that produces the m/z 323 product ion involves some form of fatty acid substituent loss. The difference between the mass of the fatty acid substituent (242 amu) and the difference between the precursor ion at m/z 547 and the m/z 323 product ion (224 amu) is 242 − 224 amu = 18 amu. The mass of 18 amu is usually attributed to the mass of H_2O, indicating a loss of water from the C15:0 fatty acid substituent. If we look at simply breaking the carbon–oxygen bond of the bottom C15:0 fatty acid substituent in the 1,3-dipentadecanoin structure, as illustrated in Figure 6.5, the masses of the resultant product structures do not match those that are involved in the product ion spectrum of Figure 6.3. Notice that one product is ionic as a lithium adduct, $C_{18}H_{35}O_3Li$ at 306.2746 Th, and one product is neutral, $C_{15}H_{29}O_2$ at 241.2168 Da. This is illustrative of product ion spectra such as for the simple case of generation of a single product ion where there is always two species produced: the charged product ion and a neutral. We now begin to see that the production of the m/z 323 product ion from the precursor m/z 547 ion involves a more complex fragmentation pathway than that illustrated in Figure 6.5. However, Figure 6.5 gives us a starting point toward a fragmentation pathway that can resonably describe the production of the m/z 323 product ion. This more complete fragmentation pathway would include loss of a C15:0 fatty acid substituent from the precursor ion that itself has lost H_2O, whose mass has been retained in the m/z 323 product ion. Figure 6.6 shows the mechanism for the fragmentation pathway that produces the m/z 323 product ion. The mechanism involves the shift of an alpha hydrogen (α-H) of the

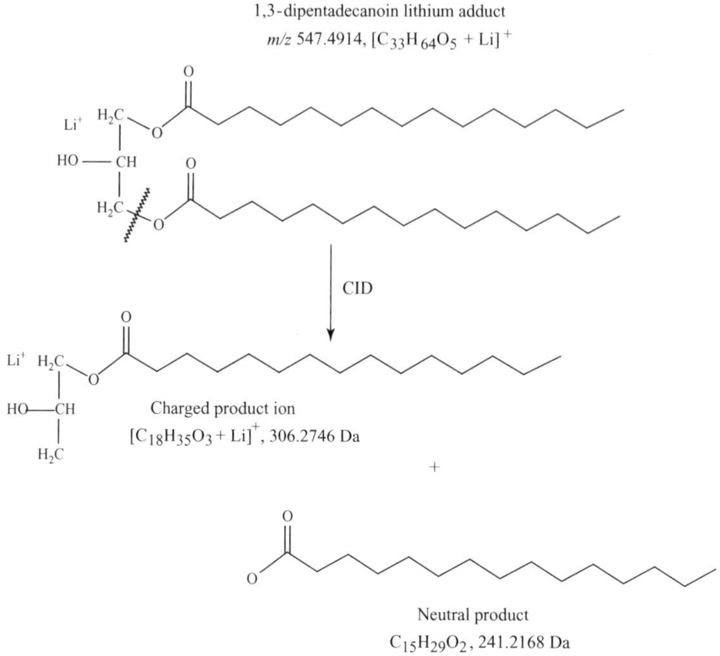

1,3-dipentadecanoin lithium adduct
m/z 547.4914, $[C_{33}H_{64}O_5 + Li]^+$

CID

Charged product ion
$[C_{18}H_{35}O_3 + Li]^+$, 306.2746 Da

+

Neutral product
$C_{15}H_{29}O_2$, 241.2168 Da

FIGURE 6.5 Products produced through direct carbon–oxygen bond breakage of the C15:0 fatty acid substituent.

C15:0 fatty acyl substituent to the glycerol backbone with the retainment of the ester linkage oxygen. This produces lithiated monopentadecanoin at m/z 323, $[C_{18}H_{36}O_4 + Li]^+$, with a concomitant neutral loss of a C15:0 fatty acyl chain as a ketene, $C_{15}H_{28}O$ at 224.2140 Da. The equation to describe the loss of the fatty acyl chain as ketene from the precursor 1,3-dipentadecanoin lithium adduct forming the m/z 323 product ion has the form of $[C_{33}H_{64}O_5 + Li - C_{15}H_{28}O]^+$. The next product ion observed in the CID spectrum of 1,3-dipentadecanoin in Figure 6.3 is the m/z 305 peak. The difference between this peak and the precursor ion at m/z 547 is 242 Da. We see that in Figure 6.2 this is the same mass as the C15:0 fatty acid substituent indicating that the mechanism for the production of the m/z 305 product ion involves neutral loss of the C15:0 fatty acid substituent. This mechanism differs from the production of the m/z 323 product ion in that the ester oxygen has been retained on the fatty acid substituent, and there is a hydrogen transfer from the glycerol backbone to the fatty acid substituent. The fragmentation pathway mechanism for the production of the m/z 305 product ion is illustrated in Figure 6.7. The overall description of the production of the m/z 305 product ion is represented as $[C_{33}H_{64}O_5 + Li - C_{15}H_{30}O_2]^+$. In this mechanism the ester oxygen has been retained on the fatty acid substituent with a hydrogen transfer from the glycerol backbone to the fatty acid substituent producing the m/z 305 product ion $[C_{18}H_{34}O_3 + Li]^+$, with a neutral loss of a C15:0 fatty acid

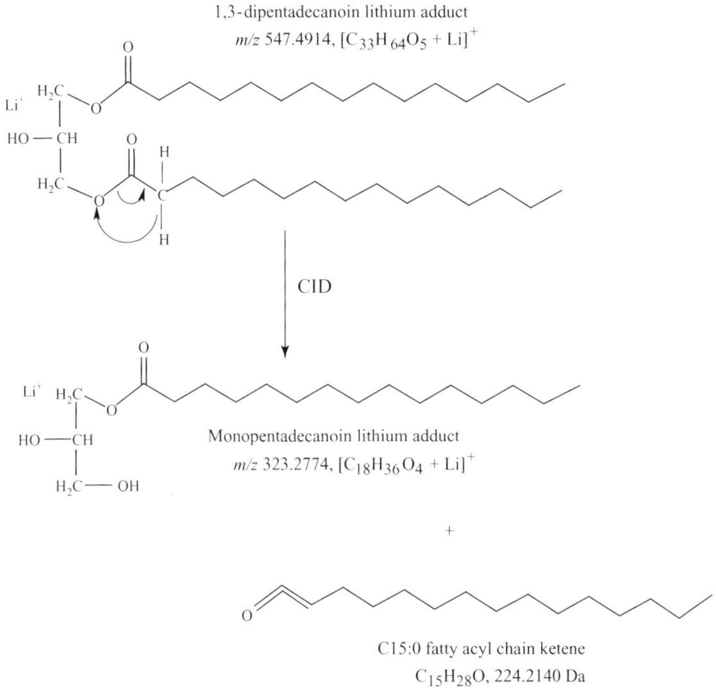

FIGURE 6.6 Fragmentation pathway mechanism for the production of the m/z 323 product ion, $[C_{33}H_{64}O_5 + Li - C_{15}H_{28}O]^+$. Alpha hydrogen shift of the C15:0 fatty acyl substituent to the glycerol backbone with the retainment of the ester linkage oxygen producing lithiated monopentadecanoin at m/z 323, $[C_{18}H_{36}O_4 + Li]^+$, with a concomitant neutral loss of a C15:0 fatty acyl chain as a ketene, $C_{15}H_{28}O$ at 224.2140 Da.

substituent, $C_{15}H_{30}O_2$ at 242.2246 Da. The next product ion in the CID spectrum of 1,3-dipentadecanoin is the m/z 299 ion. The difference between the precursor ion's mass (547 amu) and the m/z 299 product ion is 248 amu. In Figure 6.2 the C15:0 fatty acid substituent has a mass of 242 Da, so again it is likely that this lose involves the fatty acid substituent. Remember that the precursor ion of 1,3-dipentadecanoin at m/z 547 is an adduct of lithium (Li), which has an atomic weight of 7.0160 amu. If we add the weight of lithium to the weight of the C15:0 fatty acid substituent ($C_{15}H_{30}O_2$) we get 7.0160 amu + 242.2246 amu = 249.2406 amu. This is 1 amu greater than the loss we are seeing of 248 amu for the production of the m/z 299 product ion from the m/z 547 precursor ion. This 1-amu difference indicates 1 hydrogen, so we can decipher that the loss from the m/z 547 precursor ion would be C15:0 fatty acid substituent minus 1 hydrogen plus lithium: $C_{15}H_{30}O_2 - H + Li^+ = C_{15}H_{29}O_2Li^+$ (248 amu), which is 547 amu – 248 amu = 299 amu, the mass of the product ion observed. The overall description for the production of the m/z 299 product ion from the m/z 547 precursor ion would be $[C_{33}H_{64}O_5 + Li - C_{15}H_{29}O_2Li]^+$. How do we account

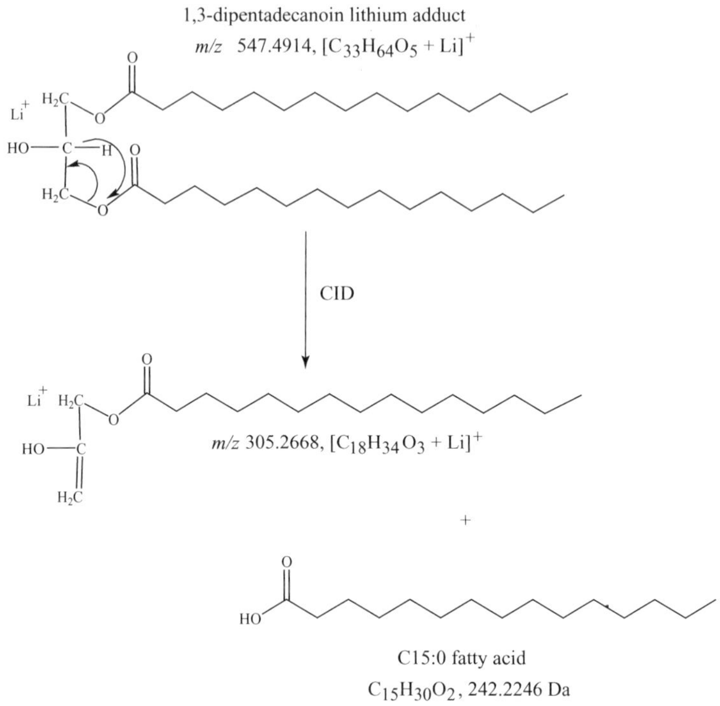

1,3-dipentadecanoin lithium adduct
m/z 547.4914, $[C_{33}H_{64}O_5 + Li]^+$

CID

m/z 305.2668, $[C_{18}H_{34}O_3 + Li]^+$

+

C15:0 fatty acid
$C_{15}H_{30}O_2$, 242.2246 Da

FIGURE 6.7 Fragmentation pathway mechanism for the production of the m/z 305 product ion, $[C_{33}H_{64}O_5 + Li - C_{15}H_{30}O_2]^+$. The ester oxygen has been retained on the fatty acid substituent with a hydrogen transfer from the glycerol backbone to the fatty acid substituent producing the m/z 305 product ion $[C_{18}H_{34}O_3 + Li]^+$, with a neutral loss of a C15:0 fatty acid substituent, $C_{15}H_{30}O_2$ at 242.2246 Da.

for this loss though? It does not involve a simple loss of a fatty acid substituent. The best way to rationalize product ion formation is to draw fragmentation pathway mechanisms such as those illustrated in Figures 6.6 and 6.7. Number one, always start with the structure of the precursor ion. From here it is often not too difficult to rationalize the fragmentation pathway mechanism. The fragmentation pathway mechanism for the production of the m/z 299 product ion from the m/z 547 precursor ion is illustrated in Figure 6.8. In this mechanism the lithium cation is transferred to the C15:0 fatty acid substituent from the glycerol backbone, producing the neutral loss of lithium pentadecanoate, $C_{15}H_{29}O_2Li$, 248.2328 amu. Notice that the charged product ion (m/z 299) is not a lithium adduct but a carbocation. In the mechanism we see first a hydroxyl shift followed by a rearrangement taking place where the charge on the primary carbocation is transferred to a secondary carbocation on the glycerol backbone. Finally, rearrangement again takes place where the charge is tranferred to the carbonyl oxygen through an electron-donating effect producing a five-membered ring. The production of a five membered ring also helps to stabilize the product ion at m/z 299.

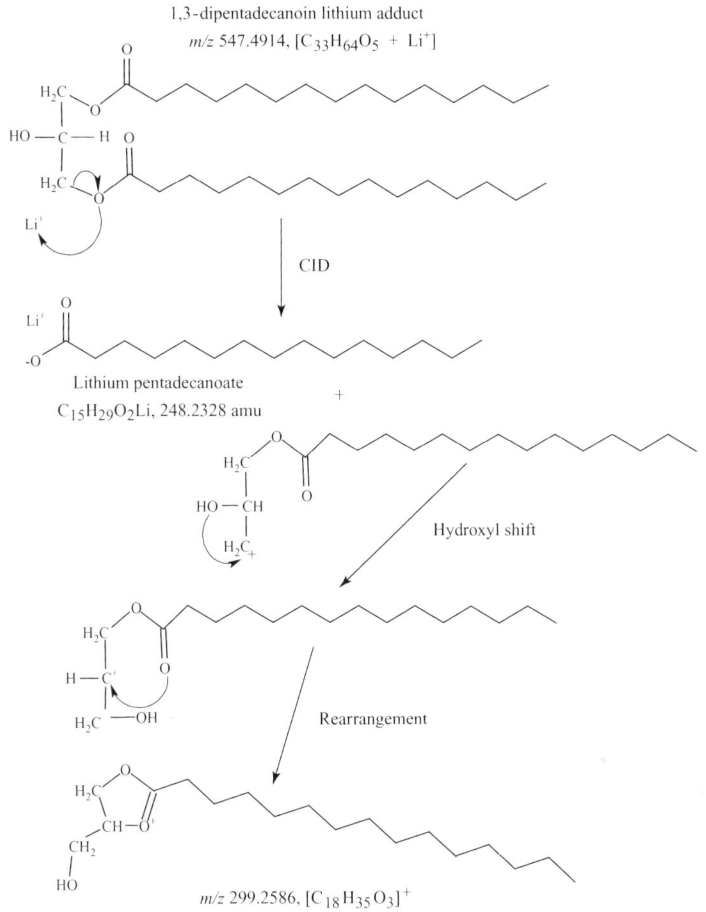

FIGURE 6.8 Fragmentation pathway mechanism for the production of the m/z 299 product ion from the m/z 547 molecular. Lithium cation is transferred to the C15:0 fatty acid substituent from the glycerol backbone, producing the neutral loss of lithium pentadecanoate, $C_{15}H_{29}O_2Li$, 248. 2328 amu, accompanied by rearrangement producing a stable five-membered ring structure and transfer of charge to the carbonyl oxygen.

Continuing in this fashion we see that the next product ion in the Figure 6.3 CID spectrum of the lithium adduct of 1,3-dipentadecanoin is an m/z 289 product ion. The difference between the m/z 289 product ion and the precursor ion at m/z 547 is 258 amu. We know that the neutral loss of the C15:0 fatty acid substituent involves a mass loss of 242 amu. This is 16 amu different from the neutral loss of 258 we have calculated for the production of the m/z 289 product ion. The value of 16 amu is readily recognizable as the difference of one oxygen. However, this is a bit difficult to rationalize upon first inspection of the structure of the m/z 547 ion in Figure 6.4. This mechanism needs to involve loss of the C15:0 fatty acid

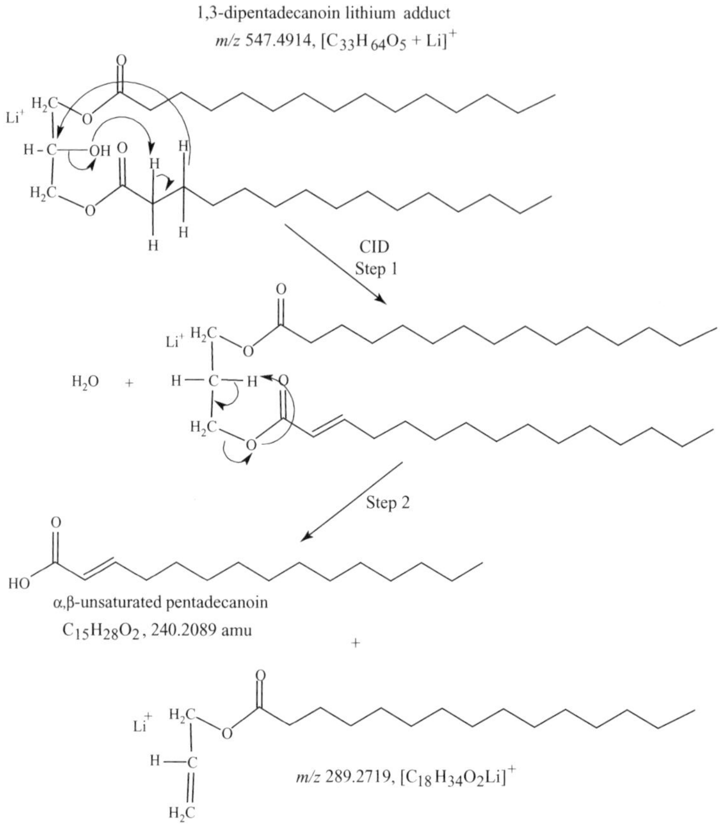

FIGURE 6.9 Two-step mechanism for the generation of the m/z 289 product ion from the m/z 547 precursor ion. Step 1 is water loss taking place from the glycerol backbone producing an α,β-unsaturated fatty acid as the substituent on the fragmentation pathway intermediate, $[C_{33}H_{64}O_5 + Li - H_2O]^+$. Step 2 involves neutral loss of the α,β-unsaturated pentadecanoic acid substituent, $C_{15}H_{28}O_2$ 240.2089 amu, to produce the m/z 289.2719 product ion, $[C_{18}H_{34}O_2Li]^+$. Overall is the neutral loss of H_2O followed by the α,β-unsaturated pentadecanoic acid substituent, $[C_{33}H_{64}O_5 + Li - H_2O - C_{15}H_{28}O_2]^+$.

substituent along with an oxygen from the glycerol backbone. However, loss of one oxygen is not likely to take place, instead water loss from the glycerol backbone is more likely. Again, at this point it is advantageous to remember that when we are rationalizing losses from structures, a resonable loss that can be drawn as a mechanism (which includes double-arrowhead 2-electron transfers) is more likely than simply considering loss of 16 amu (one oxygen) from the glycerol backbone. Now, for accounting purposes, if we are to lose H_2O (18 amu) from the glycerol backbone, then we need to subtract 2 amu (two hydrogen) from the C15:0 fatty acid

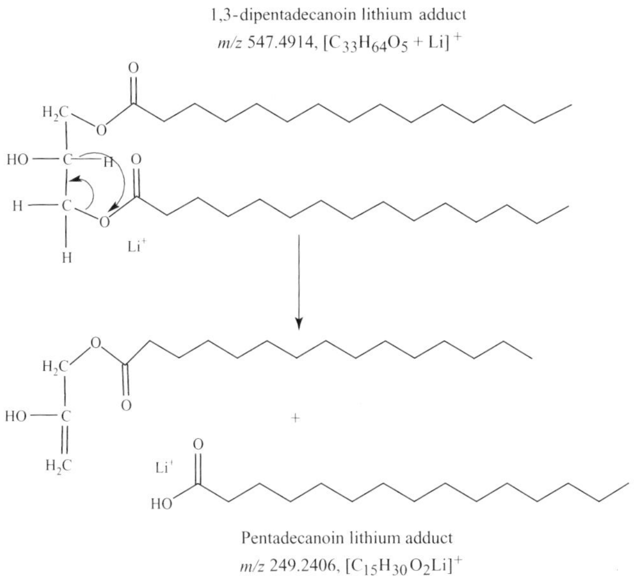

1,3-dipentadecanoin lithium adduct
m/z 547.4914, $[C_{33}H_{64}O_5 + Li]^+$

Pentadecanoin lithium adduct
m/z 249.2406, $[C_{15}H_{30}O_2Li]^+$

FIGURE 6.10 Fragmentation pathway mechanism for the generation of the m/z 249 product ion pentadecanoin lithium adduct $[C_{15}H_{30}O_2Li]^+$.

lose, keeping in mind that a charge balance also must be preserved. For the charge balance, we are postulating that a combination of neutral loss of H_2O and fatty acid substituent is taking place, thus indicating that the charge of the lithium cation is retained within the m/z 289 product ion. The fragmentation pathway mechanism for the production of the m/z 289 product ion is illustrated in Figure 6.9. This is a more complex mechanism than those illustrated in Figures 6.6 (m/z 323), 6.7 (m/z 305), and 6.8 (m/z 299), where a two-step mechanism is required for the description of the generation of the m/z 289 product ion from the m/z 547 precursor ion. In the first step there is water lose taking place from the glycerol backbone. However, we see that the C15:0 fatty acid substituent is also involved with a combination of proton extraction to form the H_2O lose and hydrogen transfer to the glycerol backbone. This effectively produces what is known as an α,β-unsaturated fatty acid as the substituent on the fragmentation pathway intermediate, $[C_{33}H_{64}O_5 + Li - H_2O]^+$. The second step in the fragmentation pathway mechanism involves neutral loss of the α,β-unsaturated pentadecanoic acid substituent, $C_{15}H_{28}O_2$, 240.2089 amu, to produce the m/z 289.2719 product ion, $[C_{18}H_{34}O_2Li]^+$. The overall neutral lose combination of H_2O followed by the α,β-unsaturated pentadecanoic acid substituent would be depicted as $[C_{33}H_{64}O_5 + Li - H_2O - C_{15}H_{28}O_2]^+$.

Lastly, in the structurally informative midrange product ion region of m/z 249 to m/z 323 illustrated in Figure 6.3 is the product ion at m/z 249. In the exploded structure illustrated in Figure 6.2 the neutral fatty acid pentadecanoin has a mass of

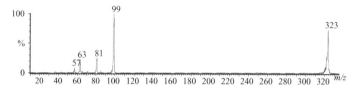

FIGURE 6.11 CID product ion spectrum for the lithium adduct of monopentadecanoin at m/z 323.2774, $[C_{15}H_{36}O_4 + Li]^+$.

242 amu for $C_{15}H_{30}O_2$. The difference in mass between the m/z 249 product ion and the neutral fatty acid is 249 amu – 242 amu = 7 amu. This, of course, is the mass of the lithium cation that is adducted to the 1,3-dipentadecanoin precursor ion at m/z 547. We therefore conclude that the m/z 249 product ion is the lithium adduct of one of the pentadecanoin fatty acid substituents, $[C_{15}H_{30}O_2Li]^+$. Figure 6.10 illustrates the fragmentation pathway mechanism for the generation of the m/z 249 product ion from the m/z 547 precursor ion, $[C_{33}H_{64}O_5 + Li - C_{18}H_{34}O_3]^+$.

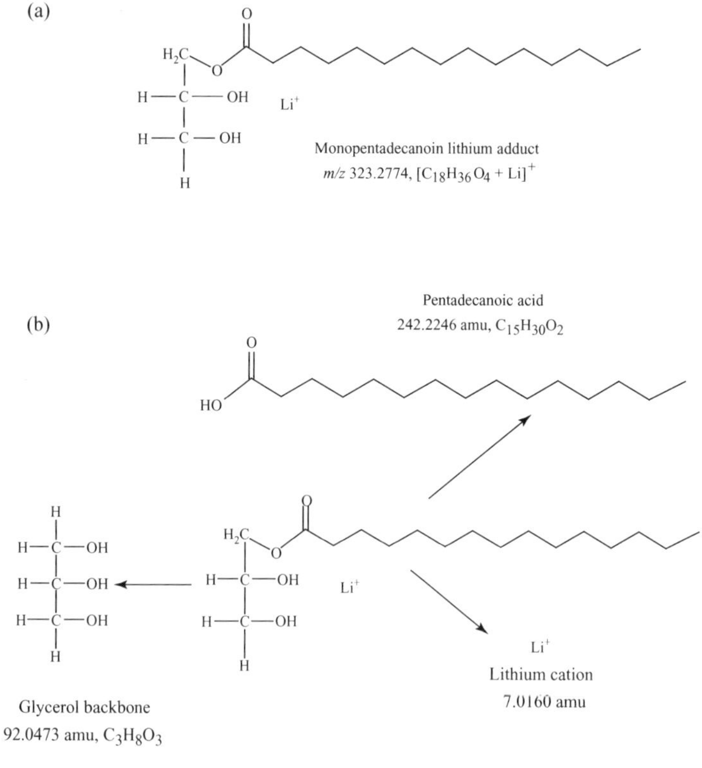

(a)

Monopentadecanoin lithium adduct
m/z 323.2774, $[C_{18}H_{36}O_4 + Li]^+$

Pentadecanoic acid
242.2246 amu, $C_{15}H_{30}O_2$

(b)

Lithium cation
7.0160 amu

Glycerol backbone
92.0473 amu, $C_3H_8O_3$

FIGURE 6.12 (*a*) structure of monopentadecanoin lithium adduct m/z 323.2774, $[C_{18}H_{36}O_4 + Li]^+$. (*b*) Exploded structure of monopentadecanoin.

6.3 PROBLEM: LITHIATED MONOPENTADECANOIN PRODUCT ION SPECTRUM

Figure 6.11 shows the product ion spectrum for the lithium adduct of monopentadecanoin at m/z 323.2774, $[C_{15}H_{36}O_4 + Li]^+$. Also included is the structure of monopentadecanoin and an exploded structure giving the major elements of the structure of monopentadecanoin and their masses (Fig. 6.12). Try to use the example we just covered of 1,3-dipentadecanoin to assign structures to the four product ions generated by CID for the lithium adduct of monopentadecanoin. The four product ions are m/z 57, m/z 63, m/z 81, and m/z 99. Appendix 2 contains a complete solution to the product ion spectrum, including the steps involved in solving the spectrum and fragmentation pathway mechanisms. However, before referring to Appendix 2 for the solutions, attempt to solve the product ion spectrum by following the steps that were used before for the lithium adduct of 1,3-dipentadecanoin:

1. Subtract the masses of the product ions from the precursor ion and compare the masses to the exploded structure masses.
2. Using 1,3-dipentadecanoin as a guide try to draw resonable fragmentation pathway mechanisms for the four product ions.

CHAPTER 7

BIOMOLECULE SPECTRAL INTERPRETATION: PROTEINS

7.1 INTRODUCTION TO PROTEOMICS

As first presented in Section 1.3.1, Proteomics, the study of a biological system's complement of proteins (e.g., from cell, tissue, or a whole organism) at any given state in time, has become a major area of focus for research and study in many different fields and applications. In proteomic studies mass spectrometry can be employed to analyze both the intact, whole protein or the resultant peptides obtained from enzyme-digested proteins. The mass spectrometric analysis of whole intact proteins is often called top-down proteomics where the measurement study starts with the analysis of the intact protein in the gas phase and subsequently investigating its identification and any possible modifications through CID measurements. The mass spectrometric analysis of enzyme-digested proteins that have been converted to peptides is known as bottom-up proteomics. Finally, mass spectrometry is also used to study posttranslation modifications (PTM) that have taken place with the proteins such as glycosylation, sulfation, and phosphorylation. We shall begin with a look at bottom-up proteomics, the most common approach, followed by top-down proteomics, which is seeing more applications and study lately, and finally the posttranslational modifications of glycosylation, sulfation, and phosphorylation. Bioinformatics has become an important tool used in the interpretation of results obtained from mass spectrometry studies. In the last part of this chapter we will briefly look at what bioinformatics is and what it can be used for in relation to mass spectrometry and proteomic studies. Due to the enormous impact of proteomics on

Even Electron Mass Spectrometry with Biomolecule Applications By Bryan M. Ham
Copyright © 2008 John Wiley & Sons, Inc.

research into biological processes, organisms, diseased states, tissues, and so on, we will begin this section starting with a brief overview of proteins, including their structure and makeup.

7.2 PROTEIN STRUCTURE AND CHEMISTRY

Of all biological molecules, proteins are one of the most important, next only to the nucleic acids. All living cells contain proteins, and their name is derived from the Greek word *proteios*, which has the meaning of "first."[1] There are two broad classifications for proteins related to their structure and functionality: water-insoluble fibrous proteins and water-soluble globular proteins. The three-dimensional configuration of a protein is described by its primary, secondary, tertiary, and quaternary structures. Figure 7.1 is a three-dimensional ribbon representation of the protein RNase. The primary structures of proteins are made up of a sequence of amino acids forming a polypeptide chain. Typically, if the chain is less than 10,000 Da, the compound is called a polypeptide, if greater than 10,000 Da, the compound is called a protein. There are 20 amino acids that make up the protein chains through carbon-to-nitrogen peptide bonds. Figure 7.2 illustrates the 20 amino acid structures that make up the polypeptide backbone chain of proteins. Amino acids posses an amino group (NH_2) and a carboxyl group (COOH), which are bonded to the same carbon atom that is α to both groups, therefore amino acids are called alpha amino acids (α amino acid). At physiological pH ($\sim$7.36) the amino acids can be subdivided into four classes according to their structure, polarity, and charge state: (1) negatively charged consisting of aspartic acid (Asp) and glutamic acid (Glu), (2) positively charged consisting of lysine (Lys), arginine (Arg), and histidine (His), (3) polar consisting of serine (Ser), threonine (Thr), tyrosine (Tyr), cysteine (Cys), glutamine (Gln), and asparagine (Asn), and (4) nonpolar consisting of glycine (Gly), leucine (Leu), isoleucine (Ile), alanine (Ala), valine (Val), proline (Pro), methionine (Met),

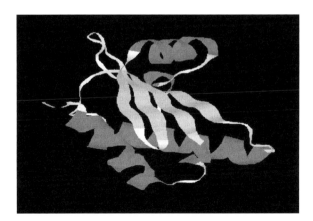

FIGURE 7.1 Ribbon structure representation of the RNase protein illustrating substructures of α helixes and β sheets.

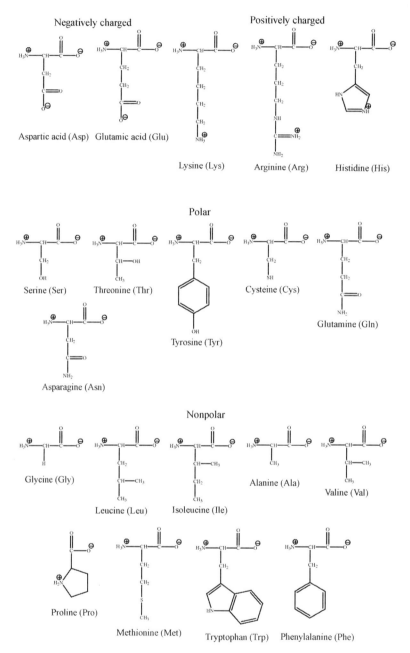

FIGURE 7.2   Structures of the 20 amino acids that make up the polypeptide backbone of proteins. Divisions include negatively charged, positively charged, polar, and nonpolar.

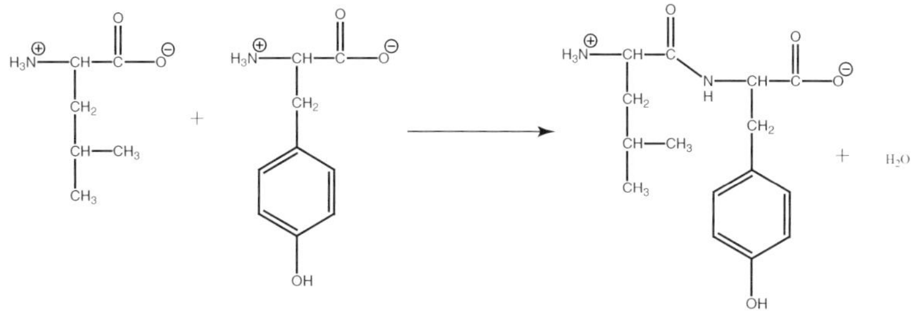

Leucine   Tyrosine

FIGURE 7.3 Condensation reaction between the amino acids leucine and tyrosine forming a peptide bond.

tryptophan (Trp), and phenylalanine (Phe). The carbon-to-nitrogen peptide bonds are formed through condensation reactions between the carboxyl and amino groups. An example condensation reaction between the amino acids leucine and tyrosine is illustrated in Figure 7.3. The peptide C–N bonds are found to be shorter than most amine C–N bonds due to a double-bond nature that contributes to 40% of the peptide bond.[2] This double-bond character lessens the free rotation of the bond, thus affecting the overall structure of the protein.[3] The secondary structure of the protein is described by two different configurations and turns. The two configurations are α helixes (first proposed by Linus Pauling and Robert B. Corey in 1951), and β sheets (parallel and antiparallel) and are illustrated in Figure 7.4. The α helix is described as a right-hand turned spiral that has hydrogen bonding between oxygen and the hydrogen of the nitrogen atoms of the chain backbone. This hydrogen bonding stabilizes the helical structure. The R-group side chains that make up the amino acid residues extrude out from the helix. The β sheet is a flat structure that also has hydrogen bonding between oxygen and the hydrogen of the nitrogen atoms, but from different β sheets (parallel and antiparallel) that run along side each other. These hydrogen bonds also work to stabilize the structure. The R-group side chains alternatively extrude out flat with the sheet, from the sides of the sheet. The third secondary structure, the turn, basically changes the direction of the polypeptide strand. The tertiary structure, which includes the disulfide bonds, is comprised of the ordering of the secondary structure, which is stabilized through side-chain interactions. The quaternary structure is the arrangement of the polypeptide chains into the final working protein. All of these structures describe what is actually a folded protein, where the apolar regions of the protein are tucked away inside the structure, away from the aqueous medium they are found in naturally, and more polar regions are on the surface.

7.3 BOTTOM-UP PROTEOMICS—MASS SPECTROMETRY OF PEPTIDES

7.3.1 History and Strategy

The proteomic approach comprised of measuring the enzymatic products of the protein digestion (after protein extraction from the biological sample), namely the

(a)

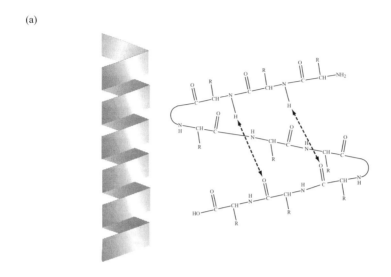

(b)   (c)

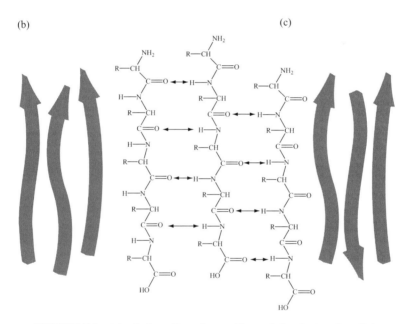

FIGURE 7.4   (*a*) α helixes. Beta sheets: (*b*) parallel and (*c*) antiparallel.

peptides, using mass spectrometry is known as bottom-up proteomics. In the bottom-up approach using nanoelectrospray high-performance liquid chromatography mass spectrometry (nano-ESI-HPLC/MS), the peptides are chromatographically separated and subjected to collision-induced dissociation in the gas phase. The product ion spectra thus obtained of the separated peptides are then used to identify the proteins present in the biological system being studied. Prior to the use of nano-ESI-HPLC/MS for peptide measurement, Edman degradation was used to sequence unknown proteins. The method of Edman sequencing involves the removal of each amino acid residue one-by-one from the polypeptide chain starting from the N-terminus of the peptide or protein.[4] The method worked well for highly purified protein samples that contained a free amino N-terminus, but the analysis was slow, usually taking a day to analyze the sequence of one protein. Mass spectrometry was first coupled with Edman sequencing in 1980 by Shimonishi et al.[5] where the products of the Edman degradation were measured using field desorption mass spectrometry. Field desorption (FD), introduced in 1969 by Beckey, is an ionization technique not commonly in use today. Field desorption consists of depositing the sample, either solid or dissolved in solvent, onto a needle and applying a high voltage. The process of desorption and ionization are obtained simultaneously. The analyte ions produced from the field desorption are then introduced into the mass spectrometer for mass analysis. Fast atom bombardment was also used as an ionization technique to measure peptides obtained from the Edman sequencing approach.[6]

Another early approach to proteomics using mass spectrometry was the application of matrix-assisted laser desorption ionization (MALDI) time-of-flight (TOF) mass spectrometry (MALDI-TOF/MS) to the measurement of peptides obtained from in-gel digestions of proteins separated by gel electrophoresis. This technique was reported by several groups and called peptide mass fingerprinting (PMF).[7–9] In the PMF approach proteins are first separated using two-dimensional gel electrophoresis (2-DE), a protein separation technique first introduced in the 1970s.[10] The gel used in electrophoresis is a rectangular gel comprised of polyacrylamide. The protein sample is loaded onto the gel, and the proteins are separated according to their isoelectric point (pH where the protein has a zero charge). This is the first dimension of the separation. The second dimension is a linear separation of the proteins according to their molecular weights. In preparation for sodium dodecylsulfate polyacrylamide gel electrophoresis, SDS-PAGE, the proteins are first denatured (usually with $8M$ urea and boiling), and sulfide bonds are cleaved, effectively unraveling the tertiary and secondary structure of the protein. Sodium dodecylsulfate, which is negatively charged, is then used to coat the protein in a fashion that is proportional to the proteins' molecular weight. The proteins are then separated within a polyacrylamide gel by placing a potential difference across the gel. Due to the potential difference across the gel, the proteins will experience an electrophoretic movement through the gel, thus separating them according to their molecular weight, with the lower molecular weight proteins having a greater mobility through the gel and the higher molecular weight proteins having a lower mobility through the gel. The resultant two-dimensional gel electrophoresis separation is a collection of spots on the gel that can be up to a few thousand in number. In 2D SDS-PAGE the proteins have been essentially separated into single protein spots. This allows the digestion of the protein within the spot (excised from the gel) using a protease with known cleavage specificity into subsequent peptides that are unique to that particular protein. The peptides extracted from the in-gel digested proteins separated by 2D SDS-PAGE are then measured by MALDI-TOF/MS, creating a spectrum of peaks that represent the molecular weight of the protein's enzymatic generated peptides. This list of measured peptides can be compared to a theoretical list according to the specificity of

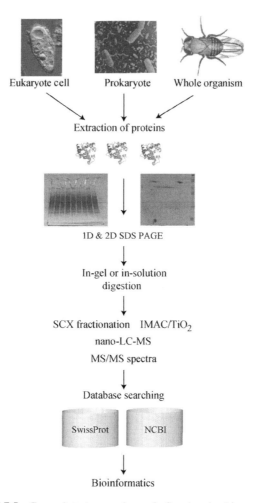

FIGURE 7.5 General strategy and sample flow involved in proteomics.

the enzyme used for digestion. There is an extensive list of references and searching software that has been introduced for the PMF approach to proteomics that has been reviewed.[11] The 2D SDS-PAGE and peptide mass fingerprint approach to proteomics is illustrated in Figure 7.5.

In bottom-up proteomics the proteins are generally extracted from the sample of interest, which can include a sample of cultured cells, bacterium, tissue, or a whole organism. A general scheme for the extraction and peptide mass fingerprint mass spectrometric analysis typically followed in early proteomic studies is illustrated in Figure 7.5. The initial sample is lysed and the proteins are extracted and solubalized. The proteins can then be separated using one-dimensional or two-dimensional sodium dodecyl sulfate polyacrylamide gel electrophoresis (1D or 2D SDS-PAGE). Proteins

TABLE 7.1   Examples of Protease Available for Polypeptide Chain Cleavage

Protease	Polypeptide Cleavage Specificity
Trypsin	At carboxyl side of arginine and lysine residues
Chymotrypsin	At carboxyl side of tryptophan, tyrosine, phenylalanine, leucine, and methionine residues
Proteinase K	At carboxyl side of aromatic, aliphatic, and hydrophobic residues
Factor Xa	At carboxyl side of Glu-Gly-Arg sequence
Carboxypeptidase Y	Sequentially cleaves residues from the carboxy (C) terminus
Submaxillary Arg-C protease	At carboxy side of arginine residues
S. aureus V-8 protease	At carboxy side of glutamate and aspartate residues
Aminopeptidase M	Sequentially cleaves residues from the amino (N) terminus
Pepsin	Nonspecifically cleaves at exposed residues favoring the aromatic residues
Ficin	Nonspecifically cleaves at exposed residues favoring the aromatic residues
Papain	Nonspecifically cleaves at exposed residues

can be digested in the gels, or the proteins in solution are digested using a protease such as trypsin. Trypsin is an endopeptidase that cleaves within the polypeptide chain of the protein at the carboxyl side of the basic amino acids arginine and lysine (the trypsin enzyme has optimal activity at a pH range of 7–10 and requires the presence of Ca^{2+}). It has been observed though that trypsin does not efficiently cleave between the residues Lys–Pro or Arg–Pro. Tryptic peptides are predominantly observed as doubly or triply charged when using electrospray as the ionization source. This is due to the amino terminal residue being basic in each peptide, except for the C-terminal peptide. There exists a number of protease that are available to the mass spectrometrist when designing a digestion of proteins into peptides. These can be used to target cleavage at specific amino acid residues within the polypeptide chain. Examples of available protease and their cleavage specificity are listed in Table 7.1. The enzymes will cleave the proteins into smaller chains of amino acids (typically from 5 amino acid residues up to 100 or so). These short-chain amino acids are mostly water soluble and can be directly analyzed by mass spectrometry. However, often a lysis and extract from a biological system will comprise a very complex mixture of proteins that requires some form of separation to decrease the complexity prior to mass spectral measurement.

7.3.2 Protein Identification Through Product Ion Spectra

More recently, nanoelectrospray high-performance mass spectrometry (nano-ESI-HPLC-MS/MS) has been employed using reverse-phase C18 columns to initially separate the peptides prior to introduction into the mass spectrometer. If a highly complex compliment of digested proteins are being analyzed such as those obtained from eukaryotic cells or tissue, a greater degree of complexity reduction is employed such as strong cation exchange (SCX) fractionation, which can separate the complex peptide mixture into up to 25 fractions or more. The coupling of online SCX with nano-ESI C18 reverse-phase HPLC-MS/MS has also been employed and is called 2D HPLC and multidimensional protein identification technology (MudPIT).[12] This is a gel-free approach that utilizes multiple HPLC-MS analysis of in-solution digestions of protein fractions. The separated peptides are introduced into the mass spectrometer and product ion spectra are obtained. The product ions within the spectra are assigned to amino acid sequences. A complete coverage of the amino acid sequence within a peptide from the product ion spectrum is known as de novo sequencing. This can unambiguously identify a protein (except for a few anomalies, which will be covered shortly) according to standard spectra stored in protein databases. Two examples of protein databases are NCBInr, a protein database consisting of a combination of most public databases compiled by the National Center for Biotechnology Information (NCBI), and Swiss Prot, a database that includes an extensive description of proteins including their functions, posttranslational modifications, domain structures, and so on. The correlation of peptide product ion spectra with theoretical peptides was introduced by Eng et al. in 1994.[13] At the same time Mann and Wilm[14] proposed a partial sequence error-tolerant database searching for protein identifications from peptide product ion spectra. There exists now a rather large choice of searching algorithms that are available for protein identifications from peptide product ion spectra. A list of identification algorithms and their associated universal resource locators (URLs) is illustrated in Table 7.2. The final step in the proteomic analysis of a biological system is the interpretation of the identified proteins, which has been called bioinformatics. Bioinformatics attempts to map and decipher interrelationships between observed proteins and the genetic description. Valuable information can be obtained in this way concerning biomarkers for diseased states, the descriptive workings of a biological system, biological interactions, and so on.

In the identification of proteins from peptides, collision-induced dissociation using mass spectrometry is performed to fragment the peptide and identify its amino acid residue sequence. In most mass spectrometers used in proteomic studies such as the ion trap, the quadrupole time of flight, the triple quadrupole, and the FTICR, the collision energy is considered low (5–50 eV), and the product ions are generally formed through cleavages of the peptide bonds. According to the widely accepted nomenclature of Roepstorff and Fohlman,[15] when the charge is retained on the N-terminal portion of the fragmented peptide, the ions are depicted as a, b, and c. When the charge is retained on the C-terminal portion, the ions are denoted as x, y, and z. The description of the dissociation associated with the peptide chain backbone and the nomenclature of the produced ions is illustrated in Figure 7.6. The ion subscript, for example the "2" in y_2, indicates the number of residues contained within the ion, two amino acid residues in this case. The weakest bond is between the carboxyl carbon and the nitrogen located directly to the right in the peptide chain. At low-energy collision-induced dissociation of the peptide in mass spectrometry, the primary breakage will take place at the weakest bond, generally along the peptide backbone chain, and produce a, b, and y fragments. Notice that the c ions and the

TABLE 7.2   List of Identification Algorithms

MS identification algorithms and URLs PMF
 Aldente http://www.expasy.org/tools/aldente/
 Mascot http://www.matrixscience.com/search_form_select.html
 MOWSE http://srs.hgmp.mrc.ac.uk/cgi-bin/mowse
 MS-Fit http://prospector.ucsf.edu/ucsfhtml4.0/msfit.htm
 PeptIdent http://www.expasy.org/tools/peptident.html
 ProFound http://65.219.84.5/service/prowl/profound.html
MS/MS identification algorithms and URLs PFF
 Phenyx http://www.phenyx-ms.com/
 Sequest http://fields.scripps.edu/sequest/index.html
 Mascot http://www.matrixscience.com/search_form_select.html
 PepFrag http://prowl.rockefeller.edu/prowl/pepfragch.html
 MS-Tag http://prospector.ucsf.edu/ucsfhtml4.0/mstagfd.htm
 ProbID http://projects.systemsbiology.net/probid/
 Sonar http://65.219.84.5/service/prowl/sonar.html
 TANDEM http://www.proteome.ca/opensource.html
 SCOPE N/A
 PEP_PROBE N/A
 VEMS http://www.bio.aau.dk/en/biotechnology/vems.htm
 PEDANTA N/A
De novo sequencing
 SeqMS http://www.protein.osaka-u.ac.jp/rcsfp/profiling/SeqMS.html
 Lutefisk http://www.hairyfatguy.com/Lutefisk
 Sherenga N/A
 PEAKS http://www.bioinformaticssolutions.com/products/peaksoverview.php
Sequence similarity search
 PeptideSearch http://www.narrador.embl-heidelberg.de/GroupPages/Homepage.html
 PepSea http://www.unb.br/cbsp/paginiciais/pepseaseqtag.htm
 MS-Seq http://prospector.ucsf.edu/ucsfhtml4.0/msseq.htm
 MS-Pattern http://prospector.ucsf.edu/ucsfhtml4.0/mspattern.htm
 Mascot http://www.matrixscience.com/search_form_select.html
 FASTS http://www.hgmp.mrc.ac.uk/Registered/Webapp/fasts/
 MS-Blast http://dove.embl-heidelberg.de/Blast2/msblast.html
 OpenSea N/A
 CIDentify http://ftp.virginia.edu/pub/fasta/CIDentify/
Congruence analysis
 MS-Shotgun N/A
 MultiTag N/A
Tag approach
 Popitam http://www.expasy.org/tools/popitam/
 GutenTag http://fields.scripps.edu/GutenTag/index.html

Source: Hernandez, P., Muller, M., and Appel, R. D. Automated protein identification by tandem mass spectrometry: Issues and strategies. *Mass Spectrom. Rev.* 2006, 25, 235–254. Reprinted with permission of John Wiley & Sons.

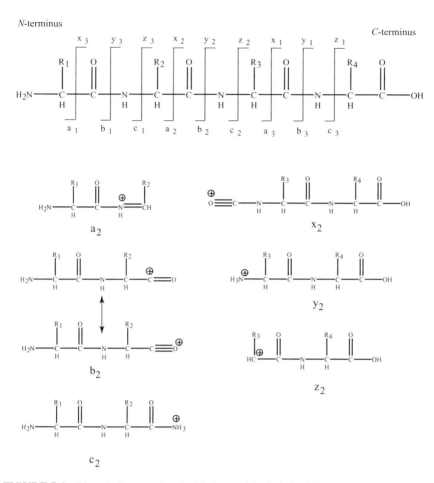

FIGURE 7.6 Dissociation associated with the peptide chain backbone and the nomenclature of the produced ions. Charge retained on the N-terminal portion of the fragmented peptide ions are depicted as a, b, and c. Charge retained on the C-terminal portion the ions are denoted as x, y, and z. Ion subscript, for example, 2 in y_2, indicates the number of residues (two) contained within the ion.

y ions contain an extra proton that they have abstracted from the precursor peptide ion. There has also been proposed a third structure for the b ion that is formed as a protonated oxazolone, which is suggested to be more stable through cyclization[16] (see b_2 ion in Fig. 7.7). The stability of the y ion can be attributed to the transfer of the proton that is producing the charge state to the terminal nitrogen, thus inducing new bond formation and a lower energy state. The model that describes the dissociation of protonated peptides during low-energy collision-induced excitation is called the *mobile proton* model.[17] Peptides fragment primarily from charge-directed reactions where protonation of the peptide can take place at side-chain groups, amide oxygen and nitrogen, and at the terminal amino acid group.

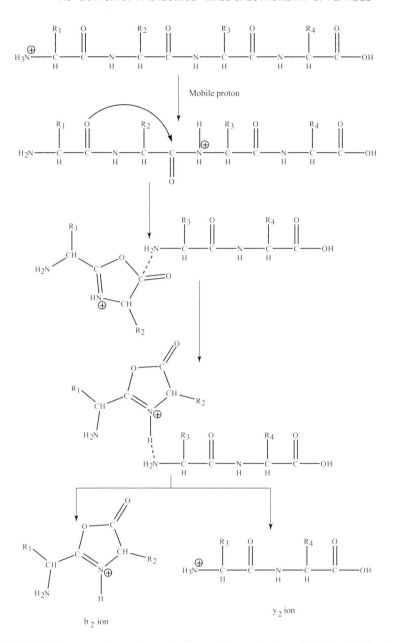

FIGURE 7.7 Fragmentation pathway leading to the production of the *b* and *y* ions from collision-induced dissociation from the polypeptide backbone chain.

On the peptide chain backbone, protonation of the amide nitrogen will lead to a weakening of the amide bond, inducing fragmentation at that point. However, it is more thermodynamically favored, as determined by molecular orbital calculations,[17,18] for protonation to take place on the amide oxygen, which also has the effect of strengthening the amide bond. Inspection of peptide product ion fragmentation spectra has demonstrated, however, that the protonating of the amide nitrogen is taking place over the protonating of the amide oxygen. This is in contrast to the expected site of protonation from a thermodynamic point of view that indicates the amide oxygen protonation and not the amide nitrogen. This discrepancy has been explained by the mobile proton model, introduced by Wysocki and co-workers[17,19] which describes that the proton(s) added to a peptide, upon excitation from CID, will migrate to various protonation sites provided they are not sequestered by a basic amino acid side chain prior to fragmentation. The fragmentation pathway leading to the production of the b and y ions is illustrated in Figure 7.7. The protonation takes place first on the N-terminus of the peptide. The next step is the mobilization of the proton to the amide nitrogen of the peptide chain backbone where cleavage is to take place. The protonated oxazolone derivative is formed from nucleophilic attack by the oxygen of the adjacent amide bond on the carbon center of the protonated amide bond. Depending upon the location of the retention of the charge either a b ion or a y ion will be produced.

Besides the amide bond cleavage producing the b and y ions that are observed in low-energy collision product ion spectra, there are also a number of other product ions that are quite useful in peptide sequence determination. Ions that have lost ammonia (–17 Da) in low-energy collision product ion spectra are denoted as a^*, b^*, and y^*. Ions that have lost water (–18 Da) are denoted as a°, b°, and y°. The a ion illustrated in Figure 7.6 is produced through loss of CO from a b ion (–28 Da). Upon careful inspection of the structures in Figure 7.6 for the product ions, it can be seen that the a ion is missing CO as compared to the structure of the b ion. When a difference of 28 is observed in product ion spectra between two m/z values, an a–b ion pair is suggested and can be useful in ion series identification. Internal cleavage ions are produced by double backbone cleavage, usually by a combination of b- and y-type cleavage. When a combination of b- and y-type cleavage takes place, an amino-acylium ion is produced. When a combination of a- and y-type internal cleavage takes place, an amino-immonium ion is produced. The structures of an amino-acylium ion and an amino-immonium ion are illustrated in Figure 7.8. These types of product ions, which are produced from internal fragmentation, are denoted with their one-letter amino acid code. Though not often observed, x-type ions can be produced using photodissociation.

7.3.3   High-Energy Product Ions

Thus far the product ions that have been discussed, the a-, b-, and y-type ions, are produced through low-energy collisions such as those observed in ion traps. The collision-induced activation in ion traps is a slow heating mechanism, produced through multiple collisions with the trap bath gas, which favors lower energy

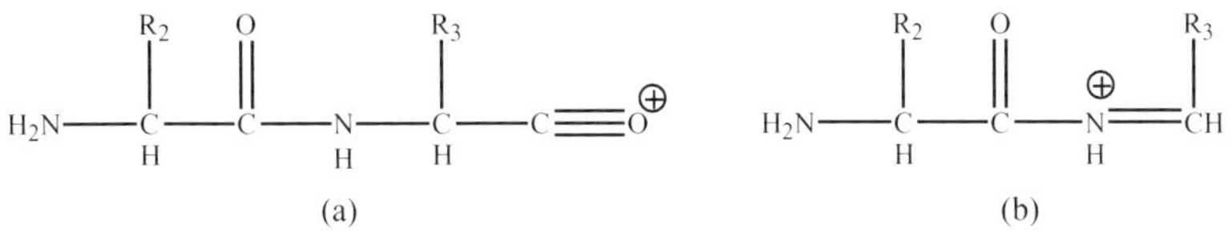

FIGURE 7.8 Structure of (*a*) an amino-acylium ion produced through a combination of *b*- and *y*-type cleavage, and (*b*) an amino-immonium ion through a combination of *a*- and *y*-type internal cleavage.

fragmentation pathways. High-energy collisions that are in the kiloelection volt range such as those produced in MALDI TOF-TOF mass spectrometry produce other product ions in addition to the types that have been discussed so far. Side-chain cleavage ions that are produced by a combination of backbone cleavage and a side-chain bond are observed in high-energy collisions and are denoted as *d*, *v*, and *w* ions. Figure 7.9 contains some illustrative structures of *d*-, *v*-, and *w*-type ions.

Immonium ions are produced through a combination of *a*- and *y*-type cleavage that results in an internal fragment that contains a single side chain. These ions are designated by the one-letter code that corresponds to the amino acid. Immonium ions are not generally observed in ion trap product ion mass spectra but are in MALDI TOF-TOF product ion mass spectra. The structure of a general immonium ion is illustrated in Figure 7.10. Immonium ions are useful in acting as confirmation of

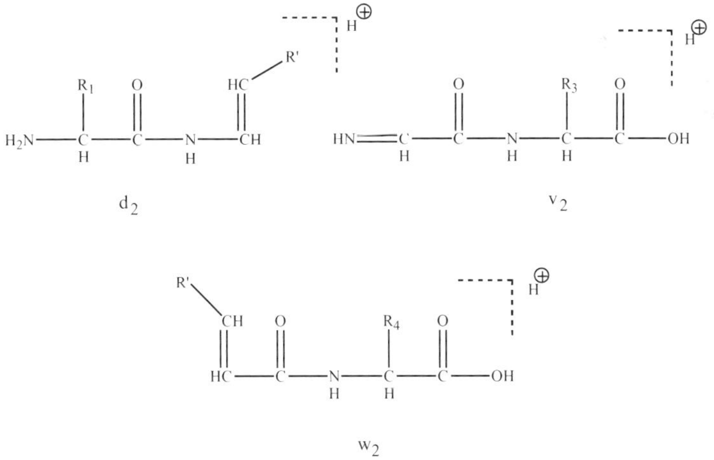

FIGURE 7.9 Structures of *d*-, *v*-, and *w*-type ions produced by a combination of backbone cleavage and a side-chain bond observed in high-energy collision product ion spectra.

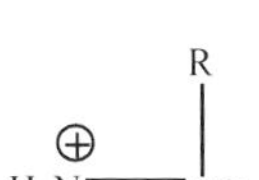

FIGURE 7.10 Structure of a general immonium ion.

TABLE 7.3   Amino Acid Residue Names, Codes, Masses, and Immonium Ion m/z Values

Residue	1-Letter Code	3-Letter Code	Residue Mass	Immonium Ion (m/z)
Alanine	A	Ala	71.04	—
Arginine	R	Arg	156.10	129
Asparagine	N	Asn	114.04	87.09
Aspartic acid	D	Asp	115.03	88.04
Cysteine	C	Cys	103.01	76
Glutamic acid	E	Glu	129.04	102.06
Glutamine	Q	Gln	128.06	101.11
Glycine	G	Gly	57.02	30
Histidine	H	His	137.06	110.07
Isoleucine	I	Ile	113.08	86.1
Leucine	L	Leu	113.08	86.1
Lysine	K	Lys	128.09	101.11
Methionine	M	Met	131.04	104.05
Phenylalanine	F	Phe	147.07	120.08
Proline	P	Pro	97.05	70.07
Serine	S	Ser	87.03	60.04
Threonine	T	Thr	101.05	74.06
Tryptophan	W	Trp	186.08	159.09
Tyrosine	Y	Tyr	163.06	136.08
Valine	V	Val	99.07	72.08

residues suspected to be contained within the peptides' backbone. Table 7.3 is a compilation of the amino acid residue information that is used in mass spectrometry analysis of peptides. The table includes the amino acid residue's name, associated codes, residue mass, and immonium ion mass.

7.3.4   De Novo Sequencing

An example of de novo sequencing is illustrated in Figure 7.11. The product ion spectrum in Figure 7.11a is for a peptide comprised of seven amino acid residues. The peptide product ion spectrum in Figure 7.11b is also comprised of seven amino acid residues, however, the serine residue (Ser, $C_3H_5NO_2$, 87.0320 amu) in Figure 7.11a has been replaced by a threonine residue (Thr, $C_4H_7NO_2$, 101.0477 amu) in 7.11b. The product ion spectra are very similar, but a difference can be discerned with the b_5 ion and the y_3 ions where a shift of 14 Da is observed due to the difference in amino acid residue composition associated with the serine and threonine.

Although the sequencing of the amino acids contained within a peptide chain can be discerned by de novo mass spectrometry as just illustrated, there is a problem associated with isomers and isobars. Isomers are species that have the same molecular formula but differ in their structural arrangement, while isobars are species with different molecular formulas that posses similar (or the same) molecular weights.

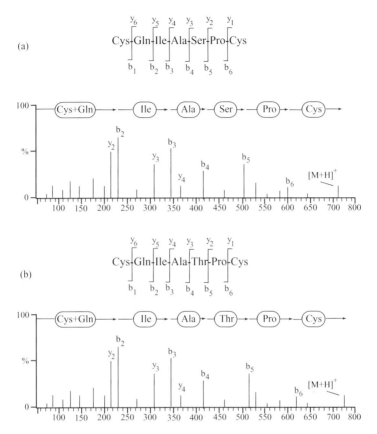

FIGURE 7.11 Example of de novo sequencing using product ion spectra collected by collision-induced dissociation mass spectrometry. (*a*) Peptide comprised of seven amino acid residues. (*b*) Peptide comprised of seven amino acid residues with the serine residue (Ser, $C_3H_5NO_2$, 87.0320 amu) replaced by a threonine residue (Thr, $C_4H_7NO_2$, 101.0477 amu). The product ion spectra are very similar, but a difference can be discerned with the b_5 ion and the y_3 ions where a shift of 14 Da is observed due to the difference in amino acid residue composition associated with the serine and threonine.

For example, it is not possible to determine whether a particular peptide contains leucine (Leu, $C_6H_{11}NO$, 113.0841 amu) or its isomer isoleucine (Ile, $C_6H_{11}NO$, 113.0841 amu), both at a residue mass of 113.0841 amu. Furthermore, even though the remaining 18 amino acid residues each contain distinctive elemental compositions and thus distinct molecular masses, some combinations of residues will actually equate to identical elemental compositions. This produces an isobaric situation where different peptides will posses either very similar or identical sequence masses. If every single peptide amide bond cleavage is not represented within the product ion spectrum, then it is not possible to discern some of these possible combinations. The use of high-resolution/high mass accuracy instrumentation such as the FTICR mass

TABLE 7.4 Examples of Combinations of Amino Acid Residues Where Isobaric Peptides Can Be Observed

Amino Acid Residue	Residue Mass (Da)	Δ Mass (Da)
Leucine	113.08406	
Isoleucine	113.08406	0
Glutamine	128.05858	
Glycine + alanine	128.05858 (57.02146 + 71.03711)	0
Asparagine	114.04293	
2 × glycine	114.04293 (2 × 57.02146)	0
Oxidized methionine	147.03540	
Phenylalanine	147.06841	0.03301
Glutamine	128.05858	
Lysine	128.09496	0.03638
Arginine	156.10111	
Glycine + valine	156.08987 (57.02146 + 99.06841)	0.01124
Asparagine	114.04293	
Ornithine	114.07931	0.03638
Leucine/Isoleucine	113.08406	
Hydroxyproline	113.04768	0.03638
2 × Valine	198.13682 (2 × 99.06841)	
Proline + threonine	198.10044 (97.05276 + 101.04768)	0.03638

spectrometer or the Orbitrap can be used to help reduce this problem when complete de novo sequencing is not possible. Table 7.4 is a listing of some of the amino acid combinations that may arise that can contribute to unknown sequence determination when complete de novo sequencing is not being obtained. For a peptide with a mass of 800 Da, the differences in the table for, for example, the glutamine versus lysine difference at 0.03638 Da, would take a mass accuracy of better than 44 ppm to distinguish the two. For the arginine versus glycine + valine at 0.01124 Da, it would require a mass accuracy of 14 ppm to distinguish the two. For the FTICR mass spectrometers and the hybrid mass spectrometers such as the LTQ-FT or the LTQ-Orbitrap, this is readily achievable, but often for ion traps this is not always achievable.

7.3.5 Electron Capture Dissociation

Other techniques such as electron capture dissociation (ECD)[20] and electron transfer dissociation (ETD, discussed in Section 7.5.3) have also been used to alleviate the problem of isobaric amino acid combinations by giving complimentary product ions (such as c and z ions) that help to obtain complete sequence coverage. The technique of ECD tends to promote extensive fragmentation along the polypeptide backbone, producing c- and z-type ions, while also preserving modifications such as glycosylation and phosphorylation. The general z-type ion that is shown in Figure 7.6 is different, however, from the z-type ion that is produced in ECD, which is a radical cation. Peptide cation radicals are produced by passing or exposing the peptides, which are already multiply protonated by electrospray ionization, through low-energy electrons. The mixing of the protonated peptides with the low-energy electrons will result in exothermic ion–electron recombinations. There are a number of dissociations that can take place after the initial peptide cation radical is formed. These include loss of ammonia, loss of H atoms, loss of side-chain fragments, cleavage of disulfide bonds, and most importantly peptide backbone cleavages. The c-type ion is produced through homolytic cleavage at the N–C peptide bond, and charges are present in the amino-terminal fragment. The $z^{\cdot}$-type ion is produced when charges are present in the carboxy-terminal fragment. The mechanism, which has been given for the promotion of fragmentation of the peptides, is due to electron attachment to the protonated sites of the peptide. The cation radical intermediate that has formed will release a hydrogen atom. A nearby carbonyl group will capture the released hydrogen atom, and the peptide will dissociate by cleavage of the adjacent N–C peptide bond. The mechanism for the production of an α-amide radical of the peptide C-terminus, a $z^{\cdot}$-type ion, and the enolamine of the N-terminus portion of the peptide, a c-type ion, is illustrated in Figure 7.12.

7.4 TOP-DOWN PROTEOMICS: MASS SPECTROMETRY OF INTACT PROTEINS

7.4.1 Background

Measuring the whole, intact protein in the gas phase using mass spectrometric methodologies is known as *top-down proteomics*. Top-down proteomics measures the intact protein's mass followed by collision-induced dissociation of the whole protein breaking it into smaller parts. A vital component of top-down proteomics is the accuracy to which the masses are measured. Often, high-resolution mass spectrometers such as the Fourier transform ion cyclotron resonance (FTICR) mass spectrometer are used to accurately measure the intact protein's mass and the product ions produced during collision-induced dissociation experiments. In early top-down experiments, however, this was not the case. Mass spectrometers such as the triple quadrupole coupled with electrospray were first used to measure intact proteins in the gas phase.[21,22] However, the triple quadrupole mass spectrometer does not allow the resolving of the isotopic distribution of the product ions being generated in the top-down approach. The use of FT-MS/MS was later reported with high enough resolution to resolve isotopic peaks.[23,24] An example of these early top-down experiments utilizing FT mass spectrometry is illustrated in Figure 7.13. Extensive initial, premass analysis sample preparation such as cleanup, digestion, desalting, enriching, and the like, all often incorporated in "bottom-up" proteomics, is not necessarily required in top-down approaches. The dynamic range in top-down proteomics can be limited by the number of analytes that can be present during analysis, but this is usually overcome by using some type of separation prior to introduction into the mass spectrometer. The separation of complex protein mixtures can be obtained using techniques such as reverse-phase high-performance liquid chromatography (RP-HPLC), gel electrophoresis, anion exchange chromatography, and capillary electrophoresis.

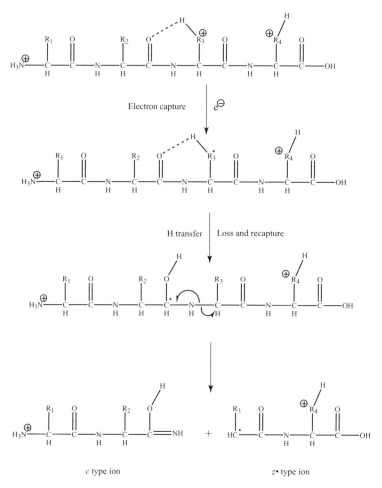

FIGURE 7.12 Mechanism for the production of c- and z'-type ions observed in ECD.

Typically, in bottom-up analysis the digested protein peptides are <3 kDa, and the complete description of the original, intact protein is not possible. With top-down analysis there is often 100% coverage of the protein being analyzed. This allows the determination of the N- and C-termini, the exact location of modifications to the protein such as phosphorylation, and the confirmation of DNA-predicted sequences.

7.4.2 Gas-Phase Basicity and Protein Charging

It is the process of electrospray ionization that allows the measurement of large molecular weight intact proteins. As a rule of thumb, for each 1000 Da of the protein, there is associated one charge state. For example, a 30-kDa protein, as an

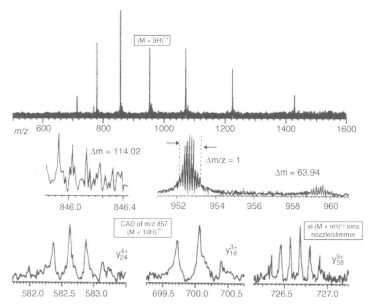

FIGURE 7.13   (*Top*) ESI mass spectrum of ubiquitin (sum of 10 scans, 64,000 data). (*Center*) Regions expanded to show the presence of impurities. (*Bottom*) MS/MS fragment ions from collisionally activated dissociation of $(M + 10H)^{10+}$ and from placing 200 V between the nozzle and skimmer of the ESI source. [Reprinted with permission from Loo, J. A., Quinn, J. P., Ryu, S. I., Henry, K. D., Senko, M. W., and McLafferty, F. W. High-resolution tandem mass spectrometry of large biomolecules (electrospray ionization/polypeptide sequencing). *Proc. Natl. Acad. Sci. USA* 1992, *89*, 286–289.]

approximation, will have a charge state of 30+. This brings the measured mass of the protein down into the range of many mass spectrometers, which typically scan between m/z 100 to m/z 4000 (m/z = 30 kDa/30+ = 1000 Th). Multiple charging of peptides and proteins is achieved during the electropsray ionization process due to the presence of amino acid residue basic sites. There is a limit to the number of charges that can be placed onto a peptide or protein during the ESI process as was demonstrated by Schnier et al. in 1995.[25] In their study the apparent gas-phase basicity as a function of charge state was measured using cytochrome c. A graphical plot of apparent gas-phase basicity versus charge state is illustrated in Figure 7.14. The curves in Figure 7.14 have a negative trend (go down) as each charge state is increased. As a new charge state is added (increasing x axis), the apparent gas-phase basicity decreases (y axis). In the graph the dashed line represents the gas-phase basicity of methanol, which is included as a reference of a species present during the electrospray ionization process that can also accept a charge. The charging of large molecules during the ionization process is thought to follow the charged residue mechanism. In this ionization process the solvent molecules evaporate from around the protein leaving charges that associate to basic sites within the protein. What

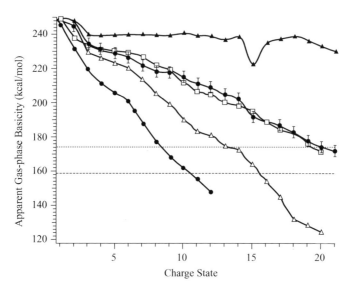

FIGURE 7.14 Apparent gas-phase basicity as a function of charge state of cytochrome c ions, measured (•) calculated, linear ($\epsilon_r = 1.0$, △; best fit $\epsilon_r = 2.0$, ○); intrinsic, ▲; calculated, X-ray crystal structure ($\epsilon_r = 2.0$, ♦); calculated, α helix ($\epsilon_r = 4.1$, □). The dashed line indicates GB of methanol (174.1 kcal/mol) and the dash-dot line indicates GB of water (159.0 kcal/mol). ϵ_r is the intrinsic dielectric polarizability. (Reprinted with permission from Schnier, D. S., Gross, E. R., and Williams, E. R. *J. Am. Chem. Soc.* 1995, *117*, 6747–6757. Copyright 1995 American Chemical Society.)

the intersection in the graph between the charging of cytochrome c and the gas-phase (GP) basicity of methanol is demonstrating is that there is a limit to the amount of charges that can be placed upon a species during ionization. At some point (intersection) it becomes thermodynamically favorable to put the next charge (proton) onto a methanol molecule than onto the protein. Here a state has been reached where Coulombic repulsion between the charges and a loss of basic sites does not allow further charging. At this point the maximum charge state has been reached for the protein molecule.

7.4.3 Calculation of Charge State and Molecular Weight

Another interesting feature of the electrospray charging of proteins is the ability to calculate the charge state and molecular weight of an unknown protein from its single-stage mass spectrum. Figure 7.15 illustrates the electrospray mass spectrum of horse heart myoglobin at a molecular weight of 16,954 Da. If this were an unknown protein species, we would only have the respective m/z values from the mass spectrum. Using the following two simultaneous equations with two unknowns, the charge states and the molecular weight of the unknown protein can be calculated using only

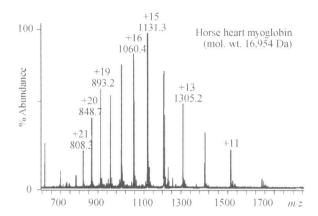

FIGURE 7.15 Electrospray mass spectrum of horse heart myoglobin at a molecular weight of 16,954 Da. The spectrum illustrates an envelope of peaks of different m/z values and charge states for the protein.

the information obtained within the mass spectrum:

$$\frac{m}{z} = \frac{m + z(1.0079)}{z} \qquad \text{(higher } m/z \text{ peak from spectrum)} \qquad (7.1)$$

$$\frac{m}{z} = \frac{m + (z + 1)(1.0079)}{z + 1} \qquad \text{(lower } m/z \text{ peak from spectrum)} \qquad (7.2)$$

If we solve the higher mass/lower charge state peak ($m/z = 1131.3$ Th) for m by taking Equation (7.1) and solving for m we obtain

$$m = 1131.3z - 1.0079z \qquad (7.3)$$

$$= 1130.2921z \qquad (7.4)$$

If we next substitute this mass into the lower mass/higher charge state ($m/z = 1060.4$ Th) [Equation (7.2)], we can calculate the charge state of the m/z 1131.3 peak:

$$\frac{m}{z} = \frac{m + (z + 1)(1.0079)}{z + 1} \qquad (7.2)$$

$$1060.4 = \frac{m + 1.0079z + 1.0079}{z + 1} \qquad (7.5)$$

$$1060.4 = \frac{1130.2921z + 1.0079z + 1.0079}{z + 1} \qquad (7.6)$$

$$z = 15 \qquad (7.7)$$

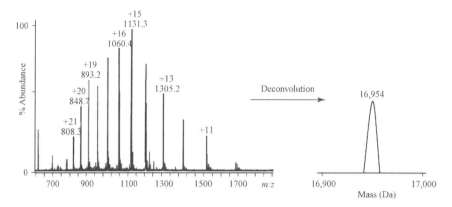

FIGURE 7.16 Deconvoluted, computer-generated spectrum of horse heart myoglobin at a molecular weight of 16,954 Da.

Therefore, we have determined that the charge state of the m/z 1131.3 peak is +15. To calculate the molecular weight of the unknown protein, we can substitute the charge state value into Equation (7.1) and solve for m:

$$\frac{m}{z} = \frac{m + z(1.0079)}{z} \tag{7.1}$$

$$1131.3 = \frac{m + 15(1.0079)}{15} \tag{7.8}$$

$$m = 16954 \tag{7.9}$$

This process is known as deconvolution where a distribution (mass spectral envelope) of a protein's m/z values and associated charge states are collapsed down to a single peak representing the molecular weight of the protein. The deconvolution of the horse heart myoglobin protein is illustrated in Figure 7.16. The width of the deconvoluted peak indicates the variability in the calculation of the molecular weight of the protein from the mass spectral peaks. While it is possible to deconvolute protein peaks by hand by solving two simultaneous equations with two unknowns, mass spectral computer software are typically used to perform this task.

7.4.4 Top-Down Protein Sequencing

In bottom-up proteomics the mass spectral identification of proteins through sequence determination of separated peptides requires the isolation of a single peptide for fragmentation experiments. This is typically done by removing from an ion trap mass spectrometer all m/z species present except for one that is of interest for fragmentation and subsequent sequencing. It has been demonstrated, however, that multiple proteins can be simultaneously fragmented and identified in top-down proteomics.

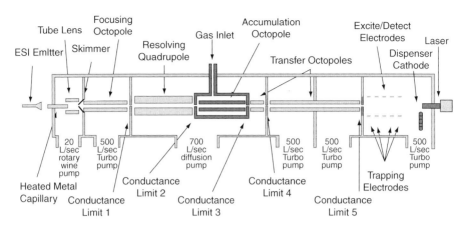

FIGURE 7.17 Schematic representation of the quadrupole/Fourier transform ion cyclotron resonance hybrid mass spectrometer for versatile MS/MS and improved dynamic range by means of m/z-selective ion accumulation external to the superconducting magnet bore. (Reprinted with permission. Patrie, S. M., Charlebois, J. P., Whipple, D., Kelleher, N. L., Hendrickson, C. L., Quinn, J. P., Marshall, A. G., and Mukhopadhyay, B. Construction of a hybrid quadrupole/Fourier transform ion cyclotron resonance mass spectrometer for versatile MS/MS above 10 kDa. *J. Am. Soc. Mass Spectrom.* 2004, *15*, 1099–1108. Copyright Elsevier 2004.)

In a study reported by Patrie et al.[26] the authors used a hybrid mass spectrometer that coupled a quadrupole mass analyzer to a Fourier transform ion cyclotron resonance mass spectrometer that uses a 9.4-T magnet. The instrumental design is illustrated in Figure 7.17. Prior to the FTICR/MS there is a resolving quadrupole that can act as either an RF-only ion guide or as a fully functional mass analyzer. Following this in the instrumental design is an accumulation octopole that was used to accumulate and store ions prior to introduction into the FTICR mass spectrometer. Nitrogen or helium gas at a pressure of approximately 1 mTorr was introduced into the accumulation octopole to help improve the accumulation. The FTICR cell located within the 9.4-T magnet is an open-ended capacity-coupled cell that is cylindrical and divided axially into five segments. At the end of the instrument (far right side) is located a laser that is used for infrared multiphoton dissociation (IRMPD) experiments. For fragmentation experiments collision-induced dissociation could be performed in the accumulation octopole, IRMPD could be performed within the ICR cell by irradiating the trapped species with the laser, or finally the instrumental design also included ECD capabilities. A top-down experiment of a mixture of proteins collected as a fraction eluting from a reverse-phase LC separation is illustrated in Figure 7.18. Figure 7.18*a* illustrates a broadband spectrum of the RPLC fraction where a very low response is observed for the proteins present. The same broadband spectrum is illustrated in Figure 7.18*b* after using the accumulation octopole to increase the amount of sample that is being introduced into the FTICR cell, and thus increase the sensitivity of the

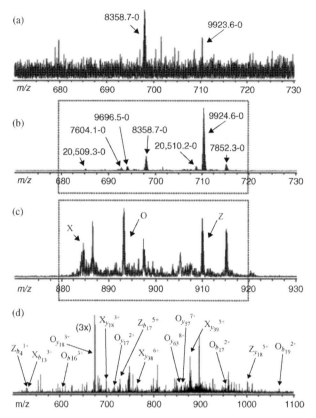

FIGURE 7.18 (*a*) Expansion of a 60-m/z section segment of the broadband spectrum from a *Methanococcus jannaschii* RPLC fraction (3.7 s scan time; 50 scans). (*b*) The same $\Delta(m/z) = 60$ segment after quadrupole-enhanced ion accumulation [$\Delta(m/z) = 40$ segment, 9.7 s scan time; 10 scans]. (*c*) $\Delta(m/z) = 60$ segment from the same sample with subsequent IRMPD fragmentation of all intact proteins in parallel [shown in (*d*), 25 scans]. Identified proteins are MJ0543 (X, Exptl. 19,431.8-0 kDa, Theoret. 19432.5-0 kDa), MJ0471 (O, Exptl. 20,511.3-0 kDa, Theoret. 20,511.1-0), and MJ0472 (asterisk, Exptl. 17.263.0-0 kDa, Theoret. 17263.6-0 kDa). (Reprinted with permission. *J. Am. Soc. Mass Spectrom.* Patrie, S. M., Charlebois, J. P., Whipple, D., Kelleher, N. L., Hendrickson, C. L., Quinn, J. P., Marshall, A. G., and Mukhopadhyay, B. Construction of a hybrid quadrupole/Fourier transform ion cyclotron resonance mass spectrometer for versatile MS/MS above 10 kDa. *J. Am. Soc. Mass Spectrom.* 2004, *15*, 1099–1108. Copyright Elsevier 2004.)

mass measurement of the 7 proteins present. Figure 7.18*c* and 7.18*d* are illustrating the fragmentation result of an IRMPD experiment of the 7 proteins that were present. Of the 7 proteins present in Figure 7.18*b*, 3 were identified by top-down proteomics, listed as X, the 19,431.8-Da protein MJ0543, O, the 20,511.3-Da protein MJ0471, and Z, the 17,263.0-Da protein MJ0472. Notice the large number of amino acid residues contained within the *b*- and *y*-type ions in Figure 7.18*d* and the associated high number of charges (e.g., the Oy_{63}^{8+} product ion that contains 63 amino acid residues and 8 charges). This is quite different from the peptides that are normally observed in bottom-up proteomics where most peptides contain between 7 and 25 amino acid residues with mostly 2 charges, but 3 or 4 charges are also observed for the longer chain peptides.

7.5 POSTTRANSLATIONAL MODIFICATION OF PROTEINS (PTM)

7.5.1 Three Main Types of PTM

In the genomic sequencing field, the use of robotic gene sequencers allowed large-scale sequencing that was essentially automated. The robotic automation of determining gene sequences is possible because the sequences involved with genes involve only four bases (see Chapter 9), and there are no variations induced in the form of postmodification. This has resulted in the well-publicized entire sequencing of the human genome (Human Genome Project, *Nature*, February 2001). This is not the case with proteins where there is not only the observance of spliced variants from alternative splicing from the messenger ribonucleic acid (mRNA); there are also post-translational modifications that can take place with the amino acids contained within the protein. There are over 200 posttranslational modifications that can take place with proteins as has been described by Wold.[27] As examples, here are 22 different types of posttranslational modifications that can take place with proteins: acetylation, amidation, biotinylation, C-mannosylation, deamidation, farnesylation, formylation, flavinylation, γ-carboxyglutamic acids, geranyl-geranylation, hydroxylation, lipoxy-lation, myristoylation, methylation, *N*-acyl diglyceride (tripalmitate), *O*-GlcNAc, palmitoylation, phosphorylation, phospho-pantetheine, pyrrolidone carboxylic acid, pyridoxyl phosphate, and sulfation.[28] There are also artifactual modifications such as oxidation of methionine. Of these, the three types of PTM that are primarily observed and studied using mass spectrometric techniques are glycosylation, sulfation, and phosphorylation. The observance of PTM is increasingly being used in expression studies where a normal-state proteome is being compared to a diseased-state pro-teome. However, the PTM of a protein during a biological of physiological change within an organism may take place without any change in the abundance of the protein involved and often is one piece of a complex puzzle. Methods that measure PTM using mass spectrometric methodologies often focus on the degree (increase or decrease, or alternatively, up-regulation or down-regulation) of PTM for any given protein or proteins. We shall briefly look at glycosylation and sulfation, which are less involved in cellular processing than phosphorylation, a major signaling cascade pathway for the response to a change in cellular condition(s).

7.5.2 Glycosylation of Proteins

Glycosylation is the covalent addition of a carbohydrate chain to amino acid side chains of a protein producing a glycoprotein. The carbohydrate side chains can be

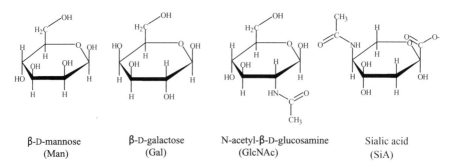

β-D-mannose   β-D-galactose   N-acetyl-β-D-glucosamine   Sialic acid
(Man)   (Gal)   (GlcNAc)   (SiA)

FIGURE 7.19   Most common four carbohydrates found in glycoproteiens: mannose (Man), galactose (Gal), *N*-acetylglucosamine (GlcNAc), and sialic acid (SiA).

anywhere from 1 to 70 sugar units in length, branched or straight chained, and are most commonly comprised of mannose, galactose, *N*-acetylglucosamine, and sialic acid. The structures and abbreviations of these four carbohydrates are illustrated in Figure 7.19. Protein glycosylation is the most common type of protein modification that is found in eukaryotes. The glycosylation of proteins helps to determine the proteins' structure, is crucial in cell–cell recognition, and may also be involved in cellular signaling events, although most likely not as important as the phosphorylation modification is in signaling. Glycoproteins are primarily membrane proteins and are abundantly found in plasma membranes where they are involved in cell-to-cell recognition processes. The carbohydrate groups of the membrane glycoproteins are positioned so that they are externally extruded from the cell membrane surface. An example of this involves erythrocytes where the externally extruded carbohydrate groups of the cell membrane possess a negative charge and thus repel each other and subsequently reduce the blood's viscosity. The two types of linkage of the carbohydrates to the amino acids are (1) to the nitrogen atom of an amino acid group called *N-linked glycosylation* and (2) to the oxygen atom of a hydroxyl group called *O-linked glycosylation*. In the N-linked glycosylation the carbohydrates are attached to the asparagine's (Asn) side-chain amino group. In the O-linked glycosylation the carbohydrates are attached to the serine (Ser) or threonine (Thr) side-chain hydroxyl group. The current specific amino acid sequences involved with glycosylation state that the N-linked carbohydrates are linked through asparagine (Asn) and N-acetylglucosamine. The associated amino acid sequence with the O-linked carbohydrate consists of a serine (Ser) or threonine (Thr) separated by one amino acid residue (any one of 19 residues excluding proline), both located toward the C-terminus of the peptide chain. The proline (Pro) amino acid residue, however, does not participate as the one amino acid between the Asn and Ser/Thr residues. The N-linked and O-linked carbohydrates are illustrated in Figure 7.20.

In applying mass spectrometry in the characterization of glycosylated proteins, there are three objectives that the researcher is attempting to achieve: (1) to get an identification of the glycosylated peptides and proteins, (2) to accurately determine

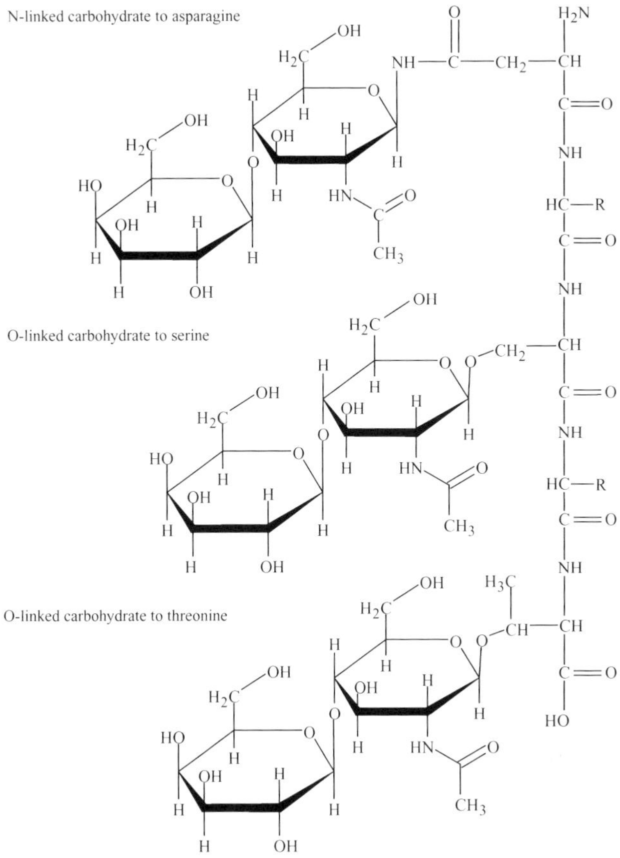

FIGURE 7.20 Covalent linking of carbohydrates to the peptide amino acid backbone. *N*-linked carbohydrate to asparagine. Notice the one amino acid residue between the next amino acid residue serine (could also be threonine). *O*-linked carbohydrate to serine. *O*-linked carbohydrate to threonine.

the sites of glycosylation, and finally (3) to determine the carbohydrates that make up the glycan and the structure of the glycan. One approach that has been employed in glycosylation characterization of proteins has been to cleave the glycans from the proteins, separate them from the proteins, and subsequently measure them by mass spectrometry. However, this approach does not give any information as to what proteins, and associated sites, belonged to what glycans. Increasingly, glycosylated proteins are being digested with an endoprotease producing glycopeptides. The glycopeptides are then analyzed by mass spectrometry. The mass spectrometric analysis of the glycopeptides is not as well defined as is the case with other modifications due to the heterogeneous nature of the oligosaccharide modifications. With glycosylation the mass shift associated with the modification is not constant as compared to

TABLE 7.5   **Common Monosaccharides and Their Associated Masses**

Monosaccharide	Formula	Mass (Da)	$[M + H]^+$ (m/z)	Residue Mass (Da)	$[M + H]^+$ (m/z) Oxonium Ion
Mannose (Man)	$C_6O_6H_{12}$	180.0634	181.0712	162.0528	163.0607
Galactose (Gal)	$C_6O_6H_{12}$	180.0634	181.0712	162.0528	163.0607
Fucose (Fuc)	$C_6O_5H_{12}$	164.0685	165.0763	146.0579	147.0657
Sialic acid (SiA)	$C_{11}O_9NH_{19}$	309.1060	310.1138	291.0954	292.1032
N-acetylglucose-amine (GlcNAc)	$C_8O_6NH_{15}$	221.0899	222.0978	203.0794	204.0872
N-acetylgalactose-amine (GalNAc)	$C_8O_6NH_{15}$	221.0899	222.0978	203.0794	204.0872

acetylation, oxidation, and phosphorylation (e.g., phosphorylation adds 80 Da to the peptide mass as HPO_3). Sometimes mass pattern recognition can be used in single-state precursor ion scans to identify glycosylation. Table 7.5 lists the monoisotopic masses of the monosaccharides commonly found in glycosylated peptides along with their associated residue masses formed through water loss. Glycosylation heterogeneity is observed in precursor mass spectra when a repeating pattern is observed. This type of repeating pattern is indicative of the subsequent addition of a monosaccharide to the glycan chain of the glycosylated peptide. The repeating pattern for a high galactose glycosylation pattern is illustrated in Figure 7.21 where repeating values of 54.0 and 40.4 Da are observed. The glycopeptide in the mass spectrum has a molecular weight of 2709.3 Da. The series associated with 54.0 Da represents the addition of a galactose moiety to the plus 3 (triply) charge state of the glycopeptide as $[M + 3H]^{3+}$. The series associated with 40.4 Da represents the addition of a galactose moiety to the plus 4 (quadruply) charge state of the glycopeptide as $[M + 4H]^{4+}$. Each addition of a monosaccharide to the glycan chain is through a condensation reaction; therefore, the series will have a difference value of the monosaccharide minus water. According to the charge state, the difference between each series will be associated with a multiple

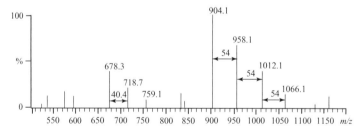

FIGURE 7.21   Repeating mass pattern for a high galactose glycosylation for a glycopeptide with molecular weight of 2709.3 Da. Series associated with a 54.0-Da difference represents the addition of a galactose moiety to the plus 3 (triply) charge state of the glycopeptide as $[M + 3H]^{3+}$. The series associated with a 40.4-Da difference represents the addition of a galactose moiety to the plus 4 (quadruply) charge state of the glycopeptide as $[M + 4H]^{4+}$.

TABLE 7.6   **Pattern Difference in Mass for Glycopeptide Heterogeneity Residue Addition**

Sugar	Formula	+1	+2	+3	+4	+5
Hexose	$C_6O_5H_{10}$	162.1	81.0	54.0	40.4	32.4
dHexose	$C_6O_4H_{10}$	146.1	73	48.7	36.5	29.2
HexNac	$C_8O_5NH_{13}$	203.2	101.6	67.7	50.8	40.6
SiA	$C_{11}O_8NH_{17}$	291.3	145.6	97.1	72.8	58.2

of the charge. Table 7.6 lists the pattern difference for glycopeptide heterogeneity residue addition. There are also diagnostic fragment ions that can be produced during collision-induced dissociation product ion spectral collection of glycated peptides known as oxonium ions. Hexose (Hex, generic name for galactose and mannose) has an oxonium ion at m/z 163, fucose (Fuc) at m/z 147, sialic acid (SiA) at m/z 292, N-acetylhexosamine (HexNAc) at m/z 204, and N-acetylglucoseamine (GlcNAc) at m/z 204. The observance of these oxonium ions in product ion spectra has been used to help in the identification of the glycan modification on a peptide. However, care must be observed in using oxonium ions as diagnostic peaks, such as in precursor ion scanning, as species other than glycans can also generate similar isobaric product ions.

Using a Finnigan LTQ-FT hybrid linear ion trap/FTICR mass spectrometer, Peterman and Mulholland[29] at Thermo Electron Corporation (Somerset, New Jersey) reported a study of high-resolution/high-mass accuracy measurement and identification of glycopeptides in bovine fetuin. Figure 7.22 is a full-scan mass spectrum of an HPLC peak eluting from a reverse-phase C18 column (retention time of 17.6 min) from a tryptic digest sample of bovine fetuin. The eluting peak was identified as T_{54-85} glycopeptide containing two glycoforms, the first as Hex5HexNAc4Neu5Ac3 with peaks at m/z 1176.7 for the +5 charge state and m/z 1470.6 for the +4 charge state, and Hex6HexNAc5Neu5Ac3 with peaks at m/z 1308.0 for the +5 charge state and m/z 1634.7 for the +4 charge state. Mass accuracies of the monoisotopic peak for each ion are labeled in the figure. Also included in the spectrum are the identified structures of the glycan modifications. The open squares are for HexNAc, the filled squares are for Hex, and the open triangles are for Neu5Ac. The inset illustrates a survey scan that was acquired directly before the full-scan mass spectrum. The survey scan is performed to look for diagnostic HexNAc fragment ions used to indicate the presence of a glycopeptide in the eluting peak.

Data-dependent product ion spectra were also collected for the glycopeptides eluting from the HPLC separation of the digested bovine fetuin. This was used for the sequencing of the oligosaccharides associated with the T_{54-85} peptide. Figure 7.23 illustrates the data-dependent MS/MS spectra for the (a) m/z 1470 +4 charge state glycopeptide and the (b) m/z 1635 +4 charge state glycopeptide. Across the top of the spectra the associated mono- and oligosaccharides for the difference between the product ions are listed. From the mass spectra the structures of the glycans in conjunction with the saccharide residue identifications can be deduced successfully.

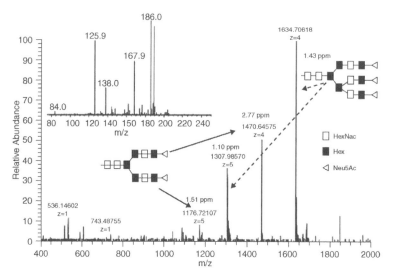

FIGURE 7.22   Full-scan mass spectrum for the retention time of 17.6 min showing two T_{54-85} glycoforms with the oligosaccharide structures and mass accuracies listed. The open squares represent HexNAc, filled squares represent Hex, and open triangles represent Neu5Ac. The inset shows the survey scan acquired directly before the high-resolution/high-mass accuracy spectrum acquisition showing the relative intensity of the characteristic fragment ions from HexNAc. (Reprinted with permission. Peterman, S. M., Mulholland, J. J. A novel approach for identification and characterization of glycoproteins using a hybrid linear ion trap/FT-ICR mass spectrometer. *J. Am. Soc. Mass Spectrom.* 2006, *17*, 168–179. Copyright Elsevier 2006.)

7.5.3   Phosphorylation of Proteins

As stated previously, the genomic DNA sequencing that has been accomplished cannot give direct information concerning posttranslational modifications such as glycosylation (see previous section) and sulfation (section to follow). Another major form of PTM that we are looking at is the phosphorylation of proteins, a significant regulatory mechanism that controls a variety of biological functions in most organisms. Examples of phosphorylation-regulating mechanisms include gene expression, cell cycle processes, apoptosis, cytoskeletal regulation, and signal transduction. It has been estimated that up to 30% of all of the proteins in humans exist in the phosphorylated form where 2% of the human genome encode for protein kinases (>2000 genes).[30] In eukaryotic cells the protein phosphorylation takes place with the serine, threonine, and tyrosine residues. The reversible protein phosphorylation of the serine, threonine, and tyrosine residues are an integral part of cellular processes involving signal transduction.[31, 32] The identification of the phosphorylation sites is important in understanding cellular signal transduction.

Protein phosphatase are enzymes responsible for the removal of phosphate groups from a target (i.e., reversible protein phosphorylation), and protein kinase are enzymes responsible for the addition of phosphate groups to a target. These two enzymes work

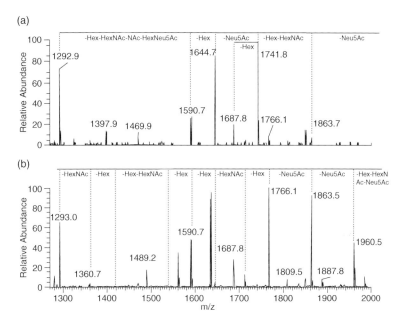

FIGURE 7.23   Full-scan data-dependent MS/MS spectra for the (*a*) *m/z* 1470 and (*b*) *m/z* 1635 glycopeptide precursor ions acquired in the linear ion trap. (Reprinted with permission. Peterman, S. M., and Mulholland, J. J. A novel approach for identification and characterization of glycoproteins using a hybrid linear ion trap/FT-ICR mass spectrometer. *J. Am. Soc. Mass Spectrom.* 2006, *17*, 168–179. Copyright Elsevier 2006.)

together to control cellular processes and signaling pathways. Greater attention has been given in the literature to the study of signaling pathways primarily involved with protein kinase as compared to specific types of phosphatase.[33–38] However, the importance of studying protein phosphatase enzymes and their targets has been demonstrated in recently reported disease-state studies where the abnormal condition has been attributed at least in part to malfunctioning protein phosphatase enzymes.[39–41] In covalent modification of proteins such as phosphorylation, the activity of the modified enzyme has been altered in the form of activated, inactivated, or to otherwise regulate its activity upwardly or downwardly. The most common mechanism for phosphorylation is the transfer of a phosphate group from adenosine triphosphate (ATP) to the hydroxyl group of either serine, threonine, or tyrosine within the protein. Figure 7.24 illustrates the general cellular mechanism involving protein kinase phosphorylation and protein phosphatase dephosphorylation. In the top portion of Figure 7.24 a cellular signal is received in a kinase/phosphatase cycle, often in the form of a messenger biomolecule such as lipid (diacylglycerol shown), which initiates the phosphorylation of the target protein by the protein kinase-ATP action. The enzyme has been phosphorylated, resulting in its participation in the signaling cycle. Usually the phosphorylated enzyme does not permanently stay in its modified form

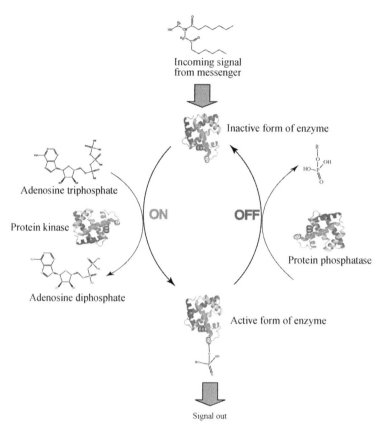

FIGURE 7.24   Example of the general cellular mechanism involving protein kinase phosphorylation and protein phosphatase dephosphorylation. (*Top*) A cellular signal is received in the form of the messenger biomolecule diacylglycerol lipid that initiates the phosphorylation of the target protein by the protein kinase-ATP action. The phosphorylated enzyme propagates the signaling cycle and then is recycled through dephosphorylation by a phosphatase.

but will undergo dephosphorylation through interaction with protein phosphatase. Figure 7.25 illustrates the nonphosphorylated structure of the serine, threonine, and tyrosine residues, each having a hydroxyl moiety, and the phosphorylated form of the amino acid residue.

While the serine, threonine, and tyrosine residues all have a side-group hydroxyl moiety available for posttranslational modification, over 99% of the phosphorylated modification takes place with the serine and threonine amino acid residues in eukaryotic cells.[42] Recently, Olsen ct al.[43] have reported a slight variation to the widely referred to study of Hunter and Sefton[44] where the relative abundances of amino acid residue phosphorylation was assigned as 0.05% for phosphotyrosine (pY), 10% for phosphothreonine (pT), and 90% for phosphoserine (pS). In Olsen et al.'s recent study this has been adjusted to 1.8% pY, 11.8% pT, and 86.4% pS where the larger

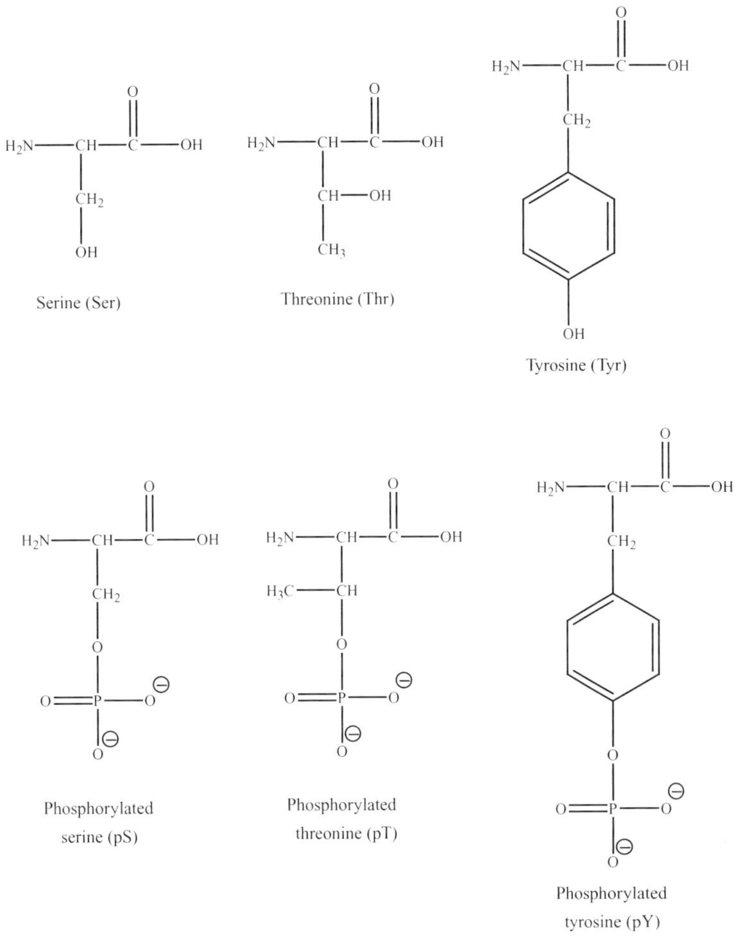

FIGURE 7.25 Structures of the nonphosphorylated (*top*) and phosphorylated (*bottom*) states of the amino acids serine, threonine, and tyrosine.

percentage value allocated to pY was attributed to more sensitive methodology being employed, thus allowing the characterization of lower abundant phosphorylated proteins. However, the stoichiometry of the phosphorylated proteome (in the form of tryptic peptides) is small (≤ 1) in relation to the nonphosphorylated proteome (in the form of tryptic peptides), thus requiring that the sensitivity of the phosphorylated proteome analysis be as optimal as possible.

Mass spectrometry is often used to identify the sites of phosphorylation in the protein backbone when studying cellular signaling pathways.[45–48] Prior to the introduction of mass spectrometric methodology for phosphorylated protein analysis, researchers studied protein phosphorylation using ^{32}P labeling followed by 2D polyacrylamide gel electrophoresis and finally Edman sequencing. This methodology was time consuming and involved the handling of radioactive isotopes. This has prompted a considerable amount of methodology development using mass spectrometric techniques that are able to measure whole proteomes from complex biological systems, and not just single proteins for which the Edman sequencing approach is most suited. There are numerous studies reported in the literature of signaling pathways primarily involved with protein kinease.[49–54] The study of protein phosphatase enzymes and their targets has also gained importance where there are recently reported numerous disease states that have been attributed at least in part to malfunctioning protein phosphatase enzymes.[55–58]

The mass spectrometric instrumentation in use today possesses the sensitivity needed for PTM analysis. Therefore, often the limiting factor in phosphorylated proteome analyses lies in the sample treatment prior to mass spectral measurement. Immobilized metal affinity chromatography (IMAC) is the methodology of choice for phosphorylated proteome cleanup and enrichment.[59–61] Labeling is also the most widely used approach for relative quantitation of the phosphorylated proteome when comparing a normal state to a perturbed state. Labeling typically involves using stable isotopes with either ^{2}D-methanol or ^{13}C-SILAC or iTRAQ reagent.[62–64] However, often labeling procedures can cause an increase in sample complexity, can be cumbersome or incomplete, and can ultimately result in sample losses.

The ideal methodology for posttranslational modification studies would entail both high sample recovery and sample specificity while avoiding any additional modifications to the proteins being studied. The first critical step in sample preparation is the lysis and solubalization of the sample's complement of proteins. This is followed by additional steps for sample cleanup and treatment that often is accompanied by protein/peptide loss at each step. Another limiting factor is the presence of nucleic acids during the IMAC enrichment step, which are known to poison or compete with the phosphorylated peptides for binding sites on the IMAC bed. Alternative approaches exist for the removal of RNA and DNA such as the addition of RNase and DNase to the whole cell lysate buffer,[65] passing the lysate repeatedly through a tuberculin syringe fitted with a 21-gauge needle to shear the RNA and DNA mechanically,[66] or using the QIAShredder (QIAgen).[67] The use of RNase and DNase is often avoided in an effort to reduce the amount of additives and the production of low-molecular-weight nucleic-acid-lysed products, which are difficult to remove and also poison the IMAC bedding. The choice of ultracentrifugation can be made over mechanical shearing for similar reasons that are associated with the production of low-molecular-weight nucleic acid species.

The use of one-dimensional sodium dodecyl sulfate polyacrylamide gel electrophoresis (1D SDS-PAGE) is another alternative approach for whole-cell lysate intact protein cleanup that includes removal of the nucleic acids. Current examples include a study by Beausoleil et al.[68] who used preparatory gels for HeLa cell nuclear protein separation followed by whole gel digestion (cut into 10 regions) where 967 proteins revealed 2002 phosphorylation sites. A recent study by de Souza et al.[69] reported that five times more proteins from tear fluid were identified after gel electrophoresis as compared to in-solution digest. Other potential advantages for the use of 1D SDS-PAGE is the ability to target specific molecular weight ranges that can be identified and excised from various band regions within the gels, and a more efficient tryptic digest due to the enhanced accessibility of the protein backbone denatured into a linear orientation locked within the gel. Recoveries from gels can be an added concern, although all steps in protein/peptide preparation can contribute to an overall loss of protein.

Mass spectrometric studies of phosphorylation posttranslational modifications typically employ tandem mass spectrometry to generate product ions spectra of phosphorylated peptides. This usually entails a premass spectrometric separation of a complex mixture of phosphorylated peptides by reverse-phase carbon-18 stationary phase high-performance liquid chromatography with an electrospray ionization source (RP C18 HPLC ESI-MS/MS). Fragmentation pathway studies using ion trap mass spectrometry of the phosphorylation of the three possible sites in peptides, serine, threonine, and tyrosine have been studied and characterized revealing different mechanisms. Peptides that contain phosphoserine generally will lose the phosphate group as the predominant product ion in the spectrum. This often results in limited information about the sequence of the peptide where other peptide backbone fragmentation is not well observed. Neutral loss of the phosphate group for the plus 1 charge state (+1 CS) is observed at 98.0 Da, $[M + H - H_3PO_4]^+$, for the plus 2 charge state (+2 CS) is observed at 49.0 Da, $[M + 2H - H_3PO_4]^{2+}$, for the plus 3 charge state (+3 CS) is observed at 32.7 Da, $[M + 3H - H_3PO_4]^{3+}$. Figure 8.26 illustrates product ion spectra for a phosphoserine containing peptide at m/z 987.5 (monoisotopic mass of 1972.98 Da). The spectrum in Figure 7.26a is the MS/MS spectrum of the doubly charged phosphorylated peptide, $[M + 2H]^{2+}$, at m/z 987.5 where the predominant product ion is observed at m/z 938.7 for the neutral loss of the phosphate group for the doubly charged precursor (−49.0 Da), $[M + 2H - H_3PO_4]^{2+}$. There is coverage in the spectrum of the peptide backbone sequence as illustrated by the b- and y-type ions; however, their response is quite low and often not observed in product ion spectra of phosphorylated peptides when using tandem ion trap mass spectrometry. The loss of the phosphate modification from the peptide is the preferred fragmentation pathway and is usually the predominant one observed. The phosphorylated peptide's sequence is shown in the spectrum as RApSVVGTTY-WMAPEVVK, where the phosphorylation is on the serine amino acid residue, which is third from the left. In the collision-induced fragmentation of the phosphorylated peptide, all of the y-type ions observed in Figure 7.26a ($y_5 - y_{14}$) do not contain the phospho group. Only one of the b-type ions contains the phospho group at b_{12}. This is due to a preferential cleavage taking place on the peptide backbone at the proline residue, while all of the other b-type ions include neutral loss of the phospho group from the serine residue and are denoted as b^{Δ}_n, which equals $b_n - H_3PO_4$.

The loss of the phosphorylation that is associated with serine is through a β-elimination mechanism producing dehydroalanine. The mechanism for dehydroalanine production through β elimination is illustrated in Figure 7.27 for phosphoserine. This would represent the structure of the serine in the peptide that is represented by the m/z 938.7 product ion in Figure 7.26a, $[M + 2H - H_3PO_4]^{2+}$.

Figure 7.26b is an example of the application of MS^3 product ion spectral collection for the enhanced fragmentation of phosphorylated peptides. In this approach, the

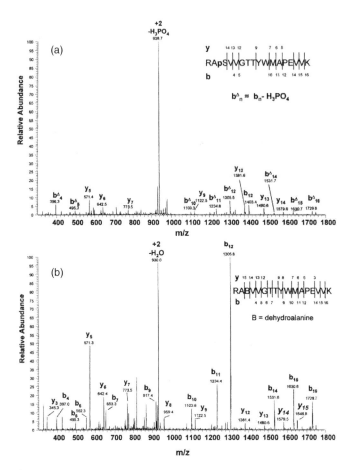

FIGURE 7.26 Phosphoserine loses predominantly H_3PO_4 through β elimination to produce dehydroalanine in tandem ion trap mass spectrometry. (*a*) MS^2 spectrum of a doubly charged phosphopeptide ion (*m/z* 987.5). A loss of 98 Da (H_3PO_4) is observed. The *bn*-D label denotes loss of 98 Da (i.e., *bn*-H_3PO_4). (*b*) MS^3 spectrum of the ion arising from loss of 98 Da [*m/z* 938.7 of the doubly charged ion in (*a*)]. The *y*14 and *y*15 fragments have a mass difference of 69 Da, which corresponds to the mass of dehydroalanine, identifying the product of phosphoserine after losing 98 Da as dehydroalanine. The "B" label denotes the dehydroalanine residue. (Reprinted with permission. DeGnore, J. P., and Qin, J. Fragmentation of phosphopeptides in an ion trap mass spectrometer. *J. Am. Soc. Mass Spectrom.* 1998, 9, 1175–1188. Copyright Elsevier 1998.)

main product ion collected from MS^2, *m/z* 938.7, is isolated and subjected to a third stage of fragmentation. In the MS^3 product ion spectrum of Figure 7.26*b* there is a predominant peak for water loss from the *m/z* 938.7 precursor peak. This also does not afford much information for the sequence of the peptide; however, there is also observed in the spectrum numerous product ion peaks of the *b* and *y* type that have

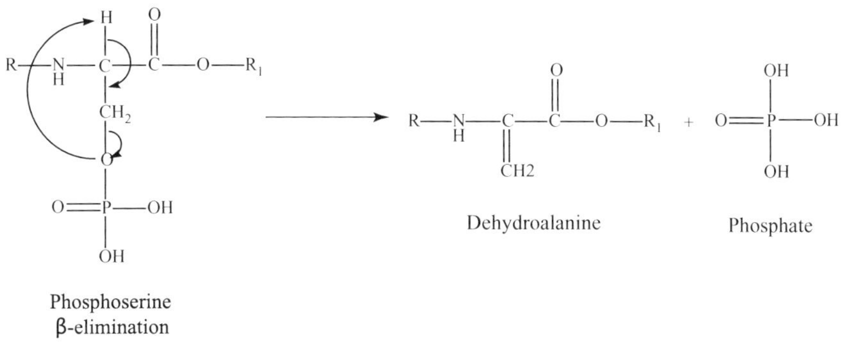

Phosphoserine
β-elimination

Dehydroalanine   Phosphate

FIGURE 7.27 *β*-elimination mechanism for the fragmentation pathway producing dehydroalanine for the phosphorylated serine amino acid residue.

a much greater response than that observed in the MS^2 product ion spectrum of Figure 7.26*a*. Notice in the fragmentation of the phosphorylated peptide in Figure 7.26*a*, product ions derived through cleavage from both sides directly adjacent to the serine residue did not take place. This means that a positive identification of the residue that contains the phosphate modification, namely the serine residue, is not possible. Notice in the third-stage product ion spectrum of Figure 7.26*b* that fragmentation on both sides of the serine residue did in fact take place. Because the precursor ion that was subjected to collision-induced dissociation in the third-stage product ion spectrum of Figure 7.26*b* was the *m/z* 938.7 product ion from the second-stage fragmentation in Figure 7.26*a*, the serine residue has been replaced in the sequence by a "B" ion, which stands for dehydroalanine. Due to the absence of the phosphate modification on the peptide chain, there is observed substantial sequence coverage in the product ions observed in Figure 7.26*b*, thus allowing specific identification of the peptide sequence and the location of the phosphorylation. Further studies concerning phosphorylated serine (pS) have demonstrated that loss of the phosphate H_3PO_4 group in product ion spectra are not dependent upon the charge state of the precursor ion. All three commonly observed charge states, +1, +2, and +3 for electrospray ionization collision-induced dissociation product ion spectral collection resulted in the loss of the phosphate group as the predominant product ion peak with other minor losses as shown in Figure 7.26*a*.

When the phosphorylation modification of a peptide takes place on the threonine (Thr) amino acid residue, the predominant product ion spectral peak derived from CID is also observed to be through *β* elimination of the phosphate group. The neutral loss of the phosphate group, –98 Da as $[M + H - H_3PO_4]^+$, from the threonine residue produces dehydroaminobutyric acid. A product ion spectrum illustrating the major species produced from a peptide containing a phosphorylated threonine is illustrated in Figure 7.28. The sequence of the phosphorylated peptide illustrated in Figure 7.28 is RASVVGTpTYWMAPEVVK, which has a monoisotopic mass of 1972.98 Da. The phosphorylation of the peptide has taken place on the threonine

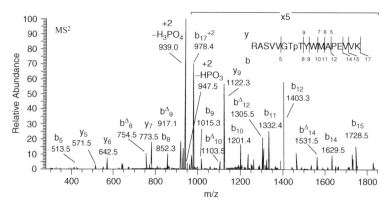

FIGURE 7.28 Product ion spectrum of phosphorylated threonine illustrating the major species produced. The sequence of the phosphorylated peptide is RASVVGTpTY-WMAPEVVK (monoisotopic mass of 1972.98 Da). Charge state of the peptide is $+2$ at m/z 987 as $[M + 2H]^{2+}$. Loss of the H_3PO_4 phosphate group is the primary fragmentation pathway at m/z 939.0 as $[M + 2H - H_3PO_4]^{2+}$. (Reprinted with permission. DeGnore, J. P., and Qin, J. Fragmentation of phosphopeptides in an ion trap mass spectrometer. *J. Am. Soc. Mass Spectrom.* 1998, *9*, 1175–1188. Copyright Elsevier 1998.)

amino acid residue (pT). The charge state of the peptide in the product ion spectrum of Figure 7.28 is $+2$ at m/z 987 as $[M + 2H]^{2+}$. Loss of the H_3PO_4 phosphate group is the primary fragmentation pathway that is observed in the product ion spectrum at m/z 939.0 as $[M + 2H - H_3PO_4]^{2+}$. In the product ions produced from CID of the m/z 987 species, there are a number of *b*-type ions that both contain the phosphate modification (b_8, b_9, b_{10}, b_{11}, b_{12}, b_{14}, b_{15}) and that do not (b^Δ_8, b^Δ_9, b^Δ_{10}, b^Δ_{12}, b^Δ_{14}). Figure 7.29 illustrates the fragmentation pathway mechanism for the production of the dehydroaminobutyric acid from the neutral loss, β elimination of the phosphate group from phosphorylated threonine amino acid. One difference in the product ion spectrum for fragmentation of phosphorylated threonine as compared to phosphorylated serine, Figure 7.26*a*, is the product ion peak observed at m/z 947.5 for the neutral loss of 80 Da. The product ion at m/z 947.5 represents dephosphorylation through neutral loss of HPO_3 from the precursor peak as $[M + 2H - HPO_3]^{2+}$. This particular loss is not observed in the product ion spectrum of phosphorylated serine. The mechanism for dephosphorylation of the threonine amino acid is illustrated in Figure 7.30. The dephosphorylation results in the structure of the original amino acid residue of threonine.

Product ion spectra of phosphorylated tyrosine (pY) containing peptides also illustrate losses associated with 80 and 98 Da, which would appear to be similar to the losses observed with phosphorylated threonine peptides. The 80-Da loss is due to dephosphorylation of the tyrosine residue, resulting in the original structure of the tyrosine residue. The mechanism is illustrated in Figure 7.31. However, due to the structure of tyrosine, a similar mechanism of β elimination for loss of the phosphate

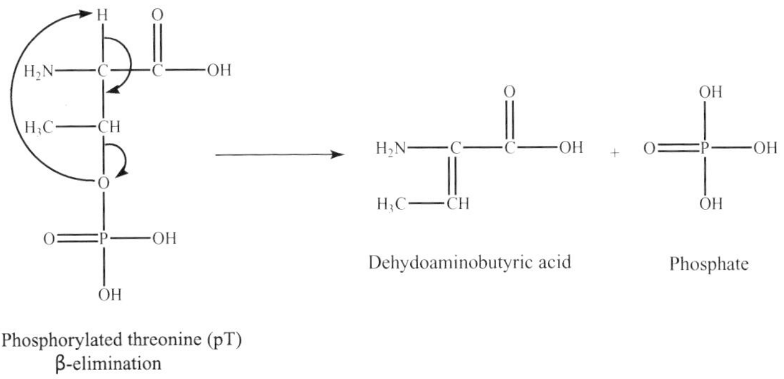

Phosphorylated threonine (pT)
β-elimination

Dehydroaminobutyric acid   Phosphate

FIGURE 7.29 Mechanism for the production of dehydroaminobutyric acid from phosphorylated threonine through neutral loss, β elimination of the phosphate group.

(−98 Da, H_3PO_4) group is most probably not likely. The neutral loss of phosphate at 98 Da has also been proposed to not happen through a two-step mechanism involving both water (H_2O) loss of 18 Da and HPO_3 loss of 80 Da, in any order. The neutral loss of 98 Da is more than likely associated with some form of rearrangement in the fragmentation pathway mechanism.

Other mass spectrometric methods are used in phosphorylated peptide analysis that have been investigated to help increase the efficiency of the fragmentation that takes place. This can allow more direct approaches for modification location without the need of third-stage fragmentation experiments that require the interpretation of two individual spectra. One example of an alternative mass spectrometric approach has been the use of the FTICR mass spectrometer using infrared multiphoton dissociation (IRMPD) and electron capture dissociation (ECD). Figure 7.32 illustrates a comparison of these two dissociation approaches. In the top figure IRMPD was used to excite and dissociate the phosphorylated peptide. The phosphorylation is located

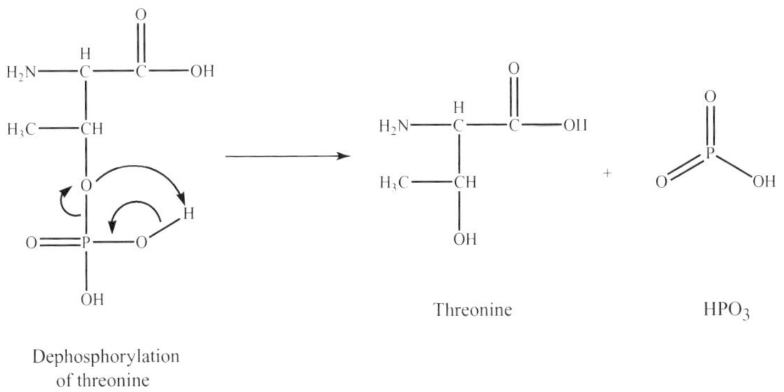

Dephosphorylation
of threonine

Threonine HPO_3

FIGURE 7.30 Mechanism for dephosphorylation of the threonine amino acid resulting in the structure of the original amino acid residue of threonine.

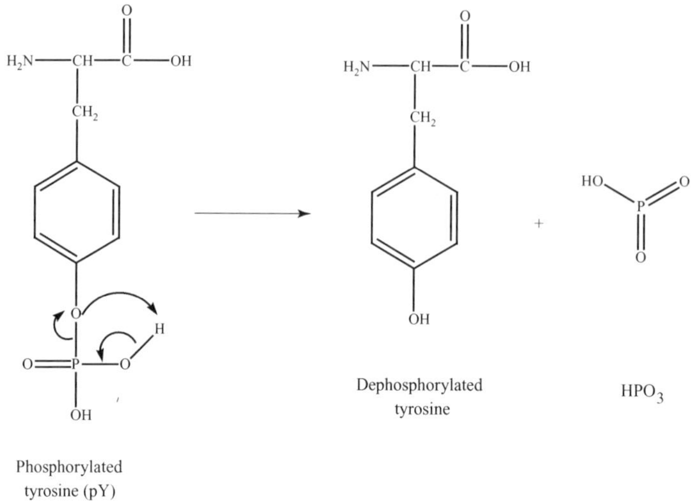

FIGURE 7.31   Mechanism for the dephosphorylation of tyrosine.

on a serine (pS) residue within the middle of the peptide chain having the sequence AKRRRL(pS)SLRASTS. In the product ion spectrum collected using IRMPD, the major product ions that were observed were all associated with the neutral loss of the phosphate group as $[M + 3H - H_3PO_4]^{3+}$, and also phosphate and water loss as $[M + 3H - H_3PO_4 - H_2O]^{3+}$. As also can be observed in the top product ion spectrum of Figure 7.32, very little information is given concerning the peptide chain's sequence. The bottom product ion spectrum of Figure 7.32 illustrates the effectiveness of using ECD for phosphorylated peptide sequence determinations. In this spectrum there are an appreciable amount of peptide backbone fragmentation ions in the form of c-type and z-type ions. The major peak in the spectrum is the triply protonated precursor ion as $[M + 3H]^{3+}$. Also notice that the phosphorylation was maintained within the structure of the product ions. The two product ion spectra are complementary where the top spectrum is diagnostic for the determination of a phosphorylation of the peptide, while the bottom spectrum gives very good sequence coverage of the peptide.

More recently, the electron transfer dissociation (ETD) capability has been incorporated into linear ion traps that also allow significant peptide backbone cleavages similar to the ECD capability usually associated with FTICR mass spectrometers. Figure 7.33 illustrates the extensive fragmentation that can be achieved with ETD for a quite long sequence of peptide. The top spectrum is the dissociation spectrum for a 35-residue phosphopeptide that has a molecular weight of 4093 Da. The phosphopeptide is in the plus 6 charge state (+6) at m/z 683.3 for $[M + 6H]^{6+}$. As can be seen in the figure for this very long sequence phosphorylated peptide, essentially complete coverage of the peptide chain has been achieved along with the determination of the phosphorylation site. The ETD product ion spectrum illustrated in Figure 7.33b is for a phosphorylated histidine (pH) peptide. Phosphorylation on the histidine residue

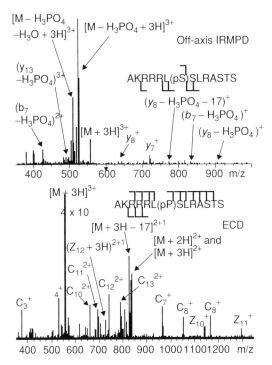

FIGURE 7.32 (*Top*) Product ion spectrum obtained from off-axis IRMPD FT-ICR MS/MS of a population of quadrupole- and SWIFT-isolated [AKRRRL(pS)SLRASTS + 3H]$^{3+}$ phosphopeptide ions. The spectrum is dominated by ions resulting from neutral losses of H_3PO_4, NH_3, and H_2O. Five (out of 13) peptide backbone bonds are broken, and the location of the phosphorylation site is identified only by observation of the ($y8–H_3PO_4–NH_3$) ion (present at very low abundance) and the singly and doubly charged ($b_7–H_3PO_4$) ions. Irradiation was for 500 ms at ~36 W laser power, and the data represent a sum of 10 scans. (*Bottom*) Product ion spectrum obtained from ECD (20 ms irradiation) FTICR MS/MS of the same quadrupole- and SWIFT-isolated phosphopeptide as in Figure 7.32 (*top*). Twelve out of 13 peptide backbone bonds are cleaved, and the location of the phosphate is readily assigned by observation of the abundant $c7$ ions. (Reprinted with permission. Chalmers, M. J., Kolch, W., Emmett, M. R., Marshall, A. G., and Mischak, H. Identification and analysis of phosphopeptides. *J. Chromatogr. B* 2004, *803*, 111–120. Copright Elsevier 2004.)

is an important type of PTM that is observed in prokaryotic proteomes and will be discussed next.

7.5.3.1 *Phosphohistidine as Posttranslational Modification*   The phosphoryl modification of serine, threonine, and tyrosine is a posttranslational modification primarily associated with regulating enzyme activity in eukaryotic systems. The phosphoester linkage resulting with the PTM of serine, threonine, and tyrosine is relatively stable at physiological pH (~7.36) and thus generally requires a phosphatase

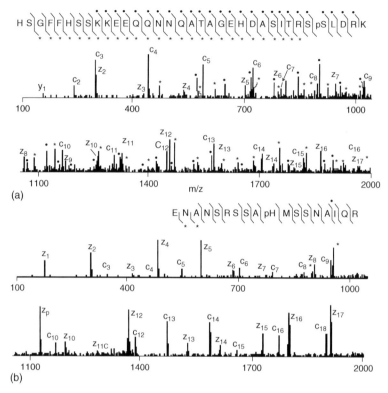

FIGURE 7.33 Phosphopeptide mass spectra. ETD mass spectra recorded on $(M + 6H)^{+6}$ ions at m/z 683.3 for a 35-residue phosphopeptide of MW 4093 (*a*), and $(M + 3H)^{+3}$ ions from a pHis containing peptide at the C-terminus of the septin protein, Cdc10 (*b*). Observed c and z ions are indicated on the peptide sequence by ⌋ and ⌈, respectively. Observed doubly charged c and z ions are indicated by an additional label, circle, and asterisk, respectively. (Reprinted with permission from Chi, A., Huttenhower, C., Geer, L. Y., Coon, J. J., Syka, J. E. P., Bai, D. L., Shabanowitz, J., Burke, D. J., Troyanskaya, O. G., Hunt, D. F. *PNAS* 2007, *104*, 2193–2198. Copyright 2007 National Academy of Sciences.)

for the removal of the modification. The phosphorylation and subsequent dephosphorylation of proteins are activities associated in biological systems with respect to cellular signaling pathways. In prokaryotic biological systems the histidine group is phosphorylated primarily for the purpose of transferring the phosphate group from one biomolecule to another. These biomolecules are known as phosphodonor and phosphoacceptor molecules where the phosphohistidine is acting as a high-energy intermediate in some type of biological process on the molecular level.[70] The phosphohistidine transfer potential ($\Delta G°$ of transfer) of the phosphate is estimated at −12 to −14 kcal/mol, reflecting a relatively high-energy system.[71] The phosphorylation of the histidine residue produces a phosphoramidate that contains a large standard free energy of hydrolysis, making them the most unstable form of phosphoamino

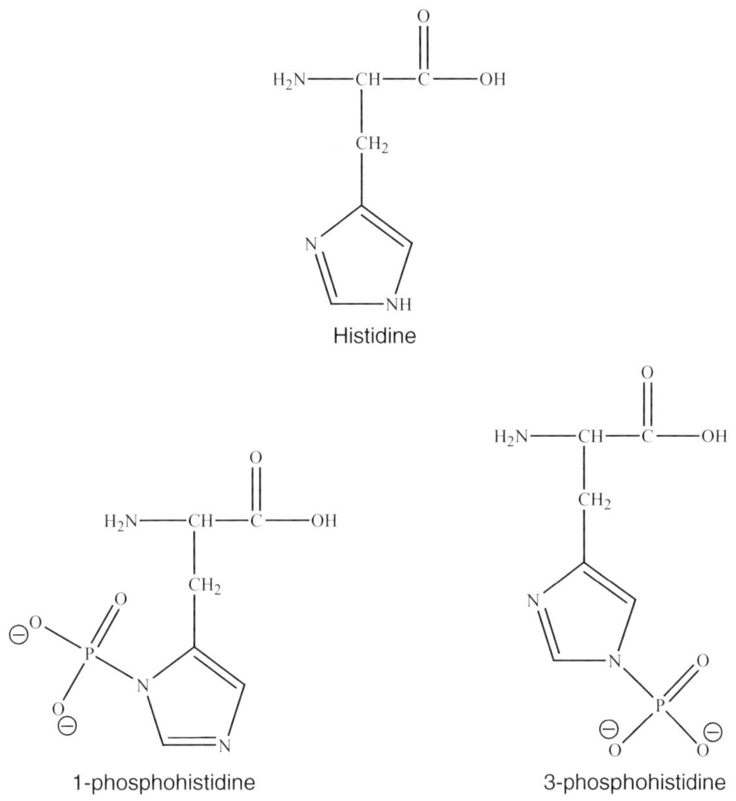

FIGURE 7.34 Structures of unmodified histidine, 1-phosphohistidine, and 3-phosphohistidine.

acids. The stability of the phosphohistidine is largely influenced by its local amino acid residues and the nature and makeup of the associated protein. It is speculated due to this instability that often a histidine phosphatase is not required. It has been estimated that up to 6% of the phosphorylation that takes place in eukaryotes[72] and prokaryotes[73] is with the histidine amino acid residue. It is also estimated that the abundance of phosphohistidine is 10- to 100-fold greater than that of phosphotyrosine but much less that phosphoserine.[72] The phosphorylation of the histidine amino acid residue can take place on either the nitrogen in the 1-position or the nitrogen in the 3-position. This is illustrated in Figure 7.34 for the phosphorylation of the histidine amino acid residue. Studies have demonstrated that the 3-phosphohistidine is more stable than the 1-phosphohistidine[74] and is therefore more likely to be the positional isomer observed in biological systems. The histidine amino acid residue is phosphorylated by kinase using the reactive intermediate ATP. This process is illustrated in Figure 7.35 where the modification produces the 3-phosphohistidine residue.

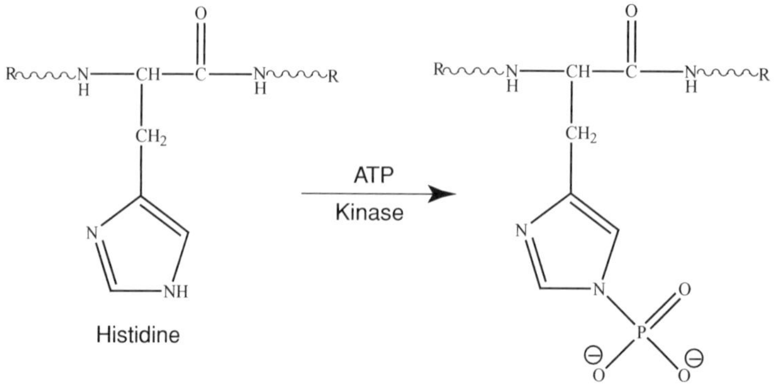

FIGURE 7.35   Kinase phosphorylation of the histidine amino acid residue producing 3-phosphohistidine.

The bacterial histidine kinases of the two-component system is an example of phosphohistidines found in prokaryotic systems. Figure 7.36 is a drawing of a two-component signaling system that involves histidine phosphorylation and dephosphorylation. The figure illustrates a membrane-bound protein that contains a carboxy-terminal histidine kinase domain. The subsequent steps involved include the binding of a ligand, which is followed by dimerization of two kinase. Phosphorylation of the histidine residue is done through the reactive intermediate ATP. The next step is the binding of a cytosolic response regulator protein that contains an aspartate residue in the amino-terminal domain. The aspartate residue in the response regulator is then phosphorylated by the first membrane-bound protein and released to perform its downstream function. There are various bacterial two-component histidine kinases signaling systems that have been observed and characterized. Table 7.7 is a listing of some of the bacteria that have been observed to have the two-component system.

The major difficulty in measuring the phosphorylation of a histidine residue in a peptide is due to the instability of the modification. The phosphorylated histidine is very unstable in an acidic environment, which is typically the matrix that the phosphorylated peptide is contained within during normal proteomic preparation steps prior to mass spectral analyses. Table 7.8 lists the acidic or alkaline stabilities of the phosphorylated amino acids. Due to the acid lability of the phosphorylated histidine residue, it is not observed during most phosphoprotome studies using mass spectrometric techniques. Current studies of histidine phosphorylated peptides are incorporating neutral level pH methodologies to preserve the modification in conjunction with enrichment approaches such as immobilized copper(II) ion affinity chromatography.[75] Figure 7.37a illustrates a nonphosphorylated peptide's product ion mass spectrum collected using electrospray ionization and a quadrupole ion trap. Very good coverage of the peptide backbone is observed with the y-series ions, allowing the sequencing of the peptide. The precursor was a doubly protonated species at m/z 642.4 as $[M + 2H]^{2+}$.

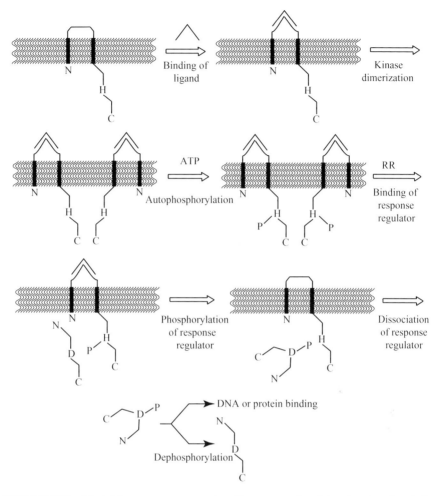

FIGURE 7.36 Model for two-component signaling systems involving the phosphorylation of the histidine amino acid residue. (Reprinted with permission. Pirrung, M. C. Histidine kinases and two-component signal transduction systems. *Chem. Biol.* 1999, *6*, R167–R175. Copyright Elsevier 1999.)

Figure 7.37*b* is the product ion spectrum of the same peptide after phosphorylation of a histidine residue contained in the peptide chain. The mass spectrum was collected on the doubly protonated species at m/z 682.0 as $[M + 2H]^{2+}$. The major product ion observed in the spectrum is for the neutral loss of the phosphate moiety and water at m/z 633.1 as $[M + 2H - HPO_3 - H_2O]^{2+}$. There are a couple of important considerations when considering product ion spectra of phosphorylated amino acid residues. When measuring these species in positive ion mode, the phosphorylation of the histidine residue, as illustrated in Figure 7.34 and 7.35, actually results in the

TABLE 7.7   Some Bacterial Two-Component Signaling System Histidine Kinases

Histidine Kinase	Control Function	Source	Receiver
CheA	Chemotaxia, flagellar motor	*E. coli*	CheY
EnvZ	Osmosensing, outer membrane proteins	*E. coli*	OmpR
VanS	Cell-wall biosynthesis	*Enterococcus faecium*	VanR
KinA	Osmosensing	*E. coli*	SpoOF
PhoR	Phosphate metabolism	*E. coli*	PhoB
FrzE	Fruiting body formation	*Myxococcus xanthus*	FrzE (internal receiver)
RscC	Cell capsule synthesis	*E. coli*	RscC (internal receiver)
VirA	Host recognition, transformation	*Agrobacter tumefaciens*	VirA (internal receiver)
ArcB	Anaerobiosis	*E. coli*	ArcB (internal receiver) and ArcA
BvgS	Virulence	*Bordetella pertusis*	BvgS (internal receiver)

Source: Pirrung, M. C. Histidine kinases and two-component signal transduction systems. *Chem. Biol.* 1999, *6*, R167–R175. Reprinted with permission. Copyright Elsevier 1999.

addition of 79.9663 Da (HPO_3) to the mass of the peptide. The ATP is contributing, in a sense, HPO_3 to the peptide, as illustrated in Figure 7.38a. In physiological conditions the addition is more likely to be $PO_3{}^-$, but measured by mass spectrometry in positive ion mode it is HPO_3. During collision-induced dissociation product ion spectral accumulation, the loss due to the phosphoryl modification is also at 79.9663 Da for HPO_3. This would suggest that in positive ion mode the gas-phase fragmentation mechanism would be something like that illustrated in Figure 7.38b. The phosphoryl group on the histidine amino acid residue is presented as being fully protonated. The channel pathway during the fragmentation process results in the neutral loss of HPO_3 and the reprotonated, original form of the histidine residue without a change in the charge state of the peptide.

TABLE 7.8   Acid and Alkaline Stabilities of the Phosphorylated Amino Acids

Phosphoamino Acid	Acid Stable	Alkali Stable
N-Phosphates		
Phosphoarginine	No	No
Phosphohistidine	No	Yes
Phospholysine	No	Yes
O-Phosphates		
Phosphoserine	Yes	No
Phsophothreonine	Yes	Partial
Phosphotyrosine	Yes	Yes
Acyl-phosphate		
Phosphoaspartate	No	No

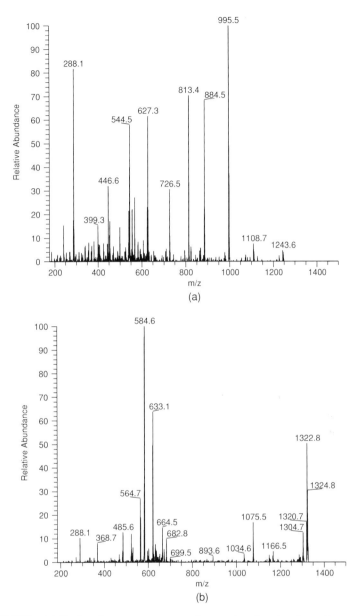

FIGURE 7.37 (*a*) Product ion mass spectrum of a nonphosphorylated peptide at *m/z* 642.4 as $[M + 2H]^{2+}$. Good coverage of the peptide backbone is observed with the *y*-series ions, allowing the sequencing of the peptide. (*b*) Product ion spectrum of the same peptide after phosphorylation of a histidine residue, at *m/z* 682.0 as $[M + 2H]^{2+}$. Major product ion observed is for the neutral loss of the phosphate moiety and water at *m/z* 633.1 as $[M + 2H - HPO_3 - H_2O]^{2+}$.

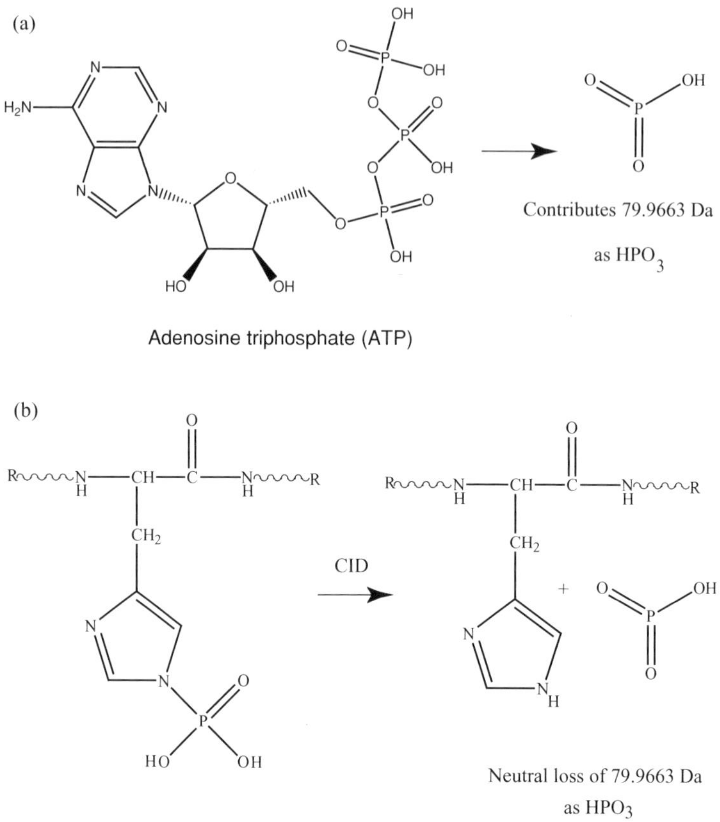

FIGURE 7.38 (*a*) ATP contributing HPO₃ to the peptide when considered mass spectrometrically in positive ion mode. (*b*) Collision-induced dissociation product ion spectra illustrate the loss due to the phosphoryl modification at 79.9663 Da for HPO₃.

Similar to the product ion spectra of the phosphorylated serine, threonine, and tyrosine (also sulfated) amino acid residues, the phosphorylated histidine product ion spectra are primarily dominated by the neutral loss of the modification. Future work utilizing mass spectrometric methodologies such as ECD and ETD will also be beneficial in the gas-phase analysis of phosphorylated histidine peptides.

7.5.4 Sulfation of Proteins

7.5.4.1 Glycosaminoglycan Sulfation There are two primary types of sulfation that occur involving protein posttranslational modification: carbohydrate sulfation of cell surface glycans and sulfonation of protein tryosine amino acid residues. We will begin with a look at carbohydrate sulfation, which is an important process that occurs in conjunction with extracellular communication. The glycosaminoglycan modification presented previously in Section 7.5.2 occurs with the portion of the membrane-bound protein that is protruding out of the surface of the cell. The predominant enzyme carbohydrate sulfotransferases is the protein responsible for adding the sulfation modification to the cell surface proteins and is found in the extra-cellular matrix. The process of glycosaminoglycan sulfation is a precise process and has been found to be associated for the activation of cytokines and growth factors and for inflammation site endothelium adhesion.[76–79] The biological modification of protein-bound carbohydrate sulfation is a process that has been associated with normal cellular processes, arthritis, cystic fibrosis, and pathogenic diseased states.[79–83] Sulfation is used for the removal of proteins and other biomolecules such as metabolic end products, steroids, and neurotransmitters[84] from the extracellular matrix or body due to the increased solubility of the sulfated species and the reduced bioactivity. Specific glycoproteins that have been observed to undergo sulfation modification include the gonadotropin hormones follitropin, lutropin, and thyrotropin,[85] mucins,[86] and erythropoietin.[87]

Previously, we observed that phosphorylation posttranslational modification of peptides predominantly takes place on the free hydroxyl moiety of the serine, threonine, and tyrosine amino acid residues by kinase enzymes. The amino acid residue phosphorylation mechanism is through the use of the activated donor molecule adenosine triphosphate (ATP; see Section 7.5.3 and Fig. 7.24). The eukaryotic cell process of sulfation is through a class of enzymes known as sulfotransferase. The sulfotransferase class of enzymes comprises two general types that either sulfate small molecules such as steroids, and metabolic products, the cytosolic sulfotransferase, or the membrane-bound, Golgi-localized sulfotransferase that will sulfate glycoproteins or the tyrosine residue of proteins (to be covered in the next section). The transfer of the sulfonate (SO_3^-) group to the glycoprotein (or amino acid hydroxyl group) is through the activated donor phosphoadenosine phosphosulfate (PAPS) that is produced through the combination of SO_4^{2-} and ATP synthesized by the protein PAPS synthase.[88] An example of a sulfated glycoprotein is illustrated in Figure 7.39. The glycan has been O-linked to a general Ser/Thr amino acid residue within the protein's peptide chain. The mechanism for the sulfation of a saccharide is illustrated in Figure 7.40.

When glycated peptides are fragmented by tandem mass spectrometry, the product ions produced are classified according to a naming nomenclature that is similar to the peptide nomenclature presented in Section 7.3.2 (see Fig. 7.6). The fragmentation nomenclature for carbohydrates is covered extensively in Chapter 9, Section 9.2.2[89] However, a brief coverage is presented here to illustrate the products that are produced upon collision-induced dissociation of sulfated and glycated mucins. Figure 7.41 illustrates the general types of fragmentation observed in tandem mass spectrometry of carbohydrates.

Glycated peptides have an enhanced response in the negative ion mode. It has been observed that different product ions are observed in the two different ion modes (positive and negative). In positive ion mode product ions of type B and Y are observed while in negative ion mode Y-, B-, C-, and Z-type ions are often observed. The product ion spectra illustrated in Figure 7.42 are examples of sulfated mucin oligosaccharides

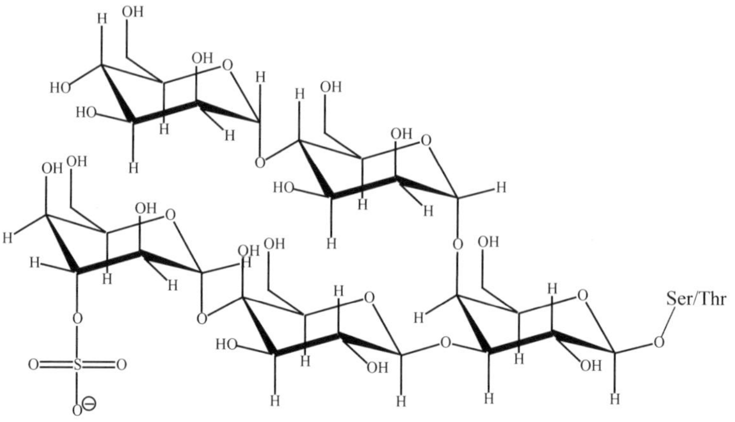

FIGURE 7.39   Sulfated glycoprotein. Glycan has been O-linked to a general Ser/Thr amino acid residue within the protein's peptide chain.

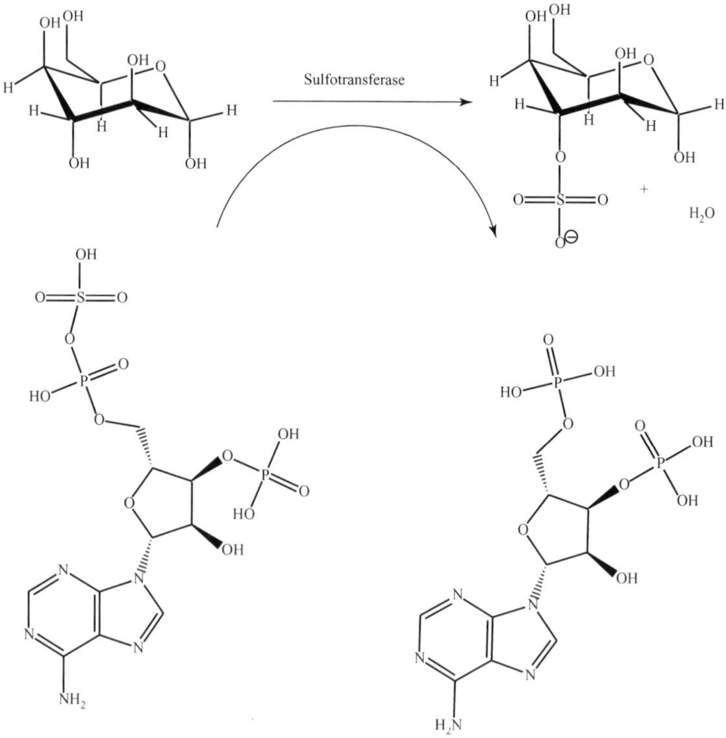

3-phosphoadenosine-5-phosphosulfate (PAPS)   3,5-ADP

FIGURE 7.40   Mechanism for biosynthesis of sulfated saccharide.

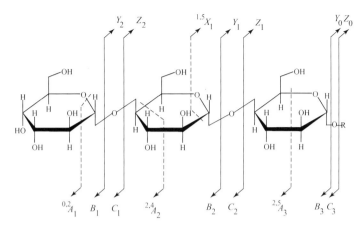

FIGURE 7.41 Carbohydrate fragmentation ion types and associated nomenclature.[91]

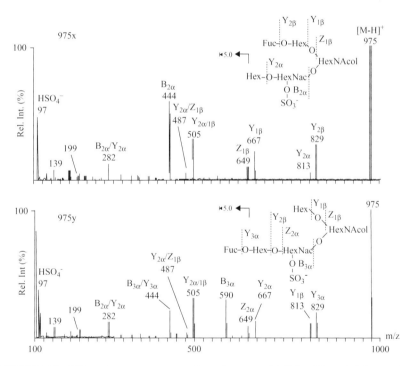

FIGURE 7.42 ESI tandem mass spectra of two isomers with $[M - H]^-$ m/z 975 that eluted at 46.5 (975x) and 49.5 min (975y) obtained from porcine stomach mucins. The fragmentation nomenclature is according to Figure 7.41. The 5 marked region has been magnified 5 times. (Reprinted with permission from Thomsson, K. A., Karlsson, H., and Hansson, G. C. *Anal. Chem.* 2000, *72*, 4543–4539. Copyright 2000 American Chemical Society.)

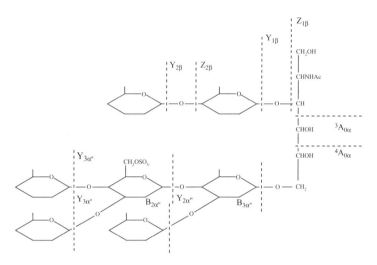

FIGURE 7.43 Fragment annotations applied in this study based upon the suggested nomenclature by Domon and Costello[91] and our previous report. (Reprinted with permission from Thomsson, K. A., Karlsson, H., and Hansson, G. C. *Anal. Chem.* 2000, *72*, 4543–4539. Copyright 2000 American Chemical Society.)

that have been chemically released from the mucins and converted to alditols prior to HPLC separation and mass spectral analysis. The two product ion spectra were obtained from isomers of an m/z 975 oligosaccharide collected in negative ion mode as deprotonated species $[M - H]^-$. The spectra illustrate the production of *B*-, *Y*-, and *Z*-type ions from the fragmentation of the oligosaccharide. Also observed in the spectra is a peak at m/z 97 produced from the neutral loss of HSO_4^- indicative of sulfation. Examples of fragmentation pathway mechanisms for the production of a select number of product ions illustrated in Figure 7.42 are presented in Figure 7.43.

7.5.4.2 *Tyrosine Sulfation* We saw previously that the tyrosine amino acid residue can be phosphorylated, although it is the least modified as compared to serine and threonine. The posttranslational modification in the form of sulfation also takes place with tyrosine, and it is the most common form of PTM that involves tyrosine. Within the total protein content of an organism it is estimated that up to 1% of all tyrosine residues may be sulfated.[90] The first reporting of the observance of the sulfate-to-tyrosine covalent modification was in 1954 by Bettelheim.[91] Sulfation takes place on trans-membrane spanning and secreted proteins. The sulfation mechanism within eukaryotic cells takes place in the trans-Golgi where the membrane-bound enzyme trosylprotein sulfotransferase catalyzes the modification (sulfation) of proteins synthesized in the rough endoplasmic reticulum (RER). The mechanism for sulfation of the tyrosine residue forming tyrosine *O*-sulfate esters (arylsulfate) is the same as for glycosaminoglycan sulfation through the activated donor 3′-phosphoadenosine-5′-phosphosulfate (PAPS). There are two different tyrosylprotein sulfotransferases that have been isolated and identified by researchers, TPST1 and TPST2.[92,93] Others have observed that *N*-sulfation and *S*-sulfation can occur,[94] and that on rare occasions

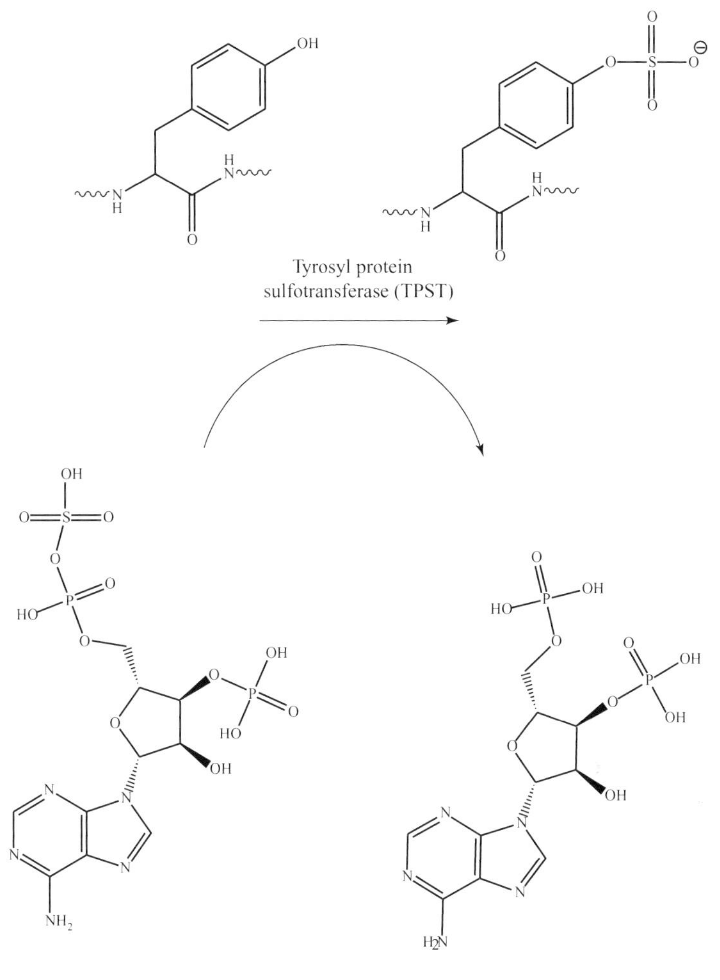

FIGURE 7.44 Tyrosine sulfation reaction mechanism.

serine and threonine can be sulfated.[95] The biosynthesis pathway for the sulfation of the tyrosine amino acid residue by tyrosylprotein sulfotransferases through the activated donor PAPS is illustrated in Figure 7.44.

Sulfation of the tyrosine amino acid residue is directly involved in the recognition processes of protein-to-protein interactions of membrane-bound and secreted proteins. A recent article from Woods et al.[96] at the National Institute of Health (NIH) gives a listing of proteins that have been observed to be sulfated that includes plasma membrane proteins, adhesion proteins, immune components, secretory proteins, and coagulation factors to name a few. A listing of O-sulfated human proteins currently included in the SWISS_Prot database are listed in Table 7.9. The UniProt database

**TABLE 7.9 Proteins Listed in SWISS-Prot Annotated as Having at Least One
O-Sulfated Amino Acid Residue**

SWISS Prot Name	Description	Sulfation Site(s)[a]	Sequence Length
1A01_HUMAN	HLA class I histocompatibility antigen, A-1 alpha chain precursor (MHC class I antigen A*1)	83	365
7B2_HUMAN	Neuroendocrine protein 7B2 precursor	—	212
A2AP_HUMAN	Alpha-2-antiplasmin precursor (Alpha-2-plasmin inhibitor) (Alpha-2-PI)	484	491
AMD_HUMAN	Peptidyl-glycine alpha-amidating monooxygenase precursor	961	973
AMPN_HUMAN	Aminopeptidase N (EC 3.4.11.2) (hAPN) (Alanyl aminopeptidase)	175, 418, 423, 912	966
C3AR_HUMAN	C3a anaphylatoxin chemotactic receptor (C3a-R) (C3AR).	174, 184, 318	482
C5AR_HUMAN	C5a anaphylatoxin chemotactic receptor (C5a-R) (C5aR) (CD88 antigen)	11, 14	350
CCKN_HUMAN	Cholecystokinins precursor (CCK)	111, 113	115
CCR2_HUMAN	C\C chemokine receptor type 2 (C\C CKR-2) (CC-CKR-2) (CCR-2) (CCR2)	26	374
CCR5_HUMAN	C\C chemokine receptor type 5 (C\C CKR-5) (CC-CKR-5) (CCR-5) (CCR5)	3, 10, 14, 15	352
CMGA_HUMAN	Chromogranin A precursor (CgA) (Pituitary secretory protein I) (SP-I)	–	457
CO4A_HUMAN	Complement C4-A precursor (Acidic complement C4)	1417, 1420, 1422	1744
CO4B_HUMAN	Complement C4-B precursor (Basic complement C4)	1417, 1420, 1422	1744
CO5A1_HUMAN	Collagen alpha-1(V) chain precursor.	234, 236, 338, 340, 346, 347, 416, 417, 420, 421, 1601,1604	1838
CXCR4_HUMAN	C–X–C chemokine receptor type 4 (CXC-R4) (CXCR-4)	21	352

TABLE 7.9 (*Continued*)

SWISS Prot Name	Description	Sulfation Site(s)[a]	Sequence Length
DERM_HUMAN	Dermatopontin precursor (Tyrosine-rich acidic matrix protein) (TRAMP).	23, 162, 164, 166, 167, 194	201
FA5_HUMAN	Coagulation factor V precursor (Activated protein C cofactor)	693, 724, 726, 1522, 1538, 1543, 1593	2224
FA8_HUMAN	Coagulation factor VIII precursor (Procoagulant component)	365, 414, 426, 737, 738, 742, 1683, 1699	2351
FA9_HUMAN	Coagulation factor IX precursor (EC 3.4.21.22) (Christmas factor)	201	461
FETA_HUMAN	Alpha-fetoprotein precursor (Alpha-fetoglobulin)	—	609
FIBG_HUMAN	Fibrinogen gamma chain precursor.	444, 448	453
FINC_HUMAN	Fibronectin precursor (FN) (Cold-insoluble globulin) (CIG).	876, 881	2386
GAST_HUMAN	Gastrin precursor [Contains: Gastrin 71 (Component I); Gastrin	52 87	101
GP1BA_HUMAN	Platelet glycoprotein Ib alpha chain precursor (Glycoprotein Ibalpha)	292, 294, 295	626
HEP2_HUMAN	Heparin cofactor 2 precursor (Heparin cofactor II) (HC-II)	79, 92	499
MFAP2_HUMAN	Microfibrillar-associated protein 2 precursor (MFAP-2)	47, 48, 50	183
MGA_HUMAN	Maltase-glucoamylase, intestinal [Includes: Maltase (EC 3.2.1.20)	415, 424, 1281	1856
NID1_HUMAN	Nidogen-1 precursor (Entactin).	289, 296	1247
OMD_HUMAN	Osteomodulin precursor (Osteoadherin)	25, 31, 39	421
OPT_HUMAN	Opticin precursor (Oculoglycan).	71	332
ROR2_HUMAN	Tyrosine-protein kinase transmembrane receptor ROR2 precursor	469, 471	943
SCG1_HUMAN	Secretogranin-1 precursor (Secretogranin I) (SgI) (Chromogranin B)	173, 341	677

TABLE 7.9   (*Continued*)

SWISS Prot Name	Description	Sulfation Site(s)[a]	Sequence Length
SCG2_HUMAN	Secretogranin-2 precursor (Secretogranin II) (SgII) (Chromogranin C)	151	617
SELPL_HUMAN	P-selectin glycoprotein ligand 1 precursor (PSGL-1) (Selectin P ligand)		
(CD162 antigen).	46, 48, 51	412	
SUIS_HUMAN	Sucrase-isomaltase, intestinal [Contains: Sucrase (EC 3.2.1.48);	236, 238, 390, 399, 666, 762, 764	1826
THYG_HUMAN	Thyroglobulin precursor.	24	2768
VTNC_HUMAN	Vitronectin precursor (Serum spreading factor) (S-protein) (V75)	75, 78, 282	478

[a] Some proteins are known to be sulfated but the sites of modification are not known. It can clearly be seen that sulfation preferentially occurs in clusters.
Source: Monigatti, F., Hekking, B., and Steen, H. Protein sulfation analysis—A primer. *Biochim. Biophy. Acta* 2006, *1764*, 1904–1913. Reprinted with permission. Copyright Elsevier 2006.

currently lists 275 proteins that are tyrosine sulfated (http://www.uniprot.org). There has not at this point been observed any type of enzymatic mechanism that would desulfate the modified proteins, and because sulfation is pH stable, proteins containing a sulfation modification can be observed in urine excretion.

Product ion mass spectra of sulfated peptides by collision-induced dissociation are similar to that observed for phosphorylated peptides in that the major product ion observed is for the neutral loss of sulfur trioxide SO_3 (-80 Da) from the precursor ion, often with little other information in the way of product ions. This is illustrated in the product ion mass spectrum of Figure 7.45. In the figure there are two predominant peaks at m/z 647 for the doubly protonated, sulfated precursor ion $[M + 2H]^{2+}$, and at m/z 607 for the neutral loss of the sulfate modification $[M + 2H - SO_3]^{2+}$. Loss of 80 Da from the precursor ion is also observed with the neutral loss of HPO_3 from the phosphorylated tyrosine and histidine amino acid residues. The exact mass of HPO_3 is 79.9663 Da while the exact mass of SO_3 is 79.9568 Da, a difference of 119 parts per million (ppm). The fragmentation pathway mechanisms for the loss of the phosphate or sulfate modifications are the same, resulting with the original, unmodified tyrosine residue, as contrasted with the production of dehydroalanine produced by loss of H_3PO_4 (98 Da). The gas-phase fragmentation pathway mechanism for the neutral loss of the sulfate modification of tyrosine during CID product ion production is illustrated in Figure 7.46.

Increasing the collision energy will generate further product ions of the m/z 647 sulfated species, which is illustrated in Figure 7.45, giving sequence information of

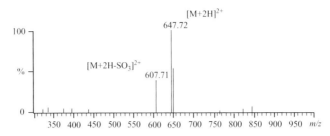

FIGURE 7.45 Product ion mass spectrum of a sulfated peptide at m/z 647 for $[M + 2H]^{2+}$. The mass spectrum is dominated by the precursor ion and the associated SO_3 neutral loss product ion at m/z 607.

the peptide in the form of b- and y-type ions. This can allow the sequencing of the peptide and subsequent identification of the protein from which it came. However, the higher energy collision-induced product ions are generated from the precursor after the loss of the labile sulfo-moeity, thus the ability to identify the modification site is often not possible. The loss of the labile sulfo-group is proton or charge driven. Conditions that allow or favor charge remote fragmentation can induce the production of product ions that still retain the sulfo-moeity. This condition will give the opportunity to determine the site of the sulfo-modification of the peptide. One approach to product ion spectral generation that has demonstrated this is the use of electron capture dissociation (ECD) in FTICR mass spectrometers. As we saw previously, the ECD fragmentation mechanism is not based on the activation of the precursor through repeated collisions but through the capture of thermal electrons that tend to promote backbone dissociations while still retaining the labile modifications (sulfation, phosphorylation, etc.). Figure 7.47 illustrates a product ion spectrum of drosulfakinin, a polypeptide known to have a sulfated tyrosine residue, in the plus 2 charge state $[M + 2H]^{2+}$. The product ion spectrum contains a considerable amount

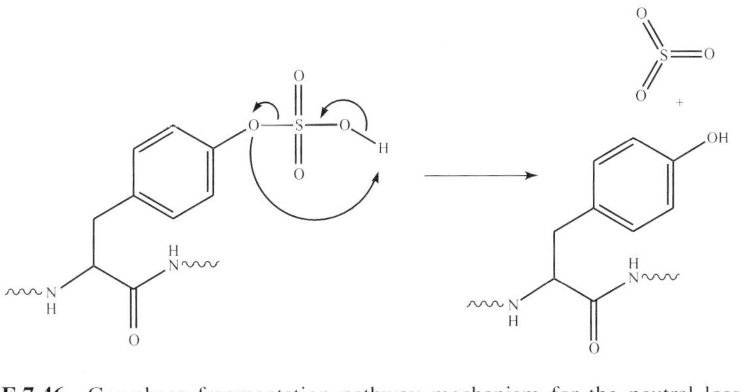

FIGURE 7.46 Gas-phase fragmentation pathway mechanism for the neutral loss of the sulfate modification of tyrosine during CID product ion production.

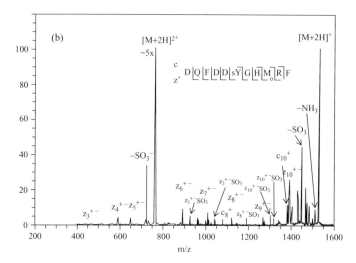

FIGURE 7.47   ECD product ion mass spectrum of drosulfakinin. (Reprinted with permission from Haselmann, K. F., Budnik, B. A., Olsen, J. V., Nielsen, M. L., Reis, C. A., Clausen, H., Johnsen, A. H., and Zubarev, R. A. *Anal. Chem.* 2001, *73*, 2998–3005. Copyright 2001 American Chemical Society.)

of backbone fragmentation, allowing sequence determination in conjunction with modification site determination. Again, one drawback to the technique of ECD is the required use of an FTICR mass spectrometer. As was presented previously, the technique of electron-transfer dissociation used in ion traps will most likely contribute to studies of sulfation posttranslational modification of peptides. A comparison of the biology, biochemistry, and mass spectrometry of sulfation versus phosphorylation is given in Table 7.10. The table gives a brief description of the subcellular protein locations associated with PTM, modification stabilities, and behaviors associated with mass spectral analysis such as ionization stabilities, and masses corresponding to the two different modifications.

7.6   SYSTEMS BIOLOGY AND BIOINFORMATICS

The application and use of mass spectrometry as an analytical tool to the field of biology is obviously apparent. The amount of information given from mass spectrometric methodologies, that is, the molecular weight, the structure, and the amount (can be relative and/or absolute) of a particular biomolecule extracted from a biological matrix is similar to the biological analytical approach traditionally used of "one gene" or "one protein" at a time study. The study of biological systems is now moving toward more encompassing analysis such as the sequencing of an organism's genome or proteome. The trend now is to study both the single components of a

TABLE 7.10 Summary of Relevant Characteristics of Phosphorylation and Sulfation as Protein Posttranslational Modifications

		Phosphorylation Intracellular	Sulfation Extracellular (Membrane Bound)
Biology	Reversibility	Reversible	Irreversible
Function	Location of modifying enzyme	Cytosolic and nuclear	Membrane bound in trans-Golgi network
	Activation, inactivation, modulation of protein interaction	Modulation of protein interaction	
Biochemistry	Radioactive isotopes	^{32}P and ^{33}P	^{35}S
	Chemical stability	pY is stable / pS/pT are alkaline labile / pE/pD/pH are acid labile	sY is acid labile
	Edman compatibility	pY compatibility limited solubility / pS undergoes β elimination to dehydroalanine / pT undergoes β elimination giving rise to many different side products	sY is hydrolyzed during TFA cleavage step
	Removal	Phosphatases	Arylsulfatases of limited use / Quantitative hydrolysis: 1 M HCl, 95°C, 5 min
	Enrichment	pS/pT: harsh alkaline treatment, many side products / α-pY and antiphosphodomain antibodies available / IMAC, TiO_2	No antibody / No affinity reagents
Mass spectrometry	Property	Acidic (2−)	Acidic (1−)
	Mass difference	+80 Da (79.9663 Da)	+80 Da (79.9568 Da)
	Stability during ionization	ESI	ESI
	Stability during CID	pS/pT: good / pY: stable / MALDI / pS/pT partial loss possible / pY stable	sY easily lost under standard conditions / MALDI / +ve: complete loss / −ve: partial loss
	Signature after CID	pS/pT: sequence dependent loss of H_3PO_4 / pS: dehydroalanine (69 Da) / pT: dehydroaminobutyric acid (83 Da)	Complete loss / Multiple sY's are more stable / None
	Characteristic neutral loss	$-H_3PO_4$ (−98 Da) $-(H_3PO_4+H_2O)$ (−116 Da)	$-SO_3$ (−80 Da)
	Characteristic fragment ion	PO_3- (−79 Da)	SO_3- (−80 Da)

Source: Monigatti, F., Hekking, B., and Steen, H. Protein sulfation analysis—A primer. *Biochim. Biophys. Acta* 2006, *1764,* 1904–1913. Reprinted with permission. Copyright Elsevier 2006.

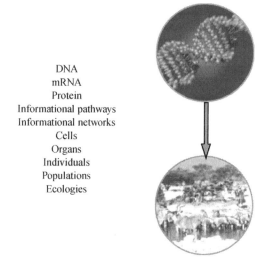

DNA
mRNA
Protein
Informational pathways
Informational networks
Cells
Organs
Individuals
Populations
Ecologies

FIGURE 7.48 Hierarchical levels of biological information. (Reprinted with permission from Hood, L. *J. Proteome Res.* 2002, *1*, 399–409. Copyright 2002 American Chemical Society.)

system in conjunction with the particular system's entire compliment of components. Of special interest is how the various components of the system interact with one another under normal conditions, and some type of perturbed condition such as a diseased state or a change in the system's environment (e.g., lack of oxygen, food, water, etc.). Thus, systems biology is the study of the processes and complex biological organizational behavior using information from its molecular constituents.[97] This is quite broad in the sense that the biological organization may go to the level of tissue up to a population or even an ecosystem. The hierarchical levels of biological information was summed up by L. Hood at the Institute for Systems Biology (Seattle, Washington) showing the progression from DNA to a complex organization as illustrated in Figure 7.48. The idea is to gather as much information about each component of a system to more accurately describe the system as a whole. This encompasses most of the disciplines of biology and incorporates analytical chemistry (mass spectrometry) through the need to measure and identify individual species on the molecular level. When attempting to describe the behavior of a biological system, often the reality of the system is that the behavior of the whole system is greater than what would be predicted from the sum of its parts.[98] The study of systems biology incorporates multiple analytical techniques and methodologies, which includes mass spectrometry, to study the components of a biological system (e.g., genes, proteins, metabolites, etc.) to better model their interactions.[99]

The experimental work flow involved in systems biology that center around mass spectrometric analysis is illustrated in Figure 7.49. Specifically, the metabolite

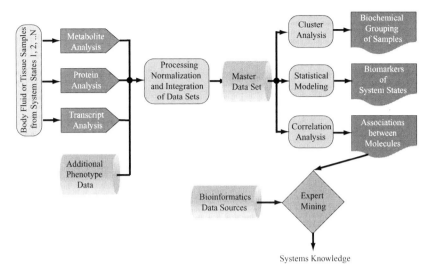

FIGURE 7.49 Systems biology work flow. Data are produced by different platforms (transcriptomics, proteomics, and metabolomics) followed by integration into a master data set. Different biostatistical strategies are pursued: clustering, modeling, and correlation analysis. Integration with extensive bioinformatics tools and expert biological knowledge is key to the creation of meaningful knowledge. (Reprinted with permission from van der Greef, J., Martin, S., Juhasz, P., Adourian, A., Plasterer, T., Verheij, E. R., and McBurney, R. N. *J. Proteome Res.* 2007, *6*, 1540–1559. Copyright 2007 American Chemical Society.)

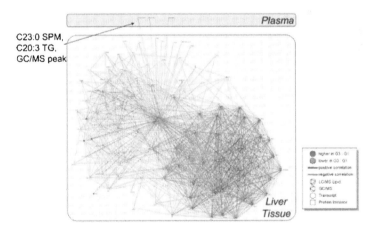

FIGURE 7.50 Correlation network of analytes across blood plasma (*top*) and liver tissue (*bottom*). Analytes include proteins, endogenous metabolites, and gene transcripts. Not only is structure evident among analytes profiled from liver tissue, but there are also a number of correlations to analytes profiled in plasma in this case. Such analytes can serve as useful circulating biomarkers for the tissue-based biochemical processes occurring in the organ. (Reprinted with permission from van der Greef, J., Martin, S., Juhasz, P., Adourian, A., Plasterer, T., Verheij, E. R., and McBurney, R. N. *J. Proteome Res.* 2007, *6*, 1540–1559. Copyright 2007 American Chemical Society.)

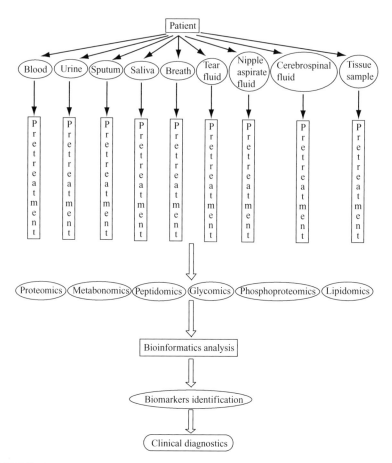

FIGURE 7.51 Scheme for mass spectrometry-based "omics" technologies in cancer diagnostics. Proteomics is the large-scale identification and functional characterization of all expressed proteins in a given cell or tissue, including all protein isoforms and modifications. Metabonomics is the quantitative measurement of metabolic responses of multicellular systems to pathophysiological stimuli or genetic modification. Peptidomics is the simultaneous visualization and identification of the whole peptidome of a cell or tissue, that is, all expressed peptides with their posttranslational modifications. Glycomics is to identify and study all the glycan molecules produced by an organism, encompassing all glycoconjugates (glycolipids, glycoproteins, lipopolysaccharides, peptidoglycans, proteoglycans). Phosphoproteomics is the characterization of phosphorylation of proteins. Lipidomics is systems-level analysis and characterization of lipids and their interacting partners. (Reprinted with permission of John Wiley & Sons, Inc. Zhang, X., Wei, D., Yap, Y., Li, L., Guo, S., and Chen, F. Mass spectrometry based "omics" technologies in cancer diagnostics. *Mass Spectrom. Rev.* 2007, *26*, 403–431.)

analysis and the protein analysis, illustrated in the second step of the flow, are increasingly performed using highly efficient separation methodology such as nano-HPLC (1D and 2D) coupled with high-resolution mass spectrometry such as FTICR-MS and LTQ-Orbitrap/MS. Much development has been done, and is still ongoing, in the processing and data extraction of the information obtained from high peak capacity nano-HPLC-ESI mass spectrometry. A tremendous amount of information is obtained from the experiments that needs to be processed, validated, and statistically evaluated. Bioinformatics for proteomics has developed with great speed and complexity with many open-source software available with statistical methods and filtering algorithms for proteomics data validation. Some examples at the present are Bioinformatics.org, sourceforge.net, Open Bioinformatics Foundation that features toolkits such as BioPerl, BioJava, and BioPython, and BioLinux, an optimized Linux operating system for Bioinformaticians.

Finally, the visualization of the data obtained from mass spectrometric analysis has also been developed significantly in the past few years. The ability to take different compliments of analyses and integrate them with the goal of correlation is quite daunting when hundreds to thousands of biomolecules are involved. Programs such as Cytoscape allow the visualization of molecular interaction networks. Figure 7.50 shows a correlation network between data obtained from proteomics, metabolomics, and transcriptomics from liver tissue and plasma. The correlation network found biomarker analytes in the plasma that may be useful in monitoring processes occurring in the organ.

7.6.1 Biomarkers in Cancer

Mass spectrometry is having a substantial impact on systems biology type of studies such as those of biomarker discovery in cancer diagnostics. Figure 7.51 illustrates a scheme for the search for biomarkers in patient-derived samples through mass spectral analyses. From the patient the figure lists nine types of patient-derived samples that includes blood, urine, sputum, saliva, breath, tear fluid, nipple aspirate fluid, and cerebrospinal fluid. Also described in the figure is the various types of "omic" studies done using mass spectrometric techniques such as proteomics, metabonomics, peptidomics, glycomics, phosphoproteomics, and lipidomics, most covered in this and Chapters 8 and 9. Indicative of the progress or status of a disease, a biomarker is a biologically derived molecule in the body that is measured along with many other species present by the omics methods. Bioinformatics analysis is performed on the mass spectral data to determine the presence of potential biomarkers through expression studies and response differentials. Once identified the biomarkers can be used for clinical diagnostics such as early detection before the onset of a serious disease such as cancer. This allows medical intervention that may have a substantial influence upon the success of an early treatment and subsequent cure. Table 7.11 lists a number of identified potential cancer biomarkers that have been discovered through mass spectrometric analyses, primarily through proteomics and metabonomics.

TABLE 7.11   Potential Cancer Biomarkers Identified by MS-Based "Omics" Technologies

Biomarkers	"Omics" Platforms	MS Methods	Sample Source	Cancer Type	References
Apolipoprotein A1, Inter-α-trypsin inhibitor Haptoglobin-a-subunit Transthyretin	Proteomics	SELDI-TOF	Serum	Ovarian	Ye et al., 2003 Zhang et al., 2004
Vitamin D-binding protein	Proteomics	SELDI-TOF	Serum	Prostate	Hlavaty et al., 2003
Stathmin (Op18), GRP 78 14-3-3 isoforms, Transthyretin Protein disulfide Isomerase	Proteomics	ESI-MS	Tissue	Lung	Chen et al., 2003
Peroxiredoxin, Enolase Protein disulfide Isomerase HSP 70, α-1-antitrypsin	Proteomics	MALDI-TOF, LC-MS	Tissue	Breast	Somiari et al., 2003
HSP 27	Proteomics	MALDI-TOF	Serum	Liver	Feng et al., 2005
Annexin I, Cofilin, GST Superoxide dismutase Peroxiredoxin, Enolase Protein disulfide Isomerase	Proteomics	MALDI-TOF, ESI-MS, Q-TOF	Tissue	Colon	Seike et al., 2003; Stierum et al., 2003
Neutrophil peptides 1-3	Proteomics	SELDI-TOF	Nipple aspirate fluid	Breast	Li et al., 2005b
PCa-24	Proteomics	MALDI-TOF	Tissue	Prostate	Zheng et al., 2003
Alkanes, Benzenes	Metabonomics	GC-MS	Breath	Lung	Phillips et al., 1999
Decanes, Heptanes	Metabonomics	GC-MS	Breath	Breast	Phillips et al., 2003
Hexanal, Heptanal	Metabonomics	LC-MS	Serum	Lung	Deng et al., 2004
Pseu, m1A, m1I	Metabonomics	HPLC, LC-MS	Urine	Liver	Yang et al., 2005b

Source: Zhang, X., Wei, D., Yap, Y., Li, L., Guo, and S., Chen, F. Mass spectrometry based "omics" technologies in cancer diagnostics. *Mass Spectrom. Rev.* 2007, 26, 403–431. Reprinted with permission of John Wiley & Sons, Inc.

7.7 PROBLEMS

7.1. What is the difference between top-down and bottom-up proteomics?

7.2. What are three forms of posttranslational modifications (PTM) of proteins that are studied using mass spectrometry.

7.3. List as many descriptive features of proteins you can.

7.4. List the four structures of proteins and their description.

7.5. What is peptide mass fingerprinting?

7.6. In Table 7.1 for what are the different proteins used? What makes one different from another?

7.7. What is de novo sequencing, and for what is it used?

7.8. What does bioinformatics attempt to do?

7.9. Describe the nomenclature of Roepstorff for fragmented peptides according to charge retainment.

7.10. What is the mobile proton model and what discrepancy does it address?

7.11. What other types of losses can be observed besides b- and y-type ions?

7.12. In high-energy collisions how are d, v, and w ions produced?

7.13. What are immonium ions and what are their usefulness?

7.14. What are isomers and isobars?

7.15. What can be used to help reduce the problem of incomplete de novo sequencing?

7.16. How does ETD differ from CID?

7.17. What is the vital component of top-down proteomics and how is it achieved?

7.18. What are two advantages of top-down proteomics versus bottom-up?

7.19. An ESI mass spectrum of a protein has an envelope of peaks at m/z 2250.7, m/z 2344.5, m/z 2446.4, and m/z 2557.5. What is the molecular weight of the protein and the associated charge states of the peaks?

7.20. What is glycosylation and what is it used for in biological systems?

7.21. What effect does glycosylation have on erythrocytes?

7.22. What three aspects of glycosylated proteins can be obtained using mass spectrometry?

7.23. If two series in a single-stage mass spectrum consisted of an increase of 145.6 Da for one series and 97.1 Da for the second, what could be ascertained from this?

7.24. What amino acid residues in eukaryotic cells can be phosphorylated?

7.25. Are the three amino acid residues phosphorylated to an even extent? If not, what are the relative percentages?

7.26. Explain why the stoichiometry of the phosphorylated proteome is so small ($\leq 1\%$).

7.27. What are known to interfere with IMAC columns, and what are some ways to remove them?

7.28. What are the neutral loss masses of the phosphate group at charge states $+1$, $+2$, and $+3$?

7.29. Neutral loss of the phosphate group from serine produces what form of the amino acid residue? What about from phosphorylated threonine?

7.30. Describe the neutral losses associated with phosphorylated tyrosine.

7.31. What sets apart ECD and ETD from CID? Is this an advantage?

7.32. Briefly describe the difference between eukaryotic and prokaryotic phosphorylation.

7.33. If up to 6% of phosphorylation in eukaryotes and prokaryotes is with the histidine residue, why is it not typically observed?

7.34. What are some stabilities of the other phosphorylated amino acid residues?

7.35. What is the structure of HPO_3?

7.36. What two types of protein sulfation occur?

7.37. Are the neutral loss masses of a phosphate modification similar to that of a sulfate modification?

7.38. Give a brief definition of systems biology.

7.39. List some of the patient samples that are used in cancer biomarker studies.

REFERENCES

1. Morrison, R. T., and Boyd, R. N. *Organic Chemistry*, 5th ed. Boston: Allyn and Bacon, 1987, p. 1345.
2. Schulz, G. E., and Schirmer, R. H. In *Principles of Protein Structure, Springer Advanced Texts in Chemistry*, Cantor, C. R., Ed. New York: Springer, 1990.
3. Franks, F. *Biophys. Chem.* 2002, *96*, 117–127.
4. Edman, P. *Acta Chem. Scand.* 1950, *4*, 283–293.
5. Shimonishi, Y., Hong, Y. M., Kitagishi, T., Matsuo, H., and Katakuse, I. *Eur. J. Biochem.* 1980, *112*, 251–264.
6. Morris, H. R., Panico, M., Barber, M., Bordoli, R. S., Sedgwick, R. D., and Tyler, A. *Biochem. Biophys. Res. Commun.* 1981, *101*, 623–631.

7. Mann, M., Hojrup, P., and Roepstorff, P. *Biol. Mass Spectrom.* 1993, *22*(6), 338–345.

8. Pappin, D. D. J., Hojrup, P., and Bleasby, A. J. *Curr. Biol.* 1993, *3*, 327–332.

9. Henzel, W. J., Billeci, T. M., Stults, J. T., Wong, S. C., Grimley, C., and Watanabe, C. *Proc. Natl. Acad. Sci. USA* 1993, *90*(11), 5011–5015.

10. Kenrick, K. G., and Margolis, J. *Anal. Biochem.* 1970, *33*, 204–207.

11. Gras, R., and Muller, M. *Curr. Opin. Mol. Ther.* 2001, *3*(6), 526–532.

12. Wolters, D. A., Washburn, M. P., and Yates, J. R. *Anal. Chem.* 2001, *73*, 5683–5690.

13. Eng, J. K., McCormack, A. L., and Yates, J. III. *J. Am. Soc. Mass Spectrom.* 1994, *5*, 976–989.

14. Mann, M., and Wilm, M. *Anal. Chem.* 1994, *66*, 4390–4399.

15. Roepstorff, P., and Fohlman, J. *Biomed. Mass Spectrom.* 1984, *11*(11), 601.

16. Yalcin, T., Csizmadia, I. G., Peterson, M. R., and Harrison, A. G. *J. Am. Soc. Mass Spectrom.* 1996, *7*, 233–242.

17. Dongre, A. R., Jones, J. L., Somogyi, A., and Wysocki, V. H. *J. Am. Chem. Soc.* 1996, *118*, 8365–8374.

18. McCormack, A. L., Somogyi, A., Dongre, A. R., and Wysocki, V. H. *Anal. Chem.* 1993, *65*, 2859–2872.

19. Somogyri, A., Wysocki, V. H., and Mayer, I. *J. Am. Soc. Mass Spectrom.* 1994, *5*, 704–717.

20. Zubarev, R. A., Kelleher, N. L., and McLafferty, F. W. *J. Am. Chem. Soc.* 1998, *120*, 3265–3266.

21. Loo, J. A., Edmonds, C. G., and Smith, R. D. *Science* 1990, *248*, 201–204.

22. Feng, R., and Konishi, Y. *Anal. Chem.* 1993, *65*, 645–649.

23. Loo, J. A., Quinn, J. P., Ryu, S. I., Henry, K. D., Senko, M. W., and McLafferty, F. W. *Proc. Natl. Acad. Sci. USA* 1992, *89*, 286–289.

24. Senko, M. W., Beu, S. C., and McLafferty, F. W. *Anal. Chem.* 1994, *66*, 415–417.

25. Schnier, D. S., Gross, E. R., and Williams, E. R. *J. Am. Chem. Soc.* 1995, *117*, 6747–6757.

26. Patrie, S. M., Charlebois, J. P., Whipple, D., Kelleher, N. L., Hendrickson, C. L., Quinn, J. P., Marshall, A. G., and Mukhopadhyay, B. *J. Am. Soc. Mass Spectrom.* 2004, *15*, 1099–1108.

27. Wold, F. *Ann. Rev. Biochem.* 1981, *50*, 783–814.

28. Wilkins, M. R., Gasteiger, E., Gooley, A. A., Herbert, B. R., Molloy, M. P., Binz, P. A., Ou, K., Sanchez, J. C., Bairoch, A., Williams, K. L., and Hochstrasser, D. F. *J. Mol. Biol.* 1999, *280*, 645–657.

29. Peterman, S. M., and Mulholland, J. J. *J. Am. Soc. Mass Spectrom.* 2006, *17*, 168–179.

30. Manning, G., Whyte, D. B., Martinez, R., Hunter, T., and Sudarsanam, S. *Science* 2002, *298*, 1912–1934.

31. Chalmers, M. J., Kolch, W., Emmett, M. R., Marshall, A. G., and Mischak, H. *J. Chromatogr. B* 2004, *803*, 111–120.

32. Olsen, J. V., Blagoev, B., Gnad, F., Macek, B., Kumar, C., Mortensen, P., and Mann, M. *Cell* 2006, *127*, 635–648.

33. Zhou, H., Watts, J. D., and Aebersold, R. *Nat. Biotechnol.* 2001, *19*, 375–378.

34. Beausoleil, S. A., Jedrychowski, M., Schwartz, D., Elias, J. E., Villen, J., Li, J., Cohn, M. A., Cantley, L. C., and Gygi, S. P. *Proc. Natl. Acad. Sci. USA* 2004, *101*, 12130–12135.

35. Amanchy, R., Kalume, D. E., Iwahori, A., Zhong, J., and Pandey, A. *J. Proteome Res.* 2005, *4*, 1661–1671.

36. Ballif, B. A., Roux, P. P., Gerber, S. A., MacKeigan, J. P., Blenis, J., and Gygi, S. P. *Proc. Natl. Acad. Sci. USA* 2005, *102*, 667–672.

37. Hoffert, J. D., Pisitkun, T., Wang, G., Shen, R. F., and Knepper, M. A. *Proc. Natl. Acad. Sci. USA* 2006, *103*, 7159–7164.

38. Yang, F., Stenoien, D. L., Strittmatter, E. F., Wang, J. H., Ding, L. H., Lipton, M. S., Monroe, M. E., Nicora, C. D., Gristenko, M. A., Tang, K. Q., Fang, R. H., Adkins, J. N., Camp, D. G., Chen, D. J., and Smith, R. D. *J. Proteome Res.* 2006, *5*, 1252–1260.

39. Arroyo, J. D., and Hahn, W. C. *Oncogene* 2005, *24*, 7746–7755.

40. Brady, M. J., and Saltiel, A. R. *Recent Prog. Horm. Res.* 2001, *56*, 157–173.

41. Oliver, C. J., and Shenolikar, S. *Front Biosci.* 1998, *3*, D961–972.

42. Brown, L., Borthwick, E. B., and Cohen, P. T. W. *Biochim. Biophys. Acta* 2000, *1492*, 470–476.

43. Olsen, J. V., Blagoev, B., Gnad, F., Macek, B., Kumar, C., Mortensen, P., and Mann, M. *Cell* 2006, *127*, 635–648.

44. Hunter, T., and Sefton, B. M. *Proc. Natl. Acad. Sci. USA* 1980, *77*, 1311–1315.

45. McLachlin, D. T., and Chait, B. T. *Curr. Opin. Chem. Biol.* 2001, *5*, 591–602.

46. Qian, W. J., Goshe, M. B., Camp II, D. G., Yu, L. R., Tang, K., and Smith, R. D. *Anal. Chem.* 2003, *75*, 5441–5450.

47. Garcia, B. A., Shabanowitz, J., and Hunt, D. F. *Methods* 2005, *35*, 256–264.

48. Salih, E. *Mass Spectrom. Rev.* 2005, *24*, 828–846.

49. Smith, R. D., Anderson, G. A., Lipton, M. S., Pasa-Tolic, L., Shen, Y., Conrads, T. P., Veenstra, T. D., and Udseth, H. R. *Proteomics* 2002, *2*, 513–523.

50. Oliver, C. J., and Shenolikar, S. *Front Biosci.* 1998, *3*, D961–972.

51. Janssens, V., and Goris, J. *Biochem. J.* 2001, *353*, 417–439.

52. Ceulemans, H., and Bollen, M. *Physiol. Rev.* 2004, *84*, 1–39.

53. Ficarro, S. B., McCleland, M. L., Stukenberg, P. T., Burke, D. J., Ross, M. M., Shabanowitz, J., Hunt, D. F., and White, F. M. *Nat. Biotechnol.* 2002, *20*, 301–305.

54. Kim, J. E., Tannenbaum, S. R., and White, F. M. *J. Proteome Res.* 2005, *4*, 1339–1346.

55. Moser, K., and White, F. M. *J. Proteome Res.* 2006, *5*, 98–104.

56. Ballif, B. A., Villen, J., Beausoleil, S. A., Schwartz, D., and Gygi, S. P. *Mol. Cell. Proteomics* 2004, *3*, 1093–1101.

57. Beausoleil, S. A., Jedrychowski, M., Schwartz, D., Elias, J. E., Villen, J., Li, J., Cohn, M. A., Cantley, L. C., and Gygi, S. P. *Proc. Natl. Acad. Sci. USA* 2004, *101*, 12130–12135.

58. Gentile, S., Darden, T., Erxleben, C., Romeo, C., Russo, A., Martin, N., Rossie, S., and Armstrong, D. L. *Proc. Natl. Acad. Sci. USA* 2006, *103*, 5202–5206.

59. Posewitz, M. C., and Tempst, P. *Anal. Chem.* 1999, *71*, 2883–2892.

60. Cao, P., and Stults, J. T. *Rapid Commun. Mass Spectrom.* 2000, *14*, 1600–1606.

61. Kocher, T., Allmaier, G., and Wilm, M. *J. Mass Spectrom.* 2003, *38*, 131–137.

62. Stover, D. R., Caldwell, J., Marto, J., Root, R., Mestan, J., Stumm, M., Ornatsky, O., Orsi, C., Radosevic, N., Liao, L., Fabbro, D., and Moran, M. F. *Clin. Proteomics* 2004, *1*, 69–80.

63. Gruhler, A., Olsen, J. V., Mohammed, S., Mortensen, P., Faergeman, N. J., Mann, M., and Jensen, O. N. *Mol. Cell. Proteomics* 2005, *4*, 310–327.

64. Zhang, Y., Wolf-Yadlin, A., Ross, P. L., Pappin, D. J., Rush, J., Lauffenburger, D. A., and White, F. M. *Mol. Cell. Proteomics* 2005, *4*, 1240–1250.

65. Garrels, J. *J. Biol. Chem.* 1979, *254*, 7961–7977.

66. Leimgruber, R. M. In *The Proteomics Protocols Handbook*, Walker, J. M. Ed. Totowa, NJ: Humana Press, 2005, p. 4.

67. Leimgruber, R. M. *Proteomics* 2002, *2*, 135–144.

68. Beausoleil, S. A., Jedrychowski, M., Schwartz, D., Elias, J. E., Villen, J., Li, J., Cohn, M. A., Cantley, L. C., and Gygi, S. P. *Proc. Natl. Acad. Sci. USA* 2004, *101*, 12130–12135.

69. de Souza, G. A., Godoy, L. M. F., and Mann, M. *Genome Biol.* 2006, 7:R72.

70. Stock, J., Ninfa, A., and Stock, A. M. *Microbiol. Rev.* 1989, *53*, 450–490.

71. Stock, J. B., Stock, A. M., and Mottonen, J. M. *Nature* 1990, *344*, 395–400.

72. Mathews, H. R. *Pharmacol. Ther.* 1995, *67*, 323–350.

73. Waygood, E. B., Mattoo, R. L., and Peri, K. G. *J. Cell Biochem.* 1984, *25*, 139–159.

74. Pirrung, M. C., James, K. D., and Rana, V. S. *J. Org. Chem.* 2000, *65*, 8448–8453.

75. Napper, S., Kindrachuk, J., Olson, D. J. H., Ambrose, S. J., Dereniwsky, C., and Ross, A. R. S. *Anal. Chem.* 2003, *75*, 1741–1747.

76. Hooper, L. V., Manzella, S. M., and Baenziger, J. U. *FASEB J.* 1996, *10*, 1137–1146.

77. Bowman, K. G., and Bertozzi, C. R. *Chem. Biol.* 1996, *6*, R9–R22.

78. Habuchi, O. *Biochim. Biophys. Acta* 2000, *1474*, 115–127.

79. Bowman, K. G., Cook, B. N., de Graffenried, C. L., and Bertozzi, C. R. *Biochemistry* 2001, *40*, 5382–5391.

80. Plaas, A. H. K., West, L. A., Wong Palms, S., and Nelson, F. R. T. *J. Biol. Chem.* 1998, *273*(20), 12642–12649.

81. Bayliss, M. T., Osborne, D., Woodhouse, S., and Davidson, C. *J. Biol. Chem.* 1999, *274*(22), 15892–15900.

82. Desaire, H., Sirich, T. L., and Leary, J. A. *Anal. Chem.* 2001, *73*(15), 3513–3520.

83. Jiang, H., Irungu, J., and Desaire, H. *J. Am. Soc. Mass Spectrom.* 2005, *16*, 340–348.

84. Falany, C. N. *FASEB J.* 1997, *22*, 206–216.

85. Green, E. D., and Baenziger, J. U. *J. Biol. Chem.* 1988, *26*, 36–44.

86. Thomsson, K. A., Karlsson, H., and Hansson, G. C. *Anal. Chem.* 2000, *72*, 4543–4539.

87. Kawasaki, N., Haishima, Y., Ohta, M., Satsuki, I., Hyuga, M., Hyuga, S., and Hayakawa, T. *Glycobiology* 2001, *11*, 1043–1049.

88. Hemmerich, S., and Rosen, S. D. *Glycobiology* 2000, *10*, 849–856.

89. Domon, B., and Costello, C. E. *Glycoconjugate J.* 1988, *5*, 397–409.

90. Baeuerle, P. A., and Huttner, W. B. *J. Biol. Chem.* 1985, *260*, 6434–6439.

91. Bettelheim, F. R. *J. Am. Chem. Soc.* 1954, *76*, 2838–2839.

92. Beisswanger, R., Corbeil, D., Vannier, C., Thiele, C., Dohrmann, U., Kellner, R., Ashman, K., Niehrs, C., and Huttner, W. B. *Proc. Natl. Acad. Sci. USA* 1998, *95*, 11134–11139.

93. Ouyang, Y. B., and Moore, K. L. *J. Biol. Chem.* 1998, *273*, 24770–24774.

94. Huxtable, R. J. *Biochemistry of Sulfur*, New York: Plenum, 1986.

95. Medzihradszky, K. F., Darula, Z., Perlson, E., Fainzilber, M., Chalkley, R. J., Ball, H., Greenbaum, D., Bogyo, M., Tyson, D. R., Bradshaw, R. A., and Burlingame, A. L. *Mol. Cell. Proteomics* 2004, *3*, 429–440.

96. Woods, A. S., Wang, H. Y. J., and Jackson, S. N. *J. Proteome Res.* 2007, *6*, 1176–1182.

97. Kirschner, M. W. *Cell* 2005, *121*, 503–504.

98. Srivastava, R., and Varner, J. *Biotechnol. Prog.* 2007, *23*, 24–27.

99. Smith, J. C., Lambert, J. P., Elisma, F., and Figeys, D. *Anal. Chem.* Accessed on the Internet May 4, 2007.

CHAPTER 8

BIOMOLECULE SPECTRAL INTERPRETATION—SMALL MOLECULES

8.1 INTRODUCTION

In this chapter we will look at the use of mass spectrometry for the analysis of biomolecules as small molecules, which has become an important and increasingly used tool in numerous fields of research and study. Today, when one considers small-molecule analysis by mass spectrometry, it is generally associated with the fields of lipidomics and metabolomics, disciplines that study the complement of small biomolecules that can be associated with biological processes, cells, tissue, and physiological fluid samples such as plasma or urine. Examples of small biomolecules are steroids, fatty acid amides, fatty acids, wax esters, fatty alcohols, phosphorylated lipids, and acylglycerols. Biomolecules referred to as small molecule are contrasted to the larger molecular weight biopolymers such as peptides, proteins, polysaccharides, and nucleic acids, all covered in Chapters 7 and 9.

Lipidomics involves the study of the lipid profile (or lipidome) of living systems and the processes involved in the organization of the protein and lipid species present,[1] including signaling processes and metabolism.[2] Mass spectrometry has lead to a dramatic increase in our understanding of lipidomics over the past several years in a variety of cellular systems. Lipids are a diverse class of physiologically important biomolecules that are often classified into three broad classes: (1) the simple lipids such as the fatty acids, the acylglycerols, and the sterols such as cholesterol, (2) the more complex polar phosphorylated lipids such as phosphatidylcholine and sphingomyelin, and (3) the isoprenoids, terpenoids, and vitamins. Recently, the Lipid

Even Electron Mass Spectrometry with Biomolecule Applications   By Bryan M. Ham
Copyright © 2008 John Wiley & Sons, Inc.

Maps Organization has proposed an eight-category lipid classification (fatty acyls, glycerolipids, glycerophospholipids, sphingolipids, sterol lipids, prenol lipids, saccharolipids, and polyketides) that is based upon the hydrophobic and hydrophilic characteristics of the lipids.[3]

8.2 IONIZATION EFFICIENCY OF LIPIDS

We will begin with a look at the ionization behavior of lipids when using electrospray as the ionization source. Ionization efficiency studies were performed on an ion trap mass spectrometer, and a Q-TOF mass spectrometer of a five-component lipid standard consisting of cis-9-octadecenamide [oleamide, a primary fatty acid amide, molecular weight (MW) of 281.3 Da], palmityl oleate (an unsaturated wax ester, MW 506.5 Da), palmityl behenate (a saturated wax ester, MW 564.6 Da), cholesteryl stearate (a saturated cholesterol ester, MW 652.6 Da), and sphingomyelin (a phosphorylated lipid, MW 730.6 Da). The ionization solutions consisted of either 10 mM LiCl in 1 : 1 chloroform/methanol or 1% acetic acid in 1 : 1 chloroform/methanol. Figure 8.1a illustrates a positive ion mode single-stage mass spectrum of the five-component lipid standard in the acidic solution. The standard was run on a Q-TOF mass spectrometer with direct infusion nanoflow electrospray as the ionization source. The acidic solution did not promote ionization of all of the lipids present in the standard. The species that are being ionized and observed in the mass spectrum include the acid adduct of oleamide at m/z 282.4 as $[M + H]^+$, the sodium adduct of oleamide at m/z 304.4 as $[M + Na]^+$, and sphingomyelin as an acid adduct at m/z 731.7 as

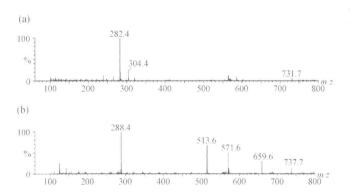

FIGURE 8.1 Single-stage positive ion mode mass spectra of a five-component lipid standard collected on a Q-TOF mass spectrometer for: (a) an acidic solution showing the ionization of oleamide at m/z 282.4 as $[M + H]^+$, the sodium adduct of oleamide at m/z 304.4 as $[M + Na]^+$, and sphingomyelin as an acid adduct at m/z 731.7, and (b) a lithium solution showing the ionization of the lithium adduct of oleamide at m/z 288.4 as $[M + Li]^+$, at m/z 513.6 for the lithium adduct of palmityl oleate as $[M + Li]^+$, at m/z 571.6 for the lithium adduct of palmityl behenate as $[M + Li]^+$, at m/z 659.6 for the lithium adduct of cholesteryl stearate as $[M + Li]^+$, and at m/z 737.7 for the lithium adduct of sphingomyelin as $[M + Li]^+$.

[M + H]$^+$, which was observed to ionize to a limited extent. Figure 8.1*b* illustrates a positive ion mode single-stage mass spectrum of the five-component lipid standard in the lithium solution. The LiCl ionization solution is observed to promote ionization of all five of the lipids in the standard at *m/z* 288.4 for the lithium adduct of oleamide as [M + Li]$^+$, at *m/z* 513.6 for the lithium adduct of palmityl oleate as [M + Li]$^+$, at *m/z* 571.6 for the lithium adduct of palmityl behenate as [M + Li]$^+$, at *m/z* 659.6 for the lithium adduct of cholesteryl stearate as [M + Li]$^+$, and at *m/z* 737.7 for the lithium adduct of sphingomyelin as [M + Li]$^+$.

The addition of Li$^+$ is the most common method to cationize lipids using electrospray in the positive ion mode.[4–6] Many lipids are not readily protonated, while sodiated ions are not preferred as sodium cations give very little fragmentation information. Lithium cations give rich fragmentation patterns, and, thus, it is fairly

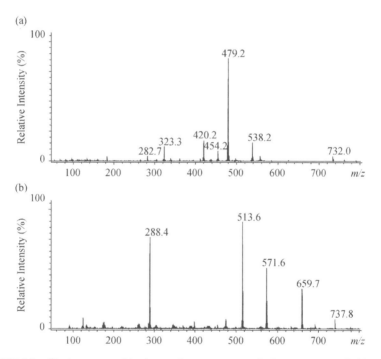

FIGURE 8.2   Single-stage positive ion mode mass spectra of a five-component lipid standard collected on a three-dimensional quadrupole ion trap mass spectrometer for: (*a*) an acidic solution showing the ionization of oleamide at *m/z* 282.4 as [M + H]$^+$, the potassium adduct of oleamide at *m/z* 323.3 as [M + K]$^+$, and sphingomyelin as an acid adduct at *m/z* 731.7, and (*b*) a lithium solution showing the ionization of the lithium adduct of oleamide at *m/z* 288.4 as [M + Li]$^+$, at *m/z* 513.6 for the lithium adduct of palmityl oleate as [M + Li]$^+$, at *m/z* 571.6 for the lithium adduct of palmityl behenate as [M + Li]$^+$, at *m/z* 659.7 for the lithium adduct of cholesteryl stearate as [M + Li]$^+$, and at *m/z* 737.8 for the lithium adduct of sphingomyelin as [M + Li]$^+$.

standard in the analytical field to use this approach. Further, adding a specific cation will often alleviate variation in the metal ion adducts. For example, a mass spectrum run without the addition of a cationizing reagent will show a distribution of M + H^+, M + Na^+, and M + K^+. When a cation is added to the sample, it will primarily shift all the species to that cation. The addition of Na^+ will shift the equilibrium to form almost exclusively M + Na^+ and makes interpretation of unknowns easier. Finally, many compounds ionize according to a cation affinity. Some compounds have a higher affinity to sodium than a proton, for example. In order to fully observe all lipid species present, researchers will use multiple cations in order to confirm the assignments of the compounds and to discover new species that might not have a high proton affinity or sodium affinity (e.g., oleamide is seen at 282 Th as M + H^+, 304 Th as M + Na^+, and 288 Th as M + Li^+).

Figure 8.2 illustrates the same effect for the five-component lipid standard when run on a three-dimensional quadrupole ion trap mass spectrometer. In Figure 8.2*a* the acidic solution shows the ionization of oleamide at m/z 282.4 as $[M + H]^+$, the potassium adduct of oleamide at m/z 323.3 as $[M + K]^+$, and sphingomyelin as an acid adduct at m/z 731.7. The other species that are present are primarily chemical noise in the form of contaminants. The spectrum in Figure 8.2*b* is in a lithium solution, showing the ionization of the lithium adduct of oleamide at m/z 288.4 as $[M + Li]^+$, the lithium adduct of palmityl oleate at m/z 513.6 as $[M + Li]^+$, the lithium adduct of palmityl behenate at m/z 571.6 as $[M + Li]^+$, the lithium adduct of cholesteryl stearate at m/z 659.7 as $[M + Li]^+$, and the lithium adduct of sphingomyelin at m/z 737.8 as $[M + Li]^+$.

8.3 FATTY ACIDS

Fatty acids are often measured in the negative mode when analyzing by electrospray mass spectrometry.[7] Studies of fatty acids by electrospray mass spectrometry include very long chain fatty acid analysis.[8] fingerprinting of vegetable oils,[9] and the analysis of fatty acid oxidation products.[10] An excellent coverage of the mass spectrometric analysis of the simple class of lipids (fatty acids, triacylglycerols, bile acids, and steroids) is given by Griffiths in a recent review.[11]

Mass spectrometric analysis performed on free fatty acid samples in acid solution do not show a good response for the free fatty acids as acid adducts, $[M + H]^+$. Figure 8.3 is a mass spectrum of a free fatty acid containing biological sample in 1 mM ammonium acetate 50 : 50 methanol/chloroform solution, collected in the negative ion mode using a three-dimensional ion trap mass spectrometer. For optimized sensitivity and resolution of the lower molecular weight lipid species, the ion trap was scanned during spectral accumulation for a mass range of m/z 50–400. The major peaks in the spectrum are comprised of deprotonated ions $[M - H]^-$ identified as myristic acid at m/z 227.5, palmitic acid at m/z 255.6, oleic acid at m/z 281.6, and stearic acid at m/z 283.6. Figure 8.4 illustrates the structures of these four common free fatty acids that are found in biological matrices. Figure 8.5 is a product ion spectrum of the oleic acid precursor ion at m/z 281.6 in the biological sample, illustrating diagnostic

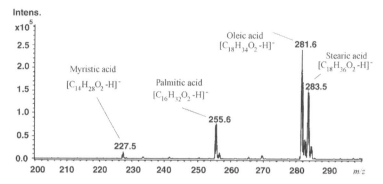

FIGURE 8.3   Negative ion mode mass spectrum of a biological extract in 1 mM ammonium acetate 50 : 50 methanol/chloroform solution. Peaks in the spectrum are deprotonated ions [M − H]⁻ of myristic acid at m/z 227.5, palmitic acid at m/z 255.6, oleic acid at m/z 281.6, and stearic acid at m/z 283.6.

peaks of fatty acids at m/z 263.6 for the production of a fatty acyl chain as ketene ion through neutral loss of water $[C_{18}H_{34}O_2 − H − H_2O]^-$, and at m/z 249.6 for the neutral loss of carbon dioxide $[C_{18}H_{34}O_2 − H − CO_2]^-$, both diagnostic losses for the identification of fatty acids. The other fatty acids (myristic at m/z 227.5, palmitic at m/z 255.6, and stearic at m/z 283.6) product ion spectra also contained these diagnostic losses, identifying them also as free fatty acids.

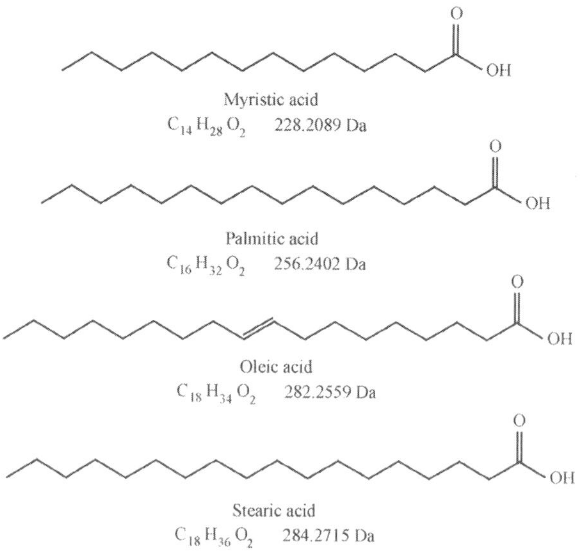

Myristic acid
$C_{14}H_{28}O_2$ 228.2089 Da

Palmitic acid
$C_{16}H_{32}O_2$ 256.2402 Da

Oleic acid
$C_{18}H_{34}O_2$ 282.2559 Da

Stearic acid
$C_{18}H_{36}O_2$ 284.2715 Da

FIGURE 8.4   Structures of four common free fatty acids found in biological matrices.

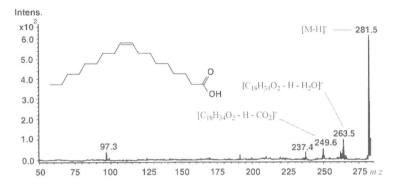

FIGURE 8.5 Product ion spectrum of the oleic acid precursor ion at m/z 281.6 in the biological extract illustrating diagnostic peaks of fatty acids at m/z 263.6 for the production of a fatty acyl chain as ketene ion through neutral loss of water $[C_{18}H_{34}O_2 - H - H_2O]^-$, and at m/z 249.6 for the neutral loss of carbon dioxide $[C_{18}H_{34}O_2 - H - CO_2]^-$.

The fragmentation pathway mechanism for the production of the fatty acyl chain ion as a ketene is illustrated in Figure 8.6. This is a three-step mechanism where step 1 involves the transfer of an α proton from the α position to the terminal fatty acid oxygen. In step 2 a fatty acyl chain proton is abstracted by the leaving hydroxyl group, resulting in neutral loss of water and the production of a negatively charged fatty acyl chain as ketene, as illustrated by step 3 in the mechanism. The fragmentation pathway mechanism for the neutral loss of carbon dioxide producing the m/z 249.6 product ion is illustrated in Figure 8.7. This is a simpler one-step mechanism where a negatively charged fatty acyl hydrocarbon chain is produced.

8.3.1 Negative Ion Mode Electrospray Behavior of Fatty Acids

The spectrum of the four free fatty acids (see Fig. 8.3), which were previously observed in the biological extract presented in Section 8.2, were collected in negative ion mode analysis by electrospray ion trap mass spectrometry (ESI-IT/MS). These were identified as the deprotonated molecules $[M - H]^-$ of myristic acid at m/z 227.5, palmitic acid at m/z 255.6, oleic acid at m/z 281.6, and stearic acid at m/z 283.6. Quantitative analysis of lipids can be difficult due to differences in ionization efficiencies when using electrospray ionization.[12–14] For the more complex phosphorylated lipids the electrospray ionization process has been shown to be directly dependent upon concentration.[15] Both unsaturation and acyl chain length affect the efficiency of the lipid ionization at higher lipid concentrations (>0.1 nmol/μL). The interactions between the lipids are greatly reduced, and a linear correlation can be achieved between concentration and ion intensities when the concentration of the lipids has been reduced.[15] Due to differences in ionization efficiencies of lipids (and subsequent suppression), internal standard spiking of lipids known to be of very low concentration is performed to directly quantitate the lipid species of interest in the

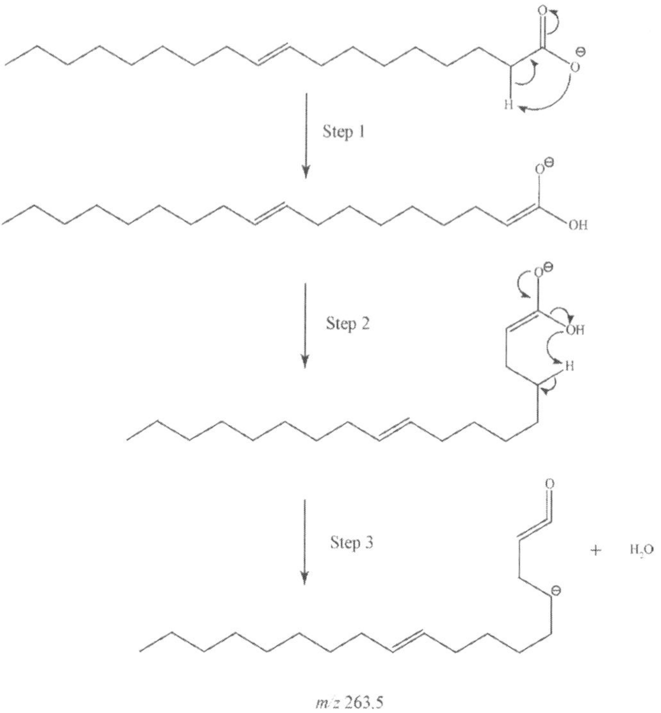

FIGURE 8.6 Fragmentation pathway mechanism for the neutral loss of water producing a negatively charged fatty acyl chain as ketene at m/z 263.5.

biological extract. This has been demonstrated to be possible due to a direct linear correlation between lipid concentration and lipid response.[12–15] Figure 8.8 contains negative ion mode spectra of a series of equal molar standards of myristic acid at m/z 227.5, palmitic acid at m/z 255.6, oleic acid at m/z 281.6, and stearic acid at m/z 283.5, at successively decreasing concentrations of the lipids from top to bottom. Figure 8.8a

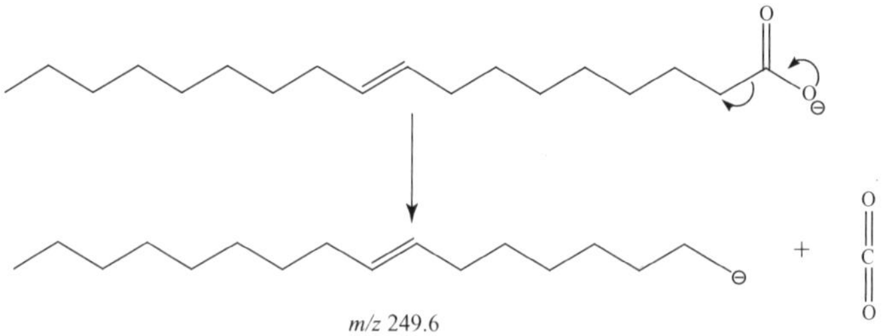

FIGURE 8.7 One-step fragmentation pathway mechanism for the production of the m/z 249.6 product ion through neutral loss of carbon dioxide.

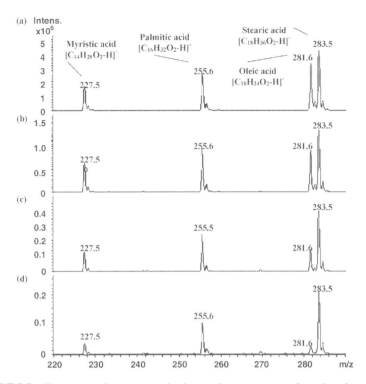

FIGURE 8.8   Electrospray ion trap negative ion mode mass spectra of a series of equal molar standards of myristic acid at m/z 227.5, palmitic acid at m/z 255.6, oleic acid at m/z 281.6, and stearic acid at m/z 283.5. The spectra represent scans of serial dilutions at successively decreasing concentrations of the lipids from top to bottom where (a) is a 0.1 mM equal molar standard with 1 mM ammonium acetate, (b) is 10 μM lipid, (c) is 1 μM lipid, and (d) is 0.1 μM lipid. Ionization efficiency increases with increasing fatty acid chain length and apparent suppression effect for oleic acid as the concentration of the equal molar standard is decreased.

is a 0.1 mM equal molar standard with 1 mM ammonium acetate, while Figure 8.8b is 10 μM lipid, Figure 8.8c is 1 μM lipid, and Figure 8.8d is 0.1 μM lipid. The spectrum illustrated in Figure 8.8a might indicate that the ionization efficiency increases with increasing fatty acid chain length as demonstrated by the lower intensity peak for myristic acid at m/z 227.5, which has a 14-carbon chain, as compared to stearic acid at m/z 283.5, which has an 18-carbon chain. However, there is also observed in the Figure 8.8 spectra an apparent suppression effect for oleic acid as the concentration of the equal molar standard is being decreased. This decrease in ionization efficiency for oleic acid at lower concentrations may be attributed to the single unsaturation in its hydrocarbon chain. It does not appear that the fatty acids measured in negative ion mode using the ion trap possess the same linear correlation that the more complex phosphorylated lipids have exhibited in the past.[15] To ascertain whether

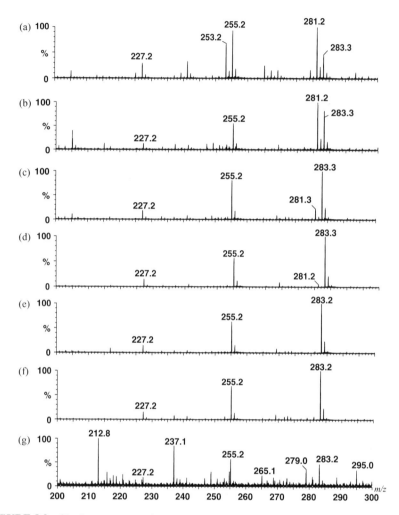

FIGURE 8.9   Single-stage negative ion mode electrospray quadrupole time-of-flight (Q-TOF/MS) mass spectra of a series of oleic acid standards at successively decreasing concentrations of: (*a*) 90 μ*M*, (*b*) 50 μ*M*, (*c*) 10 μ*M*, (*d*) 1 μ*M*, (*e*) 100 n*M*, (*f*) 10 n*M*, and (*g*) 1 n*M*, all in 1 m*M* ammonium acetate 1 : 1 CHCl₃/MeOH solutions.

the single unsaturation in the oleic fatty acyl chain has an effect upon the ionization efficiency, a series of oleic acid standards at successively decreasing concentration were analyzed using the electrospray quadrupole time-of-flight (Q-TOF/MS) mass spectrometer with the results presented in Figure 8.9*a–g*. Single-stage negative ion mode mass spectra were collected of oleic acid at concentrations of 90 μ*M* (Fig. 8.9*a*), 50 μ*M* (Fig. 8.9*b*), 10 μ*M* (Fig. 8.9*c*), 1 μ*M* (Fig. 8.9*d*), 100 n*M* (Fig. 8.9*e*), 10 n*M* (Fig. 8.9*f*), and 1 n*M* (Fig. 8.9*g*), all in 1 m*M* ammonium acetate 1 : 1 CHCl₃/MeOH solutions. In the spectra oleic acid was observed to decrease in its response while the emergence and subsequent spectral dominance of myristic acid, palmitic acid, and stearic acid were noted to take place. In the negative ion mode spectrum of Figure 8.9g for 1 n*M* oleic acid, the spectrum is being dominated primarily by chemical noise from contaminants. Adjustment of the source parameters appeared to have little effect; the emergence in the spectra of the *m/z* 227.5 (myristic acid) and *m/z* 255.6 (palmitic acid) species was also observed. The adjustment of source parameters entailed varying the sample cone voltage between approximately 20–80 V and varying the capillary voltage between approximately 2000–3500 V. This behavior of the suppression of oleic acid or the reduction of oleic acid to stearic acid and the emergence of the other two fatty acid species (myristic acid and palmitic acid) would explain the observation made in Figure 8.8. With decreasing concentration of the lipids, the oleic acid is either suffering from suppression or is being converted to stearic acid with lesser amounts of myristic and palmitic being produced. This is illustrated in Figure 8.8*d* where stearic acid at *m/z* 283.5 is the predominant peak in the spectrum, followed by palmitic acid at *m/z* 255.6, myristic acid at *m/z* 227.5, and lastly oleic acid at *m/z* 281.6.

To further investigate the suppression of oleic acid or the reduction of oleic acid during the electrospray process, a number of experiments were performed. The double-bond position was first studied by electrospray experiments, which included 6-octadecenoic and 11-octadecenoic standards. Both of the different positional C18:1 unsaturated fatty acid lipids (6- and 11-) showed similar behavior to the C18:1 9-octadecenoic fatty acid illustrated in Figure 8.8 (i.e., the C18:1 fatty acid decreased significantly with the significant increase in the C18:0 fatty acid, followed by the palmitic and myristic fatty acids). This indicates that the position of the double bond within the acyl chain does not influence this ESI behavior. A trans C18:1 fatty acid, 9-trans-octadecenoic was analyzed to determine whether a conformational effect would be observed. It was found that 9-trans-octadecenoic also had similar behavior as is illustrated in Figure 8.8 for C18:1 9-octadecenoic. This indicates that location of the double bond is not a factor nor is the conformation of the fatty acid. A C20:1 11-eicosenoic lipid standard was analyzed to determine the electrospray effect upon a longer chain fatty acid lipid. At decreasing concentrations the production of a C20:0 fatty acid by reduction of the C20:1 fatty acid was clearly not observed to any extent that was observed for the C18:1 electrospray experiments. However, the emergence of the other three fatty acids was observed, which included myristic acid at *m/z* 227.1, palmitic acid at *m/z* 255.2, and stearic acid at *m/z* 283.2. Finally, 9-hexadecenoic (C16:1 fatty acid) was analyzed by electrospray in the negative ion mode to determine whether a lower molecular weight singly unsaturated fatty acid would exhibit the same behavior as the C18:1 fatty acid. This in fact was observed where at decreasing concentrations palmitic acid (C16:0) was observed to emerge along with a small amount of myristic acid. Stearic acid was also observed at lower concentrations to become a major peak in the spectra; however, it was also observed as a contaminant in the standard. The fact that the singly unsaturated C20:1 fatty acid did not significantly produce the C20:0 saturated fatty acid during the electrospray process shows that the reduction of the double bond is not taking place. Also, that a C18:0 fatty acid was observed as a major species in the C16:1 electrospray experiments further shows that this is an ionization suppression effect that is taking place with the singly unsaturated fatty acids at decreasing concentrations. Small amounts of the fatty acids are observed in all of the standards as contaminants, except a C20:0 fatty acid. This suggests that at decreasing concentration the singly unsaturated fatty acids are suffering from significant suppression during the electrospray process.

To rule out the contribution of contaminant myristic acid, palmitic acid, and stearic acid presence (e.g., from the oleic acid standard or internal instrument carryover), influencing the spectra, a carbon-13 (^{13}C) algal fatty acid standard mix consisting of 51.0% 16:0, 8.7% 16:1, 2.3% 18:0, 18.3% 18:1, 17.7% 18:2, and 2.0% 18:3, was measured at decreasing concentrations. Single-stage negative ion mode mass spectra of dilutions of the ^{13}C fatty acid standard mixture were collected by ESI-Q-TOF/MS in solutions of 1 mM ammonium acetate 1:1 CHCl$_3$/MeOH. Figure 8.10

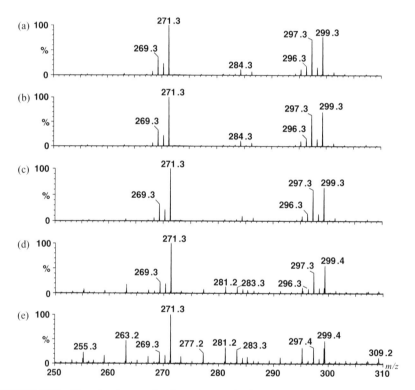

FIGURE 8.10   Negative ion mode electrospray single-stage mass spectra collected on a Q-TOF mass spectrometer illustrating the results of decreasing the concentration of the carbon 13 lipids. (*a*) Contains 940 μM ^{13}C16:0, (*b*) contains 470 μM ^{13}C16:0, (*c*) contains 240 μM ^{13}C16:0, (*d*) contains 9.4 μM ^{13}C16:0, (*e*) contains 0.94 μM ^{13}C16:0. The ^{13}C16:0 lipid is at m/z 271.3, the ^{13}C16:1 lipid is at m/z 269.3, the ^{13}C18:1 lipid is at m/z 299.3, and the ^{13}C18:2 lipid is at m/z 297.3. The ^{13}C16:0 lipid at m/z 271.3 and the ^{13}C16:1 lipid at m/z 269.3 are decreasing in response for the unsaturated ^{13}C16:1 lipid at m/z 269.3 as compared to the saturated lipid ^{13}C16:0 lipid at m/z 271.3.

illustrates the results of the negative ion mode studies where Figure 8.10a contains 940 μM ^{13}C16:0, Figure 8.10b contains 470 μM ^{13}C16:0, Figure 8.10c contains 240 μM ^{13}C16:0, Figure 8.10d contains 9.4 μM ^{13}C16:0, and Figure 8.10e contains 0.94 μM ^{13}C16:0. In the figure, the ^{13}C16:0 lipid is at m/z 271.3, the ^{13}C16:1 lipid is at m/z 269.3, the ^{13}C18:1 lipid is at m/z 299.3, and the ^{13}C18:2 lipid is at m/z 297.3. Close inspection of the ^{13}C16:0 lipid at m/z 271.3 and the ^{13}C16:1 lipid at m/z 269.3 reveals the decreasing response for the unsaturated ^{13}C16:1 lipid at m/z 269.3 as compared to the saturated lipid ^{13}C16:0 lipid at m/z 271.3. To better illustrate the observed effect visually, ratio graphs were constructed. Figure 8.11 contains graphs illustrating the behavior of the unsaturated fatty acids C16:1, C18:1, and C18:2 contained within the ^{13}C fatty acid standard mixture analyzed in Figure 8.10. Figure 8.11a is the ratio of the mass spectral response of the singly unsaturated C16:1 fatty acid to the mass spectral response to its saturated counterpart C16:0 fatty acid plotted versus concentration. The graphs are from 7 standards that cover a concentration range from 95 μM C16:0 to 950 μM C16:0, and 16 μM C16:1 to 160 μM C16:1. The graphs are presented with the x axis represented by the concentration of the C16:0 fatty acid. The graph demonstrates the change in response (suppression) of the C16:1 fatty acid as compared to the C16:0 fatty acid at successively decreasing concentrations. This behavior is also observed in the suppression of C18:1 and C18:2 at successively lower concentrations as illustrated in Figure 8.11b. Due to the relationship that the response of a lipid using ESI is linearly dependent upon concentration, it is also apparent that the suppression effect can also be linearly dependent upon concentration as is illustrated in Figures 8.11a and 8.11b. Also of note is that Figure 8.11b includes C18:1 and C18:2 as a comparison of the influence of the number of unsaturation to suppression. As stated previously the carbon-13 algal fatty acid standard mix contained 18.3% of the C18:1 fatty acid and 17.7% of the C18:2 fatty acid (giving a 3% difference in concentration). The curves in Figure 8.11b illustrate that there appears to be a reduced response for the C18:2 unsaturated lipid as compared to the C18:1 unsaturated lipid as the unsaturation increases from 1 to 2.

The significance of the observance of the suppression of the singly unsaturated fatty acids (oleic acid and palmitoleic acid) illustrates the difficulty in quantitating the amounts of the free fatty acids in biological samples using electrospray as the ionization technique. Care must be taken in assigning relative amounts to the free fatty acids, as well as to all the other lipids present in biological extracts when using an electrospray source.

8.4 QUANTITATIVE ANALYSIS BY GC/EI MASS SPECTROMETRY

Gas chromatography (GC) mass spectrometry is a useful tool for performing quantitative analysis of volatile metabolites that can be separated with a gas chromatograph. The response of the ion detectors coupled with the single quadrupole mass spectrometers are linear in mass regions allowing quantitative analysis. The method presented here is an internal standard method where a series of standards are analyzed that contain a common internal standard that is also spiked into the samples. The response

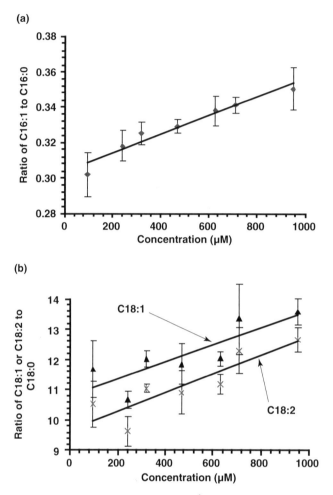

FIGURE 8.11 ESI-QTOF/MS plotted results of single-stage negative ion mode mass spectral carbon 13 fatty acid responses for decreasing concentrations. The graphs are from 7 standards that cover a concentration range from 95 to 950 μ*M*. Each standard spectrum was collected for 1 min, in triplicate, for each standard concentration (the average plotted as data points) and standard deviation (*y*-error bars). (*a*) Palmitoleic acid (C16:1) graphical results of the response of C16:1 divided by the response of C16:0 at each standard concentration. (*b*) Oleic acid (C18:1) and linoleic acid (C18:2) graphical results of the response of C18:1 and C18:2 divided by the response of stearic acid (C18:0) at each standard concentration.

of the common internal standard is used to adjust the response of the current sample being measured. In this way, variations in sample injection amount and changes in the instrumental response can be adjusted for in a normalized way.

The particular analysis presented here is the measurement of the fatty acid amides myristamide, palmitamide, oleamide, stearamide, and erucamide contained within a

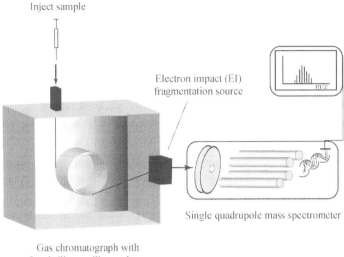

Inject sample

Electron impact (EI)
fragmentation source

Single quadrupole mass spectrometer

Gas chromatograph with
fused silica capillary column

FIGURE 8.12 Basic design and components of a gas chromatograph coupled to a single-quadrupole mass spectrometer (GC/MS) with electron ionization (EI).

biological sample extract. The internal standard method presented here consists of dissolving or diluting the biological sample within a solvent solution containing an internal standard that is then quantitated against a series of fatty acid amide standards ratioed to the internal standard (IS) methyl oleate. The separation and quantitation of the fatty acid amides were performed on a gas chromatograph coupled to a single quadrupole mass spectrometer (GC-MS) with electron ionization (EI). Figure 8.12 illustrates the basic design and components of a GC-EI/MS system. Separations were obtained using a RESTEK Rtx-5MS Crossbond 5% diphenyl–95% PDMS 0.25 mm × 30 m, 0.25 μM df column. The separation is based upon the boiling points (bp) of the analytes. A lower bp analyte will elute faster than a higher bp analyte. Raising the GC oven temperature during an analysis will help in the separation efficiency. The GC oven temperature program consisted of an initial temperature of 50°C for 2 min, ramp at 15°C/min to 300°C and hold for 10 min. The injection port temperature was 270°C and the transfer line was held at 280°C. The injection volume was 2 μL with a 1 : 5 split ratio. The electron ionization impact was at the standard 70 eV and the quadrupole was scanned for a mass range of 50–470 amu.

Figure 8.13*a* is a total ion chromatogram (TIC) of the measurement of a multicomponent fatty acid amide standard consisting of 402 μM myristamide at 13.85 min, 298 μM palmitamide at 15.13 min, 290 μM oleamide at 16.19 min, 317 μM stearamide at 16.31 min, 290 μM erucamide at 18.32 min, and an internal standard spiked in all standards and samples at a constant concentration of 100 μM methyl oleate at 14.62 min. Calibration curves were constructed for a linear standard range spanning from approximately 400–90 μM fatty acid amides. The total ion chromatograms for the serial dilutions used for construction of calibration curves are illustrated in

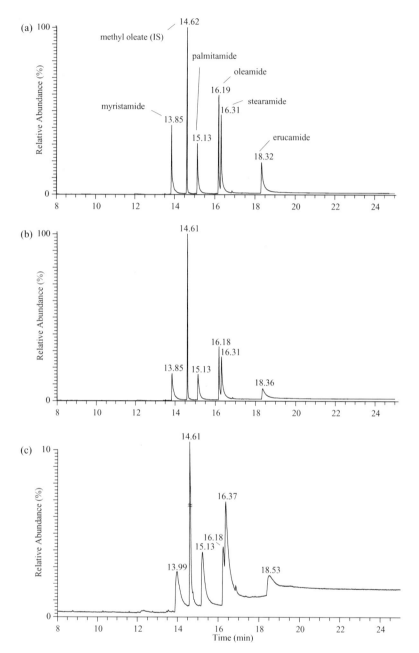

FIGURE 8.13   GC/MS TIC of (*a*) multicomponent fatty acid amide standard consisting of 4.02×10^{-4} *M* myristamide at 13.85 min, 2.98×10^{-4} *M* palmitamide at 15.13 min, 2.91×10^{-4} *M* oleamide at 16.19 min, 3.17×10^{-4} *M* stearamide at 16.31 min, 2.90×10^{-4} *M* erucamide at 18.32 min, and an internal standard spiked in all standards and samples at a constant concentration of 1×10^{-4} *M* methyl oleate at 14.62 min, (*b*) and (*c*) are serial dilutions of the multicomponent standard used for calibration curve construction.

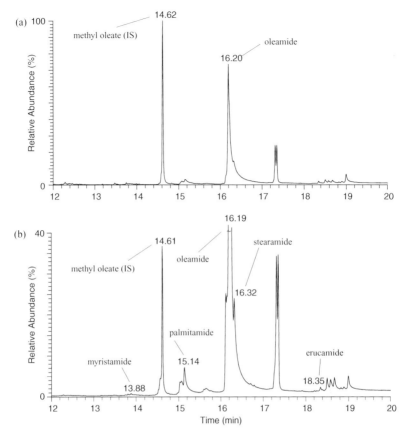

FIGURE 8.14 Total ion chromatograms of a biological sample extract at (*a*) more diluted concentration allowing the measurement of the most predominant fatty acid amide present at 16.20 min (oleamide), and (*b*) more concentrated sample containing myristamide at 13.88 min, methyl oleate (IS) at 14.61 min, palmitamide at 15.14 min, oleamide at 16.19 min, stearamide at 16.32 min, and erucamide at 18.35 min.

Figures 8.13*b* and 8.13*c*. The TICs presented in Figures 8.14*a* and 8.14*b* illustrate the process of measuring a sample at different concentrations in order to target specific analytes within the sample. This is done to measure the species within the sample at responses that are within the linear response of the associated calibration curves. Figure 8.14*a* is the total ion chromatogram of a biological sample extract at a more diluted concentration, allowing the measurement of the most predominant fatty acid amide present at 16.20 min (oleamide). Notice the internal standard methyl oleate at 14.62 min, which was spiked into the sample prior to analysis. The TIC shown in Figure 8.14*b* represents the same biological extract sample as in Figure 8.14*a*, but at a higher concentration. The biological extract sample illustrated in Figure 8.14*b* contains myristamide at 13.88 min, methyl oleate (IS) at 14.61 min, palmitamide at 15.14 min, oleamide at 16.19 min, stearamide at 16.32 min, and erucamide at

TABLE 8.1   Fatty Acid Amide Concentration in Biological Sample by GC-MS

	Myristamide (μM)	Palmitamide (μM)	Oleamide (μM)	Stearamide (μM)	Erucamide (μM)
Sample 1	1.80	0.74	3.46	1.79	0.29
Sample 2	0.55	0.92	2.24	0.55	0.29
Sample 3	1.44	0.94	7.18	1.43	0.30
Sample 4	1.51	1.56	5.79	1.51	0.54
Sample 5	1.00	0.92	4.86	1.00	0.27
Average	1.26 ± 0.49	1.02 ± 0.31	4.71 ± 1.93	1.26 ± 0.49	0.34 ± 0.11

18.35 min. Table 8.1 lists the quantitative results for the fatty acid amides determined for five biological samples. As can be seen in Table 8.1, oleamide was consistently present, with the largest quantity averaging at least four times greater quantity than the other fatty acid amide species present.

8.5 WAX ESTERS

Wax esters are lipid species that are quite common in both plants and animals and have been studied using mass spectrometric techniques. Wax ester and oxidized wax ester analysis has primarily been performed using the chromatographic coupled mass spectrometric technique of GC-MS.[16] Recently, a study was reported in the literature where the wax esters were analyzed by electrospray mass spectrometry.[17]

Figure 8.15 is the product ion spectrum of the lithium adduct of palmityl behenate, a representative fatty acid/fatty alcohol wax ester. The product ion spectrum was collected on a triple quadrupole mass spectrometer by collision-induced dissociation.

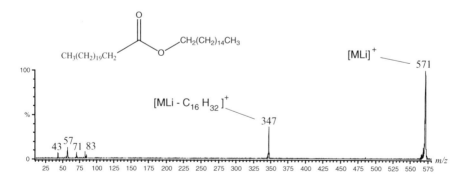

FIGURE 8.15   Collision-induced dissociation product ion spectrum of the lithium adduct of palmityl behenate at m/z 571 collected on a triple quadrupole mass spectrometer. Product ions include m/z 43, m/z 57, m/z 71, m/z 83, and m/z 347.

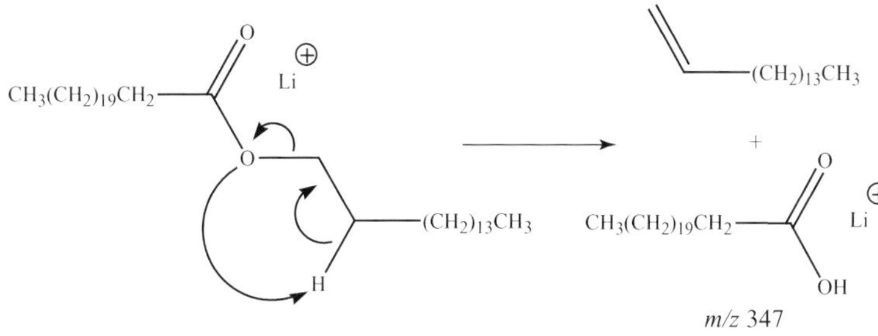

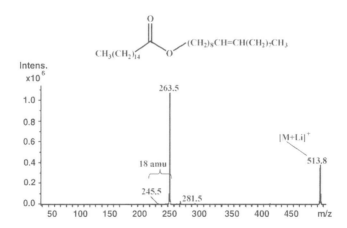

FIGURE 8.16 The *m/z* 347 product ion fragmentation pathway mechanism.

In the low mass region, only the hydrocarbon fragments are observed at *m/z* 43, *m/z* 57, *m/z* 71, *m/z* 83 (unsaturated), and *m/z* 85. In the *m/z* 350 region there is one predominant peak appearing at *m/z* 347. This represents the neutral loss of the carbon chain of the fatty alcohol [MLi − C$_{16}$H$_{32}$]$^+$, leaving the lithium adduct of the fatty acid portion of the wax ester at *m/z* 347 [C$_{22}$H$_{44}$O$_2$Li]$^+$. The fragmentation pathway mechanism for the production of the *m/z* 347 product ion is illustrated in Figure 8.16.

Figure 8.17 shows the fragmentation behavior of a standard unhydroxylated, single unsaturation containing wax ester oleyl palmitate (the saturated fatty acid palmitic acid esterified to the singly unsaturated fatty oleyl alcohol, wax ester structure also shown in figure) as the lithium cation at *m/z* 513.8 [M + Li]$^+$. The major product ion at *m/z* 263.5 is produced through neutral loss of the fatty alcohol alkyl chain forming

FIGURE 8.17 Fragmentation behavior of an unhydroxylated, single unsaturation containing wax ester oleyl palmitate (the saturated fatty acid palmitic acid esterified to the singly unsaturated fatty oleyl alcohol) as the lithium adduct at *m/z* 513.8 [M + Li]$^+$. Product ions include *m/z* 263.5 through neutral loss of the fatty alcohol alkyl chain forming the lithiated palmitic acid and *m/z* 245.5 (18 amu apart from *m/z* 263.5), a C16 ketene formed through neutral loss of the oleyl fatty alcohol.

the lithiated palmitic acid. The minor peak at m/z 245.5, which is 18 amu apart from the major product ion at m/z 263.5, is the C16 ketene formed through neutral loss of the oleyl fatty alcohol. However, this is a very minor fragmentation pathway. A mixture of wax esters such as oleyl palmitate and palmityl oleate primarily produce product ions that differ by 26 amu (spectrum not shown). In other words, a mixture of wax esters would primarily differ by C_nH_{2n} if saturated such as a difference of 56 amu or C_4H_8.

8.5.1 Oxidized Wax Esters

Oxidation of lipids occurs when oxygen attacks the double bond found in unsaturated fatty acids to form peroxide and/or epoxide linkages and can result in the destruction of the original lipid, leading to the loss of function and is believed to be a significant contributor to coronary artery disease and diabetes.[18]

8.5.2 Oxidation of Monounsaturated Wax Esters by Fenton Reaction

Fenton reactions were performed on the monounsaturated wax esters oleyl palmitate and palmityl oleate for product ion fragmentation pathway determinations. The Fenton reaction is used for *in vitro* studies of the oxidation of lipids where the addition of hydrogen peroxide in the presence of Fe^{2+} induces oxidative conditions.[19] The oxidation of oleyl palmitate and palmityl oleate were performed using the following procedure.[19] Approximately 5 nmol of lipid in 50 μL of chloroform was pipetted into a 1.5-mL Eppendorf tube and brought just to dryness with a stream of dry nitrogen. The lipid was then reconstituted into 332 μL of a 0.1 M ammonium bicarbonate buffer solution (pH 7.4). For the Fenton reaction, 15 μmol of $FeCl_2$ and 158 μmol of hydrogen peroxide (H_2O_2) were added to the reconstituted lipid in ammonium bicarbonate. The mixture was then left to react for periods of time (20, 22, and 26 h) coupled with sonication by immersing the Eppendorf tube in a sonicator bath. The oxidized and unoxidized lipid species in the reaction mixture were then extracted using a modified Folch method consisting of 2 : 1 chloroform : methanol (vol/vol).[20,21] The extracted lipid species were brought just to dryness with a stream of dry nitrogen and reconstituted in a 10 mM LiCl chloroform/methanol solution (1 : 1, vol/vol) for electrospray ionization mass spectrometric analysis.

The Fenton reaction for monounsaturated wax esters was found to produce the desired oxidation effect with extended reaction periods. The molecular oxygen addition to the monounsaturated wax esters was observed and is illustrated in the single-stage mass spectrum of Figure 8.18 where the m/z 529.7 peak represents the oxidized form of oleyl palmitate as either an epoxide[22] or hydroxylated. Figure 8.19 illustrates the production of the epoxidated lipid species using the Fenton reaction and the major product ions produced from the epoxide state, or rearrangement to the hydroxide followed by neutral loss of water forming the m/z 511 product ion. The location of the hydroxide can be either at positions 8, 9, 10, or 11 with the unsaturation either remaining at the $\Delta 9$–10 original position or either at the two adjacent positions ($\Delta 8$–9 or $\Delta 10$–11).[23]

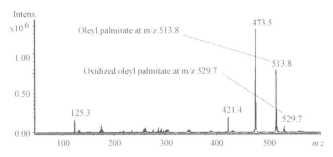

FIGURE 8.18 Results of Fenton reaction of monounsaturated wax esters. Fenton reaction of oleyl palmitate where the m/z 529.7 peak represents the oxidized form of oleyl palmitate as an epoxide.

The product ion spectrum of the oxidized wax ester at m/z 529.7 is illustrated in Figure 8.20. Important product ion peaks include the m/z 511.7 peak formed through loss of H_2O producing a diene of the wax ester $[M + Li - H_2O]^+$, the m/z 403.6 peak formed through neutral loss of C_nH_{2n} from the fatty alcohol alkyl chain producing an aldehyde at the C_9 position $[M + Li - C_9H_{18}]^+$ (diagnostic for location of double bond),[24,25] and the m/z 387.6 product ion formed through neutral loss of a 9-carbon aldehyde from the fatty alcohol substituent $[M + Li - C_9H_{18}O]^+$

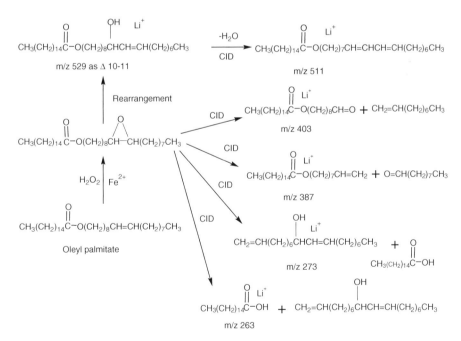

FIGURE 8.19 Major product ions produced upon collision-induced dissociation of the oxidized form of oleyl palmitate.

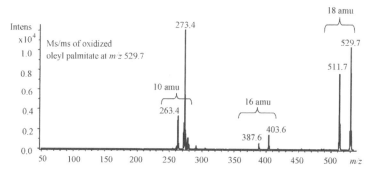

FIGURE 8.20   ESI–ion/trap product ion spectrum of the oxidized wax ester at m/z 529.7. Important product ion peaks include the m/z 511.7 peak formed through loss of H_2O, m/z 403.6 peak formed through neutral loss of C_nH_{2n} [M + Li − C_9H_{18}]$^+$ (diagnostic for location of double bond), m/z 387.6 product ion formed through neutral loss of a nine-carbon aldehyde from the fatty alcohol substituent [M + Li − $C_9H_{18}O$]$^+$ (diagnostic for location of double bond), m/z 273.4 peak formed through neutral loss of a C16:0 fatty acid forming a lithiated C18:1 fatty alcohol [M + Li − $C_{16}H_{32}O_2$]$^+$, and m/z 263.4 product ion (10 amu difference from m/z 273.4) formed through-neutral loss of the fatty alcohol alkyl chain [M + Li − $C_{18}H_{36}$]$^+$.

(diagnostic for location of double bond).[26,27] A predominant pair of product ions in the middle of the spectrum are the m/z 273.4 peak formed through neutral loss of a C16:0 fatty acid forming a lithiated C18:1 fatty alcohol [M + Li − $C_{16}H_{32}O_2$]$^+$, and the m/z 263.4 product ion (10 amu difference from m/z 273.4) formed through neutral loss of the fatty alcohol alkyl chain [M + Li − $C_{18}H_{36}$]$^+$. The m/z 263.4 product ion is the predominant pathway observed for the unoxidized oleyl palmitate wax ester where a loss analogous to m/z 273.4 is not observed. The presence of the epoxide is thus apparently allowing these two pathways to take place with a much higher efficiency for the m/z 273.4 pathway as compared to the m/z 263.4 pathway. In comparison, Figure 8.21 is the product ion spectrum of the oxidized epoxide of palmityl oleate where the unsaturation is now located on the fatty acid substituent of

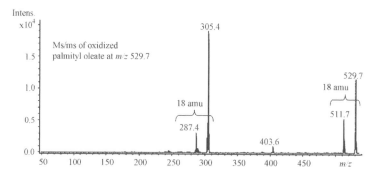

FIGURE 8.21   Product ion spectrum of the oxidized epoxide of palmityl oleate.

the wax ester as opposed to the fatty alcohol substituent as is the case with the oleyl palmitate wax ester of Figure 8.20. The same loss of H_2O at m/z 511.7 and the loss of C_9H_{18} at m/z 403.6 indicative of the location of the double bond are also observed as discussed above for the oxidized epoxide of oleyl palmitate. The major product ions in the middle of the spectrum, however, differ from that of oleyl palmitate due to the location of the epoxide. The m/z 305.4 product ion is formed through neutral loss of the fatty alcohol chain, which is also the major product ion formed with unoxidized wax esters. The product ion at m/z 287.4 is formed through neutral loss of the C14:0 fatty alcohol, producing a C18 ketene with a difference of 18 amu from the m/z 305.4 product ion. These types of losses can be used in diagnostic identification of wax ester unknowns, whether saturated or unsaturated, oxidized or unoxidized. Low levels of the peroxides and diols of the wax esters were observed, and the product ion spectra were similar to the epoxides but slightly more complex. Loss of H_2O is observed along with double-bond location, and the major substituent losses differing by either 10 amu for oleyl palmitate or 18 amu for palmityl oleate were also contained in the product ion spectra. A number of biological sample extracted lipids observed in the electrospray ionization mass spectra produce product ion spectra that indicated that they are oxidized lipid species by a characteristic neutral loss of either 16 or 18 amu from the precursor ion. Water loss from the precursor often indicates an oxidized lipid species.[22] Often in lipid work researchers have observed the 16 amu and/or 18 amu neutral loss associated with oxidation. As an example, in the study of Reis et al.[28] of linoleic acid oxidation, the inspection of their product ion spectra illustrated in Figures 8.22 and 8.23 all show spectra containing a neutral loss of 18 amu associated with water loss. Also included in the product ion spectra of Figures 8.22*d* and 8.23*d* is a peak derived through a neutral loss of 16 amu. The loss of atomic oxygen at 16 amu is probably attributable as a loss from an epoxide that reestablishes the double bond.

8.6 STEROLS

In metabolomic studies often an unknown biomolecule is observed in a system under study by mass spectrometry. The structural elucidation and subsequent identification of an unknown small biomolecule is a process that often involves a number of steps. Figure 8.24 is a single-stage ESI Q-TOF mass spectrum of a biological extract containing unknown biomolecules in a solution of 1:1 chloroform/methanol with 10 mM LiCl. The spectrum contains primarily lithium adducts of the lipid species in the biological extract as $[M + Li]^+$. The predominant biomolecule observed in the biological extract at m/z 473.5 (also observed at m/z 489 as the sodiated species) will be the subject of study concerning its identification in the following example of identifying an unknown biomolecule.

Figure 8.25 is an ESI Q-TOF single-stage negative ion mode mass spectrum of the same biological extract as illustrated in Figure 8.24 in a 1:1 chloroform/methanol solution containing 1 mM ammonium acetate. The spectrum is comprised of deprotonated biomolecule species as $[M - H]^-$. In the lower molecular weight region of

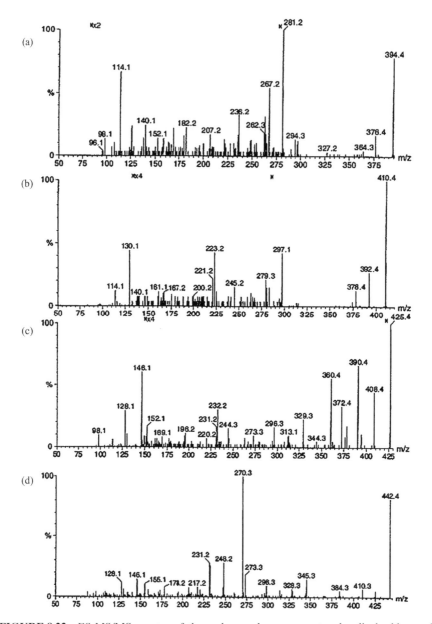

FIGURE 8.22 ES-MS/MS spectra of the carbon and oxygen-centered radical adducts of linoleic acid at (a) m/z 394, (b) m/z 410, (c) m/z 426, and (d) m/z 442. (Reprinted with permission. Reis, A., Domingues, R. M., Amado, F. M. L., Ferrer-Carreia, A. J. V., and Domingues, P. Detection and characterization by mass spectrometry of radical adducts produced by linoleic acid oxidation. *J. Am. Soc. Mass Spectrom.* 2003, *14*, 1250–1261. Copyright Elsevier 2003.)

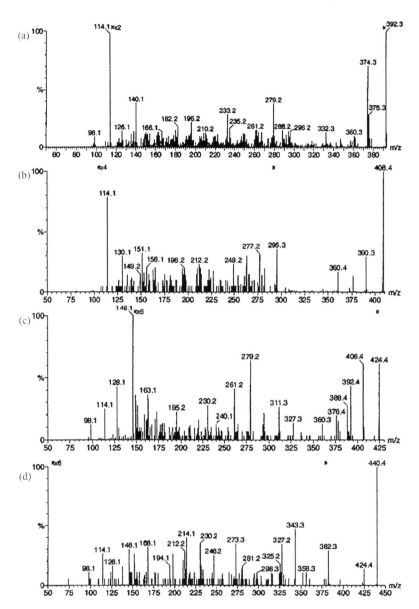

FIGURE 8.23 ES-MS/MS spectra of the carbon and oxygen-centered radical adducts of linoleic acid of (*a*) *m/z* 392, (*b*) *m/z* 408, (*c*) *m/z* 424, and (*d*) *m/z* 440. (Reprinted with permission. Reis, A., Domingues, R. M., Amado, F. M. L., Ferrer-Carreia, A. J. V., Domingues, P. Detection and characterization by mass spectrometry of radical adducts produced by linoleic acid oxidation. *J. Am. Soc. Mass Spectrom.* 2003, *14*, 1250–1261. Copyright Elsevier 2003.)

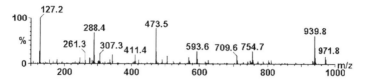

FIGURE 8.24 Single-stage ESI Q-TOF mass spectrum of a biological extract in a solution of 1 : 1 chloroform/methanol with 10 mM LiCl containing primarily lithium adducts as [M + Li]$^+$. M/z 473.5 (also observed at m/z 489 as the sodiated species) is the subject of study concerning its identification.

the spectrum there are four peaks observed representing the free fatty acids myristic acid at m/z 227.3, palmitic acid at m/z 255.4, oleic acid at m/z 281.4, and stearic acid at m/z 283.4. The peak at m/z 465.5 is suspected to be the same biomolecule as that of m/z 473.5 observed in positive ion mode. The unknown species was suspected to be cholesteryl phosphate. To verify this, the cholesteryl phosphate species was synthesized, characterized, and compared to the unknown for confirmation of its identification.

8.6.1 Synthesis of Cholesteryl Phosphate

The first step was to synthesize the cholesteryl phosphate species. Cholesteryl phosphate (CP) was synthesized following the methodology outlined by Sedaghat et al.[29] and Gotoh et al.[30] where the major modification consisted of changing the methodology from a gram scale to a milligram scale. Briefly, 0.67 g of cholesterol was dissolved in 3.5 mL of dry pyridine. The solution was then cooled in ice water and phosphorus oxychloride (0.175 mL dissolved in 3.5 mL acetone) was added slowly with stirring. A precipitate was formed immediately, and the solution was allowed to cool for 10 min in the ice water. The precipitate (cholesteryl phosphochloridate) was filtered and washed with 15 mL cold dry acetone. The solid was dissolved in tetrahydrofuran (THF) and refluxed for 2 h after adding aqueous NaOH (2.1 equivalent). The resulting cholesteryl phosphate precipitate was filtered and washed with dry cold acetone and dried in a Vacufuge (Eppendorf). High-resolution mass

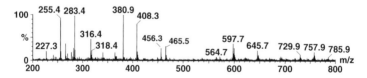

FIGURE 8.25 ESI/Q-TOF single-stage mass spectrum of the biological extract collected in negative ion mode in a 1 : 1 chloroform/methanol solution containing 1 mM ammonium acetate. Spectrum is comprised of deprotonated lipid species as [M – H]$^-$ illustrating the free fatty acids myristic acid at m/z 227.3, palmitic acid at m/z 255.4, oleic acid at m/z 281.4, and stearic acid at m/z 283.4, and an unknown species at m/z 465.5.

measurements (including both the sodium adduct and the lithium adduct) of the synthesized cholesteryl phosphate and proton nuclear magnetic resonance (^{1}H-NMR) results are presented for verification of its identification in the next two sections.

8.6.2 Single-Stage and High-Resolution Mass Spectrometry

The synthesized cholesteryl phosphate (CP) was next characterized using high-resolution mass spectrometry. Figure 8.26a is a single-stage mass spectrum in positive ion mode acquired on the Q-TOF mass spectrometer of the synthesized cholesteryl phosphate in an acidic solution containing NaI (1% acetic acid in 1 : 1 CHCl$_3$/MeOH).

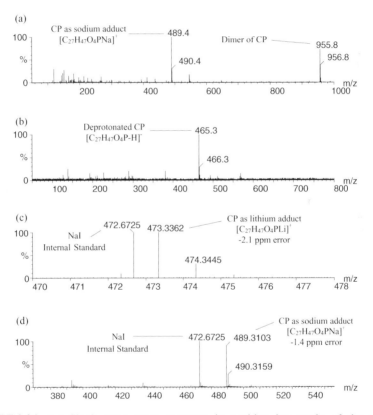

FIGURE 8.26 (*a*) Single-stage mass spectrum in positive ion mode of the synthesized cholesteryl phosphate in an acidic solution containing NaI (1% acetic acid in 1 : 1 CHCl$_3$/MeOH). (*b*) Single-stage mass spectrum in negative ion mode of the synthesized cholesteryl phosphate in a 10 mM ammonium hydroxide 1 : 1 CHCl$_3$/MeOH solution. (*c*) High-resolution mass measurement of the synthesized CP as the lithium adduct [C$_{27}$H$_{47}$O$_4$PLi]$^+$ at m/z 473.3362 (calculated -2.1 ppm error). (*d*) High-resolution mass measurement of the synthesized CP as the sodium adduct [C$_{27}$H$_{47}$O$_4$PNa]$^+$ at m/z 489.3103 (calculated -1.4 ppm error).

The two major peaks in the spectrum are at m/z 489.4 for cholesteryl phosphate as the sodium adduct and m/z 955.8 for a cholesteryl phosphate dimer. Figure 8.26b is a single-stage mass spectrum in negative ion mode acquired on the Q-TOF mass spectrometer of the synthesized cholesteryl phosphate in a 10 mM ammonium hydroxide 1 : 1 CHCl$_3$/MeOH solution. The major peak in the spectrum is at m/z 465.3 for the deprotonated form of cholesteryl phosphate as $[C_{27}H_{47}O_4P-H]^-$. Figure 8.26c is the high-resolution mass measurement of the synthesized CP as the lithium adduct $[C_{27}H_{47}O_4PLi]^+$ at m/z 473.3362 (m/z 472.6725 is the internal standard NaI peak used for the exact mass measurement). The theoretical mass of CP as the lithium adduct is m/z 473.3372, equating to a calculated -2.1-ppm error. Figure 8.26d is the high-resolution mass measurement of the synthesized CP as the sodium adduct $[C_{27}H_{47}O_4PNa]^+$ at m/z 489.3103 (m/z 472.6725 is the internal standard NaI peak used for the exact mass measurement). The theoretical mass of CP as the sodium adduct is m/z 489.3110, equating to a calculated -1.4-ppm error.

8.6.3 Proton Nuclear Magnetic Resonance

The next step was to characterize the synthesized cholesteryl phosphate using NMR to confirm its synthesis. Single-stage proton (^{1}H-NMR) spectra were obtained on a Bruker DRX 800 (800 MHz) nuclear magnetic resonance spectrometer (Bruker, Bremen, Germany). The spectra of the synthesized cholesteryl phosphate were obtained at 800 MHz (shifts in ppm) with deuterated chloroform (CDCl$_3$) used as solvent. Figure 8.27 contains the structure of cholesteryl phosphate with the carbon atoms numbered and the experimental proton NMR results of the synthesized cholesteryl phosphate.

8.6.4 Theoretical NMR Spectroscopy

The NMR proton shifts were calculated using ChemDraw (Cambridge Soft Corporation, Cambridge, Massachusetts). The theoretical proton shifts for the structure are illustrated in Figure 8.28, where Figure 8.28a represents the structure of cholesteryl phosphate with the proton shifts labeled, and Figure 8.28b is the theoretical NMR spectrum generated by ChemDraw using the ChemNMR feature. Table 8.2 illustrates a comparison of the theoretically generated proton NMR shifts versus the experimentally obtained proton NMR shifts. Good agreement was observed between the theoretical and the experimental proton NMR shifts.

8.6.5 Structure Elucidation

Product ion mass spectra were collected of the unknown biomolecule to compare to the product ion spectra of the synthesized cholesteryl phosphate. Figure 8.29 is a product ion spectrum of the m/z 473.5 biomolecule, as $[M + Li]^+$, in the biological extract collected on a Q-TOF mass spectrometer in a solution of 1 : 1 chloroform/methanol with 10 mM LiCl. The major products produced and the associated fragmentation pathways are illustrated in Figure 8.30. The m/z 255.3 product ion, $[C_{11}H_{21}O_4PLi]^+$,

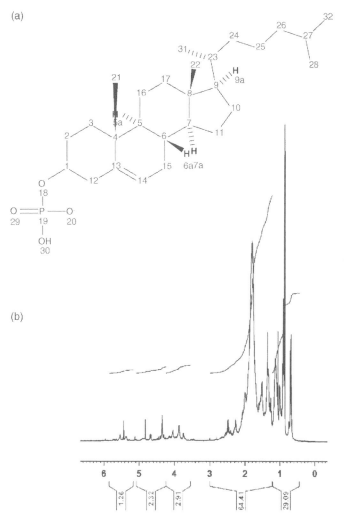

FIGURE 8.27 (*a*) Structure of cholesteryl phosphate with the carbon atoms numbered. (*b*) Experimental proton NMR results of the synthesized cholesteryl phosphate.

is formed through cleavage of the cholesteryl backbone as illustrated in Figure 8.30, $[C_{27}H_{47}O_4P + Li - C_{16}H_{26}]^+$. The m/z 237.2 product ion is derived through neutral water loss from the m/z 255.3 product ion, $[C_{11}H_{21}O_4PLi - H_2O]^+$. The m/z 293.3 and m/z 311.3 product ions, which differ by 18 amu (H_2O), are formed in an analogous fashion with their respective structures illustrated in Figure 8.30. The m/z 311.3 product ion, $[C_{15}H_{29}O_4PLi]^+$, is formed through cleavage of the cholesteryl backbone as $[C_{27}H_{47}O_4P + Li - C_{12}H_{18}]^+$. The m/z 293.3 product ion is derived through neutral water loss from the m/z 311.3 product ion, $[C_{15}H_{29}O_4PLi - H_2O]^+$.

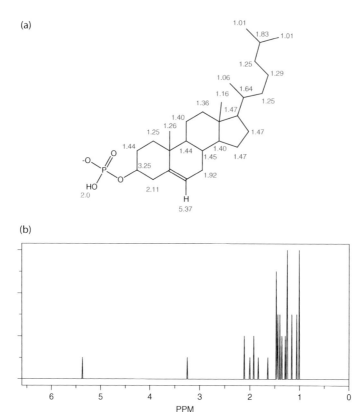

FIGURE 8.28 (*a*) Structure of cholesteryl phosphate with the proton shifts labeled. (*b*) Theoretical NMR spectrum of cholesteryl phosphate generated by ChemDraw using the ChemNMR feature.

The m/z 473.4 lipid species has been identified as the lithium adduct of cholesteryl phosphate with the ionic formula of $[C_{27}H_{47}O_4P + Li]^+$ and a theoretical mass of 473.3372 Da. High-resolution mass measurements of the m/z 473.4 lipid species in meibum was determined at m/z 473.3361 equating to a mass error of -2.3 ppm. The m/z 489.3 lipid species has been identified as the sodium adduct of cholesteryl phosphate with the ionic formula of $[C_{27}H_{47}O_4P + Na]^+$ and a theoretical mass of 489.3110 Da. High-resolution mass measurements of the m/z 489.3 lipid species in meibum was determined at m/z 489.3099 equating to a mass error of -2.2 ppm.

Figure 8.31 illustrates the product ion spectrum of the m/z 465.5 species observed in the biological extract obtained using the Q-TOF mass spectrometer with an electrospray ionization source collected in negative ion mode. In Figure 8.31 there are two major product ion peaks observed at m/z 79.0 and 97.0. The m/z 465.5 lipid species has been identified as the deprotonated form of cholesteryl phosphate

TABLE 8.2 Proton NMR Spectral Results

Node	Theoretical Shift (ppm)	Experimental Shift (ppm)
CH3(18)		0.68–0.73 (0.69)[29]
CH3(26), CH3(27)		0.88–0.89 (0.86—0.87)[29]
CH3(21)		0.92 (0.91)[29]
CH3	1.01	1.01 (1.0)[29]
CH3	1.01	1.02
CH3	1.06	1.08
CH3	1.16	1.15–1.18
CH2	1.25	
CH2	1.25	
CH2	1.25	
CH3	1.26	1.275
CH2	1.29	1.29
CH2	1.36	1.36
CH	1.40	1.40
CH2	1.40	1.40
CH2	1.44	
CH	1.44	
CH	1.45	
CH2	1.47	1.46
CH2	1.47	1.47
CH	1.47	1.48
CH	1.64	1.64
CH	1.83	1.83
CH2	1.92	1.90
OH	2.0	2.01
CH2	2.11	2.11
CH	3.25	
C—C—H	5.37	5.46 (5.3)[29]

with the ionic formula $[M - H]^-$ of $[C_{27}H_{47}O_4P - H]^-$ and a theoretical mass of 465.3134 Da. High-resolution mass measurements of the m/z 465.5 lipid species in the biological extract was determined at m/z 465.3140, equating to a mass error of 1.3 ppm. The structure of the m/z 465.5 lipid species is illustrated in Figure 8.32 along with proposed fragmentation pathways that describe the production of the m/z 97.0 and m/z 79.0 product ions. Both the m/z 97.0 and 79.0 product ions represent the H_2PO_4 phosphate portion of the headgroup. This same H_2PO_4 phosphate product ion at m/z 97.0 is also observed in the product ion spectra of the phosphorylated lipid standards 1-palmitoyl-2-oleoyl-*sn*-glycero-3-phosphate (16 : 0—18 : 1 PA or POPA, spectrum not shown), and 1-palmitoyl-2-oleoyl-*sn*-glycero-3-[phospho-*rac*-(1-glycerol)] (16 : 0—18 : 1 PG or POPG, spectrum not shown). The product ion spectrum of the synthesized cholesteryl phosphate is identical to the m/z 465.5 species in the biological extract and confirms its identification as cholesteryl phosphate.

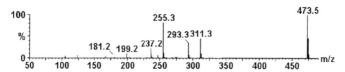

FIGURE 8.29   Product ion spectrum of m/z 473.7 identified as cholesteryl phosphate. Product ion spectrum of synthesized cholesteryl phosphate as the lithium adduct $[C_{27}H_{47}O_4P + Li]^+$ at m/z 473.4. The spectrum is identical to the product ion spectrum of the unknown m/z 473.5 species observed in biological extract.

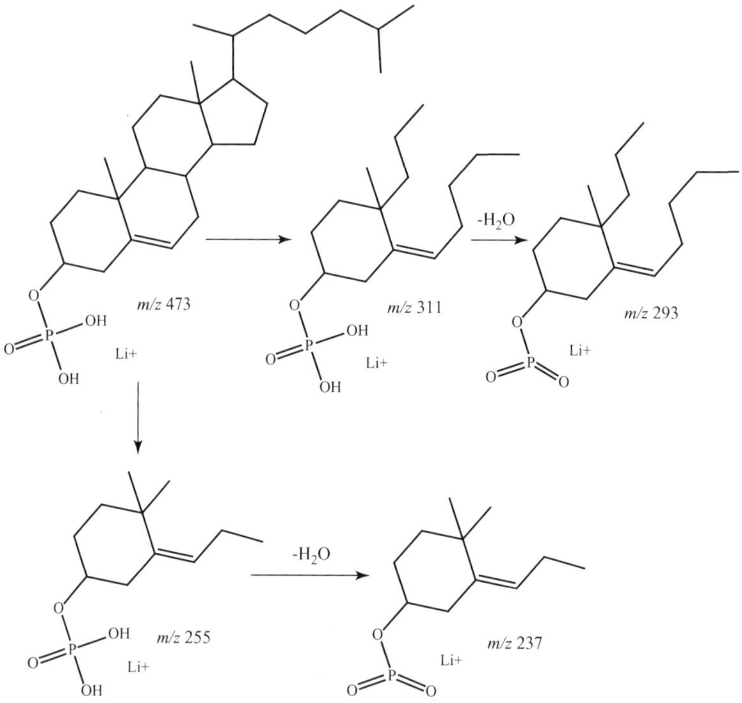

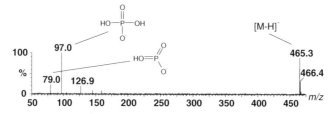

FIGURE 8.30   Structure of cholesteryl phosphate and fragmentation pathways from product ion spectrum in Figure 8.29.

FIGURE 8.31   Q-TOF MS negative ion mode product ion spectrum of the m/z 465.5 species in the biological extract identified as the deprotonated form of cholesteryl phosphate with the ionic, deprotonated formula $[M - H]^-$ of $[C_{27}H_{47}O_4P - H]^-$.

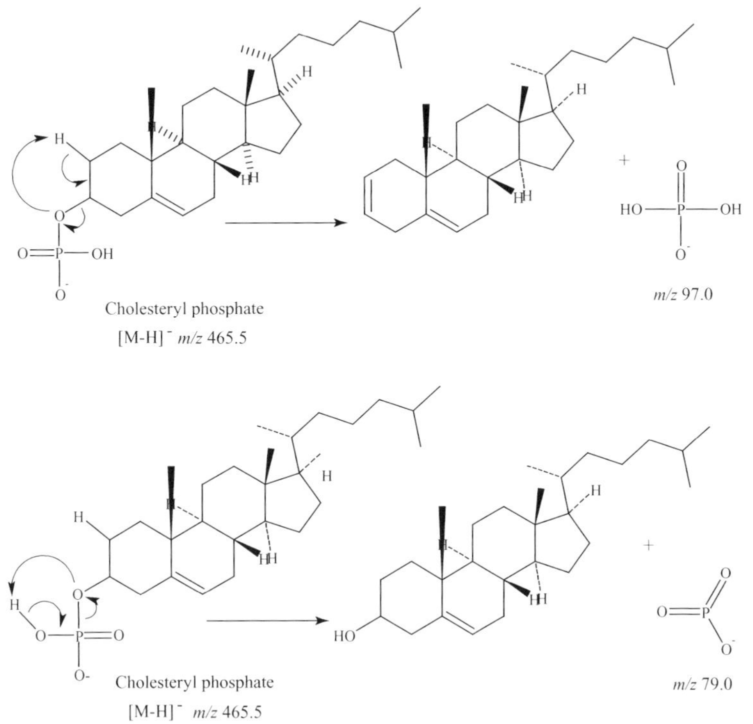

FIGURE 8.32 Structure of the *m/z* 465.5 lipid species and fragmentation pathways.

8.7 ACYLGLYCEROLS

Lipids are very important biomolecules that are found in all living species. These include nonpolar lipids such as the acylglycerols and the more polar phosphory-lated lipids. Of the nonpolar lipids, triacylglycerols are thought to be a storage form of energy in the cells,[31] while diacylglycerols are of special importance for their role as physiological activators of protein kinase C (PKC).[32,33] The acyl-glycerols are a subclass of lipids and are comprised of mono-, di-, and triacyl-glycerols. Various approaches have been reported for the mass spectrometric anal-ysis of acylglycerols such as derivative gas chromatography–electron ionization mass spectrometry GC-EI/MS,[34–36] positive chemical ionization,[34] negative chemi-cal ionization[37–39] (NCI) of chloride adducts, ES-MS/MS[40,41] including ES-MS/MS of ammonium adducts,[42,43] MALDI-TOF,[44,45] fast atom bombardment (FAB),[46] at-mospheric pressure chemical ionization (APCI) mass spectrometry,[42,47] and APCI liquid chromatography/mass spectrometry[26,27,48] (LC/MS). The nonpolar lipids, in general, have also been studied by ESI-MS/MS and include the structural character-ization of acylglycerols,[49] direct qualitative analysis,[50] regiospecific determination of triacylglycerols,[51] analysis of triacylglycerol oxidation products,[52] quantitative analysis,[41] and acylglycerol mixtures.[53] Using ratios of diacylglycerol fragment ions versus precursor ions for quantifying triacylglycerols and fragmentation schemes has been reported for single-stage APCI/MS experiments.[54–57] Even in acidified solutions saturated acylglycerols do not readily form protonated molecules; therefore, alkali metal salt (Na^+) and ammonium acetate (NH_4^+) have been used as additives in chloroform:methanol for both ESI and APCI LC/MS.

In the identification of lipid fractions in biological samples using mass spectrometry, lithium, (Li^+) is sometimes chosen as the alkali metal for adduct formation for wax esters, the nonpolar lipids such as the acylglycerols, and for the more polar phosphorylated lipids, which have previously been reported.[58] As examples of the use of lithium adducts for lipid analysis, it has been reported for FAB-MS/MS, and ES/MS studies of oligosaccharides,[59, 60] for ionization of aryl 1,2-diols by fast atom bombardment,[46] for FAB ionization and structural identification of fatty acids,[61] and for the ES ionization and structural determination of triacylglycerols.[40, 62] In this section we will take a detailed look at the major fragments produced from ES-MS/MS studies, employing collision-induced dissociation (CID) of lithium adducts of monoacylglycerols, diacylglycerols, and triacylglycerols.

8.7.1 Analysis of Monopentadecanoin

We will first have a look at the simplest form of an acylglycerol, the monoacylglycerol. This lipid biomolecule has one fatty acyl substituent on the glycerol backbone. Figure 8.33 is the collision-induced dissociation product ion spectrum of lithiated monopentadecanoin at m/z 323 as $[M + Li]^+$. At the lower molecular weight region there are observed three peaks characteristic of the fragmentation of lithiated monoglycerols. The m/z 99 major product ion is the glycerol backbone derived from the neutral loss of the C15:0 fatty acyl chain as ketene from the parent ion $[MLi - C_{15}H_{28}O]^+$. The next two product ion peaks observed in the spectrum, m/z 81 and m/z 63, are the subsequent loss of one and then two H_2O molecules from the glycerol backbone. The associated product ion fragmentation pathways are illustrated in Figure 8.34.

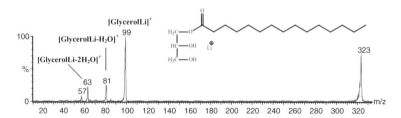

FIGURE 8.33   Positive ion mode ESI-MS/MS product ion spectra of the lithium adducts of monopentadecanoin at m/z 323 as $[MLi]^+$.

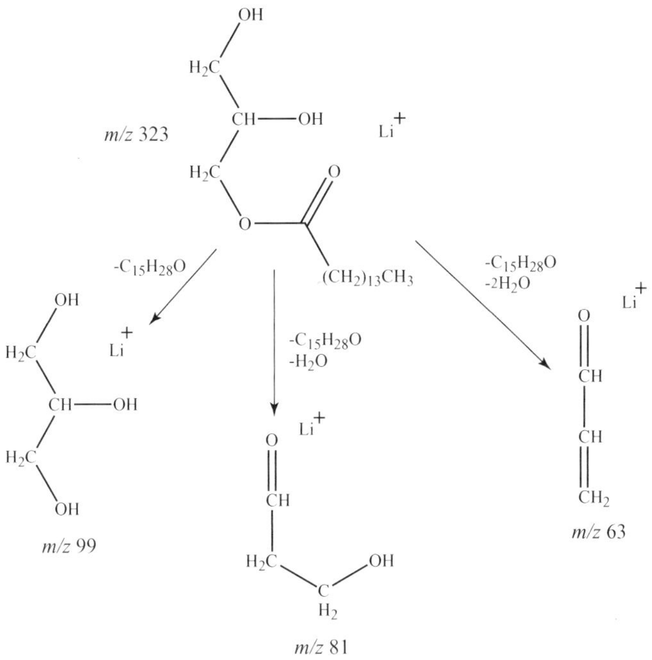

FIGURE 8.34 Major fragment ions produced from the collision-induced dissociation of monopentadecanoin m/z 323 $[MLi]^+$. Ion at m/z 99 is the glycerol backbone derived from the neutral loss of the C15:0 fatty acyl chain as a ketene from the precursor ion $[MLi - C_{15}H_{28}O]^+$. The m/z 81 and m/z 63 are the subsequent loss of one, and then two H_2O molecules from the lithiated glycerol backbone.

8.7.2 Analysis of 1,3-Dipentadecanoin

The diacylglycerol 1,3-dipentadecanoin is a disubstituted glycerol where the glycerol backbone contains two fatty acyl chains. The product ion spectrum of 1,3-dipentadecanoin is illustrated in Figure 8.35. Similar to the monoacylglycerol presented above, the 1,3-diacylglycerol also contains hydrocarbon fragment ions, two water losses from the glycerol backbone at m/z 63, and also the glycerol backbone at m/z 99 and m/z 81. Also included in the product ion spectrum is lithiated pentadecanoic acid at m/z 249 $[C_{15}H_{30}O_2Li]^+$, and a fragment ion that has undergone neutral loss of C15:0 fatty acyl chain as a ketene at m/z 323 $[MLi - C_{15}H_{28}O]^+$. There are also three losses that involve the fatty acyl substituent at m/z 289 from the consecutive neutral loss of C15:0 fatty acyl ketene, followed by loss of H_2O $[MLi - C_{15}H_{28}O - H_2O]^+$, at m/z 299 for the neutral loss of C15:0 lithium fatty acetate $[MLi - C_{15}H_{29}O_2Li]^+$, and the neutral loss of C15:0 fatty acid at m/z 305 $[MLi - C_{15}H_{30}O_2]^+$. The loss of NH_4OH, then ketene, has been observed and reported by LC-APCI/MS,[35] which is analogous to the m/z 299 peak formed through neutral loss of C15:0 lithium fatty acetate if the process occurs in two steps, involving initial

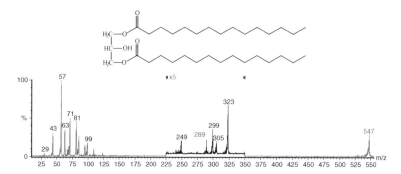

FIGURE 8.35   Positive ion mode ESI-MS/MS product ion spectra of the lithium adducts of 1,3-dipentadecanoin m/z 547.

loss of C:15 fatty ketene, followed by loss of LiOH. An experiment was performed where the loss of the C:15 fatty ketene from the m/z 547 precursor was produced in the source using high cone voltage (i.e., "in-source" or "up-front" collision-induced dissociation, CID) to distinguish which fragment pathway was responsible for the m/z 299 product ion. The m/z 323 was isolated by the first quadrupole and then subjected to CID in the central hexapole. The product ions generated were scanned by the third quadrupole for the loss of LiOH at m/z 299. A small peak was observed at m/z 299, indicating that this fragmentation pathway is possible. The proposed loss of a fatty ketene followed by loss of LiOH is much less favored than the direct neutral loss of the C:15 lithium fatty acetate from the m/z 547 precursor. Figure 8.36 illustrates fragmentation pathways concerning the major and minor losses associated with CID of the lithium adduct of 1,3-dipentadecanoin.

8.7.3   Analysis of Triheptadecanoin

Figure 8.37 is the product ion spectrum of the lithiated adduct of triheptadecanoin as $[M + Li]^+$ at m/z 855. As was observed with the mono- and diacylglycerols, the spectrum contains hydrocarbon fragment ions at m/z 43, m/z 57, m/z 71, and m/z 85. These product ions are derived from σ-bond cleavage of the fatty acid hydrocarbon chains. Also included are the product ions at m/z 99 for the lithium adduct of the glycerol backbone, and at m/z 81 derived from water loss from this glycerol backbone. The product ion spectrum includes peaks that are indicative of the triacylglycerol at m/z 253 for the C17:0 acylium ion $[C_{17}H_{33}O]^+$, at m/z 277 for the lithium adduct of heptadecanoic acid $[C_{17}H_{34}O_2Li]^+$, the m/z 317 product ion from the neutral loss of heptadecanoic acid, then the neutral loss of a C17:1 α-β unsaturated fatty acid, to give $[MLi - C_{17}H_{34}O_2 - C_{17}H_{32}O_2]^+$, at m/z 579 for the neutral loss of lithiated heptadecanoate $[MLi - C_{17}H_{33}O_2Li]^+$, and the precursor ion minus the neutral loss of heptadecanoic acid at m/z 585 $[MLi - C_{17}H_{34}O_2]^+$. The product ions of m/z 317 and m/z 585, which both involve the neutral loss of a fatty acid, have not been observed

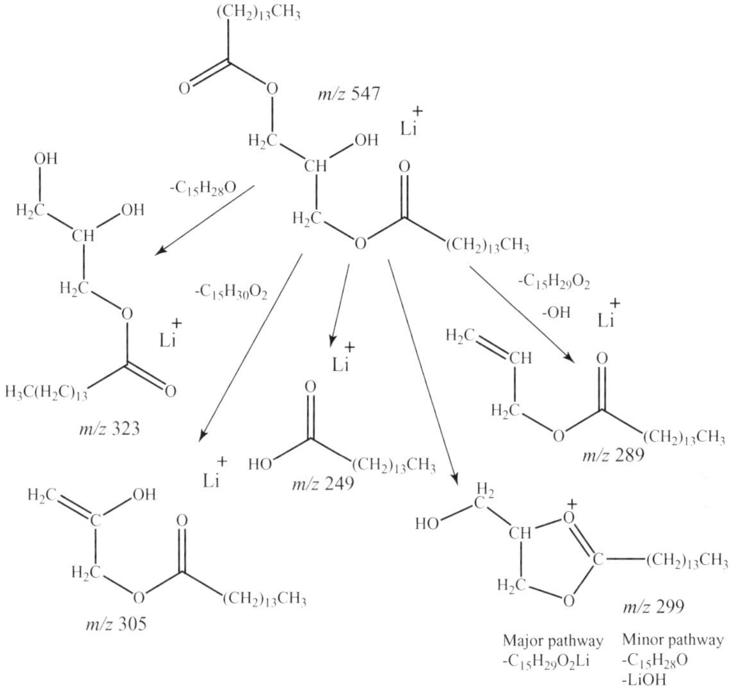

FIGURE 8.36 Listing of the collision-induced dissociation of 1,3-dipentadecanoin at m/z 547 [MLi]$^+$ major fragment ions generated. The major path of neutral loss of C15:0 lithium fatty acetate at m/z 299 [MLi – $C_{15}H_{29}O_2Li$]$^+$. The minor path of the subsequent loss of C15:0 fatty acyl chain as ketene from the precursor ion followed by LiOH at m/z 299 [MLi – $C_{15}H_{28}O$ – LiOH]$^+$. The precursor ion peak minus the neutral loss of C15:0 fatty acyl chain as ketene at m/z 323 [MLi – $C_{15}H_{28}O$]$^+$. The neutral loss of C15:0 fatty acid at m/z 305 [MLi – $C_{15}H_{30}O_2$]$^+$, the consecutive loss of C15 fatty acyl ketene, followed by loss of H_2O at m/z 289 [MLi – $C_{15}H_{28}O$ – H_2O]$^+$, and pentadecanoic acid at m/z 249 [$C_{15}H_{30}O_2Li$]$^+$.

in ESI tandem mass spectrometry studies with ammonium adducts,[45] but the m/z 579 product ion from the neutral loss of a catonized fatty acetate has. Figure 8.38 illustrates fragmentation pathways concerning the major and minor losses associated with CID of the lithium adduct of triheptadecanoin.

8.8 ESI/MS OF PHOSPHORYLATED LIPIDS

Phosphorylated lipids are also very important biomolecules that are found in all living species. Of the phosphorylated lipids that are polar lipids, those containing phosphoryl headgroups, such as the phosphatidylcholines (PC), the phosphatidylethanolamines (PE), and the phosphatidylserines (PS) (see Fig. 8.39 for structures), constitute the

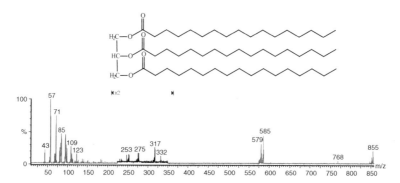

FIGURE 8.37   Positive ion mode ESI-MS/MS product ion spectra of the lithium adducts as [M + Li]$^+$ of triheptadecanoin at m/z 855.

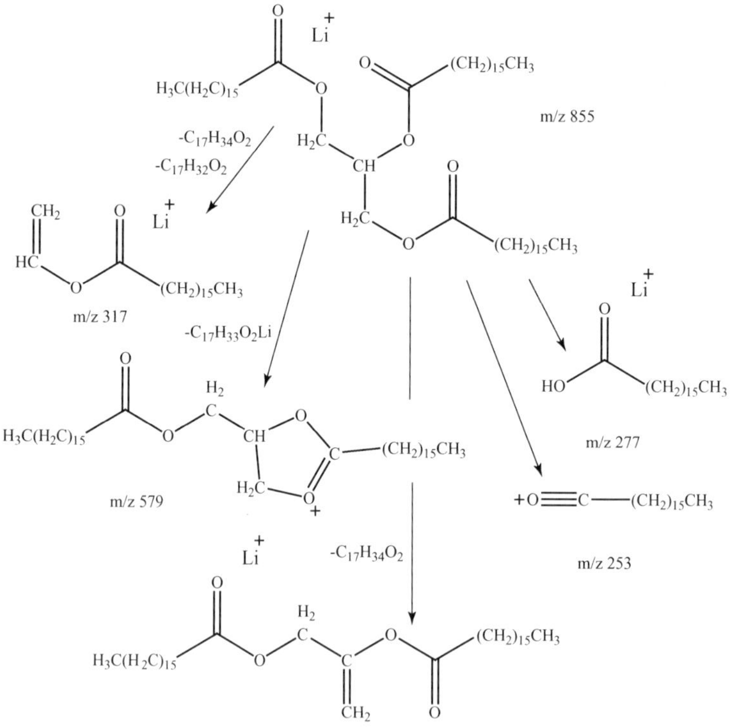

FIGURE 8.38   Major fragment ions produced from the collision-induced dissociation of triheptadecanoin at m/z 855 [MLi]$^+$. The C17:0 acylium ion at m/z 253 [C$_{17}$H$_{33}$O]$^+$, the lithium adduct of heptadecanoic acid at m/z 277 [C$_{17}$H$_{34}$O$_2$Li]$^+$, the neutral loss of heptadecanoic acid, then the neutral loss of a C17:1 α-β unsaturated fatty acid, to give the m/z 317 product ion [MLi − C$_{17}$H$_{34}$O$_2$ − C$_{17}$H$_{32}$O$_2$]$^+$, the precursor ion minus lithiated heptadecanoate at m/z 579 [MLi − C$_{17}$H$_{33}$O$_2$Li]$^+$, and the precursor ion minus the neutral loss of heptadecanoic acid at m/z 585 [MLi − C$_{17}$H$_{34}$O$_2$]$^+$.

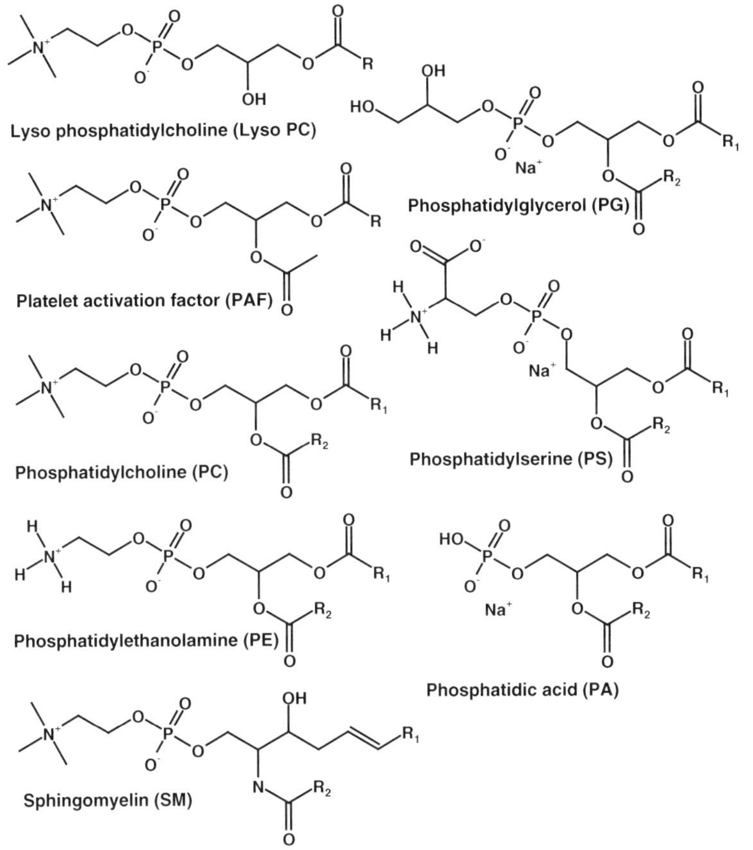

FIGURE 8.39 Structures of typical phosphorylated lipids observed in meibum classified as: (*left*) zwitterionic including lyso phosphatidylcholine (lyso PC), phosphatidylcholine (PC), phosphatidylethanolamine (PE), and sphingomyelin (SM), and (*right*) anionic including phosphatidylglycerol (PG), phosphatidylserine (PS), and phosphatidic acid (PA).

bilayer components of biological membranes, determine the physical properties of these membranes, and directly participate in membrane protein regulation and function. The phosphatidylcholines and phosphatidylethanolamines are zwitterionic lipid species that contain a polar headgroup, although the lipid is neutral overall. Other types of phosphorylated lipids are the sphingomyelins and the glycosphingolipids, which have been reported to be involved in different biological processes such as growth and morphogenesis of cells.[63] It is postulated that sphingomyelins, similar to the other zwitterionic phosphaorylated lipids, are directly involved in the anchoring of the tear lipid layer to the aqueous portion of the tear film. The polar headgroup of sphingomyelins may be associated with the aqueous layer of the tear film while the nonpolar portion of the lipid (the fatty acyl chains of the substituents) are probably associated with the other nonpolar lipids, creating an interface between the two. Phosphatidylserine is a membrane phospholipid in prokaryotic and eukaryotic cells and is found most abundantly in brain tissue. Phosphatidylserine is also a precursor biological molecule in the biosynthesis of phosphatidylethanolamine through a reaction that is catalyzed by PS decarboxylase. At physiological pH (7.38) phosphatidylserine carries a full negative charge and is often referred to as an anionic phosphorylated lipid. Thus, analysis of this special class of biomolecules has been of great interest to medical researchers, biologists, and chemists.

Recent analyses of phospholipids have employed HPLC,[64] MEKC,[64,65] CE-ES/MS,[66] ES-MS/MS,[67–74] and to a lesser extent MALDI-FTICR[75] and MALDI-TOF.[76–80] The mass spectrometric analyses of nonvolatile lipids has been reviewed by Murphy et al.,[81] and it is pointed out that the advent of soft ionization techniques such as ES/MS and MALDI/MS, has obviated the need for derivatization of lipids. Biological samples are complex mixtures of a wide variety of compounds that can have a negative influence on the ability to identify and quantify lipids by mass spectrometry; effects include signal suppression and interferences from isomeric and isobaric compounds.

8.8.1 Electrospray Ionization Behavior of Phosphorylated Lipids

The phosphorylated lipids will ionize differently depending on the ion mode that the analysis is performed in and upon the ionic species (Na^+, H^+, NH_4^+, and Li^+ in positive ion mode, and $-H$ or Cl^- in negative ion mode) that are present during the ionization process. Figure 8.40 illustrates the structure of the phosphatidic acid 1-palmitoyl-2-oleyl-*sn*-glycero-3-phosphate (POPA, 16:0–18:1 PA) as a neutral sodium salt adduct. There are two acidic hydrogens on the phosphate headgroup of POPA. The pK_a for the removal of the first acidic hydrogen (pK_{a1}) is 3.0; therefore, at a pH of 3.0 the phosphate headgroup is 50% ionized. At a pH of approximately 5 the headgroup is 100% ionized. The second acidic hydrogen has a pK_a of 8.0 (pK_{a2}), thus at physiological pH (7.36), the POPA species is ionized. In general, the common phosphorylated lipids all have a pK_{a1} that is ≤ 3 and are subsequently ionized in solutions that are not highly acidic. The common phosphorylated lipids also tend to associate with sodium during the ionization process. The neutral sodium salt adduct

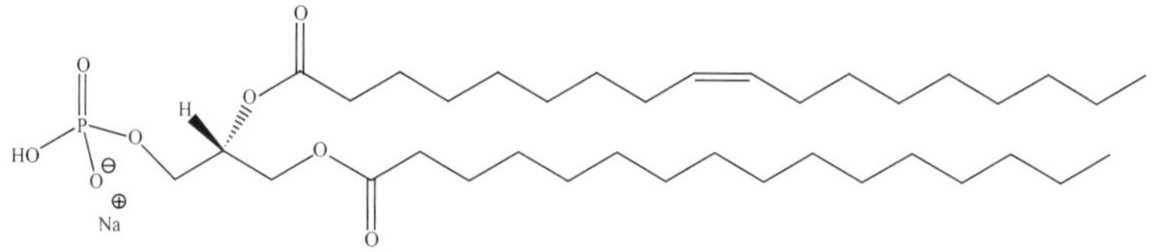

Palmityl, oleyl phosphatidic acid (POPA)

FIGURE 8.40 Structure of 1-palmitoyl-2-oleyl-*sn*-glycero-3-phosphate (POPA, 16:0-18:1 PA) as a neutral sodium salt adduct.

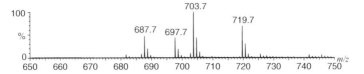

FIGURE 8.41   Nanoelectrospray Q-TOF single-stage mass spectrum of POPA in a solution of 0.1% acetic acid (H^+), 1 mM lithium (Li^+), and 1 mM sodium (Na^+). Four major peaks observed for the deprotonated dilithium adduct at m/z 687.7 as $[C_{37}H_{70}O_8P + 2Li]^+$, the protonated lithium adduct at m/z 697.7 as $[C_{37}H_{70}O_8P + H + Li]^+$, the deprotonated lithium sodium adduct at m/z 703.7 as $[C_{37}H_{70}O_8P + Li + Na]^+$, and the deprotonated disodium adduct at m/z 719.7 as $[C_{37}H_{70}O_8P + 2Na]^+$.

molecular formula for POPA is $C_{37}H_{70}O_8PNa$, and has a mass of 696.4706 Da. In negative ion mode analysis the POPA lipid is in the desodiated, deprotonated form of $[M - H]^-$, as $C_{37}H_{70}O_8P$ with a mass of 673.4808 Da. In positive ion mode analysis the phosphorylated lipids can exhibit a more complex ionization behavior. For example, let's look at the positive ion mode electrospray mass spectrum of the POPA phosphorylated lipid in an ionization solution containing 0.1% acetic acid (H^+), 1 mM lithium (Li^+), and 1 mM sodium (Na^+) in a 1 : 1 methanol/chloroform solution. The single-stage mass spectrum of the POPA lipid in the multiionic solution was collected using nanoelectrospray with a Q-TOF mass spectrometer and is illustrated in Figure 8.41. In the mass spectrum is observed four major peaks for the deprotonated dilithium adduct at m/z 687.7 as $[C_{37}H_{70}O_8P + 2Li]^+$, the protonated (this term is used loosely here as the protonated form is actually the neutral molecule) lithium adduct at m/z 697.7 as $[C_{37}H_{70}O_8P + H + Li]^+$, the deprotonated lithium sodium adduct at m/z 703.7 as $[C_{37}H_{70}O_8P + Li + Na]^+$, and the deprotonated disodium adduct at m/z 719.7 as $[C_{37}H_{70}O_8P + 2Na]^+$. This ionization behavior of the phosphorylated lipids can be used to help determine whether an unknown species is in fact a phosphorylated lipid.

8.8.2   Positive Ion Mode ESI of Phosphorylated Lipids

The product ion mass spectra can be used to determine the fatty acid substituents of the phosphorylated lipids. Figure 8.42 illustrates the product ion mass spectrum for the deprotonated dilithium adduct of 1-palmitoyl-2-oleyl-*sn*-glycero-3-phosphate at m/z 687.7 as $[C_{37}H_{70}O_8P + 2Li]^+$ collected using nanoelectrospray on a Q-TOF mass spectrometer. The two major peaks in the midmass range of the spectrum are associated with the neutral loss of the fatty acid substituents. The m/z 431.5 product ion is generated from the neutral loss of the C16:0 fatty acid substituent, and the m/z 405.4 product ion is generated from the neutral loss of the C18:1 fatty acid substituent. These two product ions can be used to identify the fatty acid substituents of the phosphorylated lipid. Another observation of interest is that the fatty acid substituent neutral losses that produce the m/z 431.5 and m/z 405.4 product ions do not involve either of the lithium cations. This directly indicates that both of the

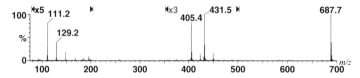

FIGURE 8.42 Nanoelectrospray Q-TOF/MS/MS product ion mass spectrum for the deprotonated dilithium adduct of 1-palmitoyl-2-oleyl-*sn*-glycero-3-phosphate at m/z 687.7 as $[C_{37}H_{70}O_8P + 2Li]^+$. The m/z 431.5 product ion is generated from the neutral loss of the C16:0 fatty acid substituent and the m/z 405.4 product ion is generated from the neutral loss of the C18:1 fatty acid substituent.

lithium cations are associated with the phosphatidic headgroup rather than with the fatty acid substituent portion of the phosphorylated lipid. This is also seen with the m/z 111.2 product ion that represents the dilithium adduct of the phosphate headgroup as $[H_2PO_4Li_2]^+$. The m/z 129.2 product ion also represents the phosphatidic headgroup as $[C_2H_3PO_4Li]^+$; however, one of the lithium ions has been transferred to one of the fatty acid substituents.

This same behavior is also observed for the product ion mass spectrum for the deprotonated lithium sodium adduct of 1-palmitoyl-2-oleyl-*sn*-glycero-3-phosphate at m/z 703.7 as $[C_{37}H_{70}O_8P + Li + Na]^+$ collected using nanoelectrospray on a Q-TOF mass spectrometer illustrated in Figure 8.43. The two major peaks in the midmass range of the spectrum are associated with the neutral loss of the fatty acid substituents. The m/z 447.4 product ion is generated from the neutral loss of the C16:0 fatty acid substituent, and the m/z 421.4 product ion is generated from the neutral loss of the C18:1 fatty acid substituent. Here too the lithium and sodium cations are not observed to be lost with the fatty acid substituents but stay associated with the phosphatidic headgroup. The m/z 127.2 product ion represents the lithium sodium adduct of the phosphate headgroup as $[H_2PO_4LiNa]^+$, similar to what was observed above. Interestingly, though, the product ion at m/z 145.2 represents the phosphatidic headgroup as $[C_2H_3PO_4Na]^+$; however, the lithium ion has been transferred to one of the fatty acid substituents instead of the sodium ion.

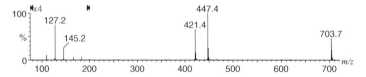

FIGURE 8.43 Nanoelectrospray Q-TOF/MS/MS product ion mass spectrum for the deprotonated lithium sodium adduct of 1-palmitoyl-2-oleyl-*sn*-glycero-3-phosphate at m/z 703.7 as $[C_{37}H_{70}O_8P + Li + Na]^+$. The m/z 447.4 product ion is generated from the neutral loss of the C16:0 fatty acid substituent and the m/z 421.4 product ion is generated from the neutral loss of the C18:1 fatty acid substituent.

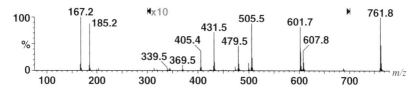

FIGURE 8.44 Nanoelectrospray Q-TOF/MS/MS product ion mass spectrum of the depro-tonated dilithium adduct of 1-palmitoyl-2-oleoyl-*sn*-glycero-3-[phospho-*rac*-(1-glycerol)] at *m/z* 761.8 as $[C_{40}H_{76}O_{10}P + 2Li]^+$.

The product ion mass spectrum of 1-palmitoyl-2-oleoyl-*sn*-glycero-3-[phospho-*rac*-(1-glycerol)] (POPG, 16:0–18:1 PG), however, is more complex that that of POPA. Figure 8.44 illustrates the nanoelectrospray Q-TOF/MS/MS product ion mass spectrum for the deprotonated dilithium adduct of 1-palmitoyl-2-oleoyl-*sn*-glycero-3-[phospho-*rac*-(1-glycerol)] at *m/z* 761.8 as $[C_{40}H_{76}O_{10}P + 2Li]^+$. There are a number of product ions in this spectrum that are associated with the headgroup as compared to that of POPA above.

The *m/z* 167 product ion is the dilithiated headgroup, and its structure is illustrated in Figure 8.45 along with the structure for *m/z* 185.2. The *m/z* 601.7 product ion is formed through the neutral loss of the headgroup and one lithium as $[C_{40}H_{76}O_{10}P + Li - C_4H_{10}PO_4Li]^+$ and is illustrated in Figure 8.45. The *m/z* 505.5 product ion is generated from the neutral loss of the C16:0 fatty acid substituent, and the *m/z* 479.5 product ion is generated from the neutral loss of the C18:1 fatty acid substituent. The

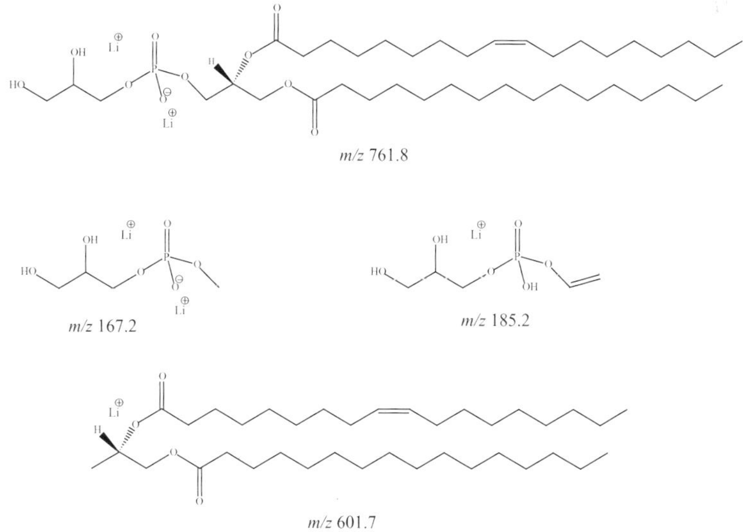

m/z 761.8

m/z 167.2

m/z 185.2

m/z 601.7

FIGURE 8.45 Some of the major product ions formed from the dissociation of POPG.

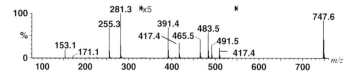

FIGURE 8.46 Product ion mass spectrum for POPG standard reference at m/z 747.6. The product ion mass spectrum contain product ions at m/z 253.4 (deprotonated form of a C16:1 palmitic acid $[C_{16}H_{30}O_2 - H]^-$) and m/z 281.4 (deprotonated form of a C18:1 oleic acid $[C_{18}H_{34}O_2 - H]^-$) that are diagnostic for the identification of the fatty acid substituents of the lipids.

m/z 431.5 and m/z 405.4 are formed through a combination of fatty acid substituent loss and partial head group loss.

8.8.3 Negative Ion Mode ESI of Phosphorylated Lipids

Figure 8.46 is the product ion spectrum of a 1-palmitoyl-2-oleoyl-*sn*-glycero-3-[phospho-*rac*-(1-glycerol)] (POPG) standard at m/z 747.6 collected in negative ion mode as the deprotonated ion $[M - H]^-$. The formula for the deprotonated ion is $[C_{40}H_{76}O_{10}P]^-$ with an exact mass of 747.5176 Da. The product ion spectrum illustrated in Figure 8.46 contains a number of diagnostic peaks used for unknown identification. These include the m/z 153.1 product ion, which is the phosphatidylglycerol headgroup as $[C_3H_6PO_5]^-$, and m/z 171.1, which is also a phosphatidylglycerol headgroup represented as $[C_3H_8PO_6]^-$. The fatty acid substituents are represented by m/z 255.3, which is a C16:0 deprotonated fatty acid as $[C_{16}H_{32}O_2 - H]^-$, and the m/z 281.3 product ion, which is a C18:1 deprotonated fatty acid as $[C_{18}H_{34}O_2 - H]^-$. The m/z 391.4 product ion is derived through neutral loss of the glycerol portion of the headgroup followed by neutral loss of the C18:1 fatty acid substituent $[C_{40}H_{77}O_{10}P - H - C_3H_6O_2 - C_{18}H_{34}O_2]^-$. The m/z 417.4 product ion is derived through neutral loss of the glycerol portion of the headgroup followed by neutral loss of the C16:0 fatty acid substituent $[C_{40}H_{77}O_{10}P - H - C_3H_6O_2 - C_{16}H_{32}O_2]^-$. The m/z 465.5 product ion is derived through neutral loss of the C18:1 fatty acid substituent, $[C_{40}H_{77}O_{10}P - H - C_{18}H_{34}O_2]^-$, while m/z 483.5 is derived through neutral loss of the C18:1 fatty acyl chain as a ketene $[C_{40}H_{77}O_{10}P -H - C_{18}H_{32}O]^-$. The C16:0 fatty acid substituent also produces product ions that are analogous to those produced by the C18:1 fatty acid substituent, such as the m/z 491.5 product ion, which is derived through neutral loss of the C16:0 fatty acid substituent, $[C_{40}H_{77}O_{10}P - H - C_{16}H_{32}O_2]^-$, and the m/z 509.5 product ion, which is derived through neutral loss of the C16:0 fatty acyl chain as a ketene $[C_{40}H_{77}O_{10}P - H - C_{16}H_{30}O]^-$. The difference between the product ions for neutral loss of fatty acid substituent versus fatty acyl chain as ketene is 18 atomic mass units (amu). When this pattern is observed in product ion spectra such as Figures 8.46, it can be used as an indication of ester linkages as fatty acid substituents in the structures and can be used in unknown lipid species identification.

8.9 PROBLEMS

8.1. List some examples of small biomolecules that are measured using mass spectrometry.

8.2. What are some benefits of using a metal adduct such as lithium for lipid analysis?

8.3. In what mode are fatty acids often measured during mass spectral analysis? Why do you think this is the case?

8.4. In electrospray mass spectrometry can lipids be quantitatively measured? Explain why or why not.

8.5. During the negative ion mode ESI analysis of fatty acids, it was suspected that reduction of unsaturation may be taking place. Is this possible?

8.6. In the negative ion mode ESI behavior of fatty acids was the location of the double bond or the structural conformation a factor?

8.7. Was reduction of the fatty acids in ESI negative ion mode analysis being observed or was it suppression?

8.8. What do the graphs in Figure 8.11 demonstrate?

8.9. How was contamination ruled out in the ESI negative ion mode behavior study of the fatty acids?

8.10. What is the significance of the ESI negative ion mode behavior study?

8.11. What is found to allow quantitative analysis by GC/EI mass spectrometry?

8.12. Briefly describe the internal standard method.

8.13. Why are different sample dilutions measured when quantitating species by mass spectrometry?

8.14. In Figure 8.16 what fragment is the lithium cation associated with and why?

8.15. What are a couple of neutral losses in the CID of wax esters that are used as diagnostic for location of a double bond?

8.16. What form of loss from a lipid often indicates an oxidized species?

8.17. In the sterols discussed in Section 8.7, what steps were performed to identify the unknown biomolecule?

8.18. List some approaches that have been used for mass spectrometric analysis of acylglycerols.

8.19. What are zwitterionic lipid species?

8.20. Describe the types of ions that can be produced in an acidic solution that also contain metal salts such as Li^+, Na^+, and K^+.

8.21. What can product ion spectra be used for in analysis of lipids?

8.22. With the phosphorylated lipids with what part is the metal cation generally found associated?

8.23. What is the difference in neutral loss for loss of fatty acid substituents versus fatty acyl chain as ketene? What can this be useful in determining?

REFERENCES

1. Han, X., and Gross, R. W. *Expert Rev. Proteomics* 2005, *2*, 253–264.
2. Han, X., and Gross, R. W. *J. Lipid Res.* 2003, *44*, 1071–1079.
3. Fahy, E., Subramaniam, S., and Brown, H. A. *J. Lipid Res.* 2005, *46*, 839–862.
4. Adams, J., and Gross, M. L. *Anal. Chem.* 1987, *59*, 1576–1582.
5. Hsu, F. F., Bohrer, A., and Turk, J. *J. Am. Soc. Mass Spectrom.* 1998, *9*, 516–526.
6. Hsu, F. F., and Turk, J. *J. Am. Soc. Mass Spectrom.* 1999, *10*, 587–599.
7. Kerwin, J. L., Wiens, A. M., and Ericsson, L. H. *J. Mass Spectrom.* 1996, *31*, 184.
8. Valianpour, F., Selhorst, J. J. M., van Lint, L. E. M., van Gennip, A. H., Wanders, R. J. A., and Kemp, S. *Mol. Genetd. Metab.* 2003, *79*, 189.
9. Catharino, R. R., Haddad, R., Cabrini, L. G., Cunha, I. B. S., Sawaya, A. C. H. F., and Eberlin, M. N. *Anal. Chem.* 2005, *77*, 7429.
10. Moe, M. K., Strom, M. B., Jensen, E., and Claeys, M. *Rapid Commun. Mass Spectrom.* 2004, *18*, 1731.
11. Griffiths, W. J. *Mass Spectrom. Rev.* 2003, *22*, 81.
12. Han, X., and Gross, R. W. *Anal. Biochem.* 2001, *295*, 88.
13. Gross, M. L. *Int. J. Mass Spectrom.* 2000, *200*, 611.
14. Brugger, B., Erben, G., Sandhoff, R., Wieland, F. T., and Lehmann, W. D. *Proc. Natl. Acad. Sci. USA* 1997, *94*, 2339.
15. Han, X., and Gross, R. W. *Mass Spectrom. Rev.* 2005, *24*, 367.
16. Wakeham, S. G., and Frew, N. M. *Lipids* 1982, *17*, 831.
17. Alveraz, H. M., Luftmann, H., Silva, R. A., Cesari, A. C., Viale, A., Waltermann, M., and Steinbuchel, A. *Microbiology* 2002, *148*, 1407.
18. Gurr, M. I., Harwood, J. L., and Frayn, K. N. In *Lipid Biochemistry: An Introduction* 5th ed. Weimer, Tx: Culinary and Hospitality Industry Publications Services, 2002.
19. Reis, A., Domingues, R. M., Amado, F. M. L., Ferrer-Carreia, A. J. V., and Domingues, P. *J. Am. Soc. Mass Spectrom.* 2003, *14*, 1250.
20. Folch, J., Lees, M., and Stanly, G. H. S. *J. Biol. Chem.* 1957, *226*, 497.
21. Bligh, E. G., and Dyer, W. J. *Can. J. Biochem. Physiol.* 1959, *37*, 911.
22. Wilcox, A. L., and Marnett, L. J. *Chem. Res. Toxicol.* 1993, *6*, 413.
23. Porter, N. A., Caldwell, S. E., and Mills, K. A. *Lipids* 1995, *30*, 277.
24. Bierl-Leonhardt, B. A., DeVilbuss, E. D., and Plimmer, J. R. *J. Chromatogr. Sci.* 1980, *18*, 364.
25. Tomer, K. B., Crow, F. W., and Gross, M. L. *J. Am. Chem. Soc.* 1983, *105*, 5487.
26. Holcapek, M., Jandera, P., and Fischer, J. *Crit. Rev. Anal. Chem.* 2001, *31*, 53.
27. Mottram, H. R., Woodbury, S. E., and Evershed, R. P. *Rapid Commun. Mass Spectrom.* 1997, *11*, 1240.
28. Reis, A., Domingues, R. M., Amado, F. M. L., Ferrer-Carreia, A. J. V., and Domingues, P. *J. Am. Soc. Mass Spectrom.* 2003, *14*, 1250.

29. Sedaghat, S., Désaubry, L., Streiff, S., Ribeiro, N., Michels, B., Nakatani, Y., and Ourisson, G. *Chem. Biodiv.* 2004, *1*, 124–128.

30. Gotoh, M., Ribeiro, N., Michels, B., Elhabiri, M., Albrecht-Gary, A. M., Yamashita, J., Hato, M., Ouisson, G., and Nakatani, Y. *Chem. Biodiv.* 2006, *3*, 198–209.

31. Ohlrogge, J., and Browse, J. *Plant Cell* 1995, *7*, 957–970.

32. Hodgkin, M. N., Pettitt, T. R., Martin, A., and Wakelam, M. J. O. *Biochem. Soc. Trans.* 1996, *24*, 991–994.

33. Pettitt, T. R., Martin, A., Horton, T., Liassis, C., Lord, J. M., and Wakelam, M. J. O. *J. Biol. Chem.* 1997, *272*, 17354–17359.

34. Murphy, R. C. In *Mass Spectrometry of Lipids*. New York: Plenum, 1993, p. 189.

35. Liu, Q. T., and Kinderlerer, J. L. *J. Chromatogr. A.* 1999, *855*, 617.

36. Harvey, D. J., and Tiffany, J. M. *J. Chromatogr.* 1984, *301*, 173.

37. Kuksis, A., Marai, L., and Myher, J. J. *J. Chromatogr.* 1991, *588*, 73.

38. Marai, L., Kuksis, A., Myher, J. J., and Itabashi, Y. *Biol. Mass Spectrom.* 1992, *21*, 541.

39. Cole, R. B., and Zhu, J. *Rapid Commun. Mass Spectrom.* 1999, *13*, 607.

40. Duffin, K. L., and Henion, J. D. *Anal. Chem.* 1991, *63*, 1781.

41. Han, X., and Gross, R. W. *Anal. Biochem.* 2001, *295*, 88.

42. Byrdwell, W. C., and Neff, W. E. *Rapid Commun. Mass Spectrom.* 2002, *16*, 300.

43. Marzilli, L. A., Fay, L. B., Dionisi, F., and Vouros, P. *J. Am. Chem. Oil Soc.* 2003, *80*, 195.

44. Waltermann, M., Luftmann, H., Baumeister, D., Kalscheuer, R., and Steinbuchel, A. *Microbiology* 2000, *146*, 1143.

45. Schiller, J., Arnhold, J., Benard, S., Muller, M., Reichl, S., and Arnold, K. *Anal. Biochem.* 1999, *267*, 46.

46. Leary, J. A., and Pederson, S. F. *J. Org. Chem.* 1989, *54*, 5650.

47. Byrdwell, W. C. *Lipids* 2001, *36*, 327.

48. Mu, H., Sillen, H., and Hoy, C. E. *J. Am. Oil Chem. Soc.* 2000, *77*, 1049.

49. Marzilli, L. A., Fay, L. B., Dionisi, F., and Vouros, P. *J. Am. Oil. Chem. Soc.* 2003, *80*, 195.

50. McAnoy, A. M., Wu, C. C., and Murphy, R. C. *J. Am. Soc. Mass Spectrom.* 2005, *16*, 1498.

51. Kalo, P., Kemppinen, A., Ollilainen, V., and Kuksis, A. *Lipids* 2004, *39*, 915.

52. Giuffrida, F., Destaillats, F., Skibsted, L. H., and Dionisi, F. *Chem. Phys. Lipids* 2004, *131*, 41.

53. Ham, B. M., Jacob, J. T., Keese, M. M., and Cole, R. B. *J. Mass Spectrom.* 2004, *39*, 1321.

54. Holcapek, M., Jandera, P., Fischer, J., and Prokes, B. *J. Chromatogr. A* 1999, *858*, 13.

55. Holcapek, M., Jandera, P., Zderadicka, P., and Hruba, L. *J. Chromatogr. A* 2003, *1010*, 195.

56. Jakab, A., Jablonkai, I., and Forgacs, E. *Rapid Commun. Mass Spectrom.* 2003, *17*, 2295.

57. Fauconnot, L., Hau, J., Aeschlimann, J. M., Fay, L. B., and Dionisi, F. *Rapid Commun. Mass Spectrom.* 2004, *18*, 218.

58. Hsu, F. F., Bohrer, A., and Turk, J. *J. Am. Soc. Mass Spectrom.* 1998, *9*, 516.

59. Zhou, Z., Ogden, S., and Leary, J. A. *J. Org. Chem.* 1990, *55*, 5444.

60. Striegel, A. M., Timpa, J. D., Piotrowiak, P., and Cole, R. B. *Int. J. Mass Spectrom. Ion Processes* 1997, *162*, 45.

61. Adams, J., and Gross, M. L. *Anal. Chem.* 1987, *59*, 1576.

62. Hsu, F. F., and Turk, J. *J. Am. Soc. Mass Spectrom.* 1999, *10*, 587.

63. Gu, M., Kerwin, J. L., Watts, J. D., and Aebersold, R. *Anal. Biochem.* 1997, *244*, 347–356.

64. Szucs, R., Verleysen, K., Duchateau, G. S. M. J. E., Sandra, P., and Vandeginste, B. G. M. *J. Chromatogr. A* 1996, *738*, 25–29.

65. Verleysen, K., and Sandra, P. *J. High. Resol. Chromatogr.* 1997, *20*, 337–339.
66. Raith, K., Wolf, R., Wagner, J., and Neubert, R. H. H. *J. Chromatogr. A.* 1998, *802*, 185–188.
67. Han, X., and Gross, R. W. *J. Am. Chem. Soc.* 1996, *118*, 451–457.
68. Hoischen, C., Ihn, W., Gura, K., and Gumpert, J. *J. Bacteriol.* 1997, *179*, 3437–3442.
69. Hsu, F. F., Bohrer, A., and Turk, J. *J. Am. Soc. Mass Spectrom.* 1998, *9*, 516–526.
70. Khaselev, N., and Murphy, R. C. *J. Am. Soc. Mass Spectrom.* 2000, *11*, 283–291.
71. Hsu, F. F., and Turk, J. *J. Mass Spectrom.* 2000, *35*, 596–606.
72. Liebisch, G., Drobnik, W., Lieser, B., and Schmitz, G. *Clin Chem.* 2002, *48*, 2217–2224.
73. Ho, Y. P., Huang, P. C., and Deng, K. H. *Rapid Commun. Mass Spectrom.* 2003, *17*, 114–121.
74. Hsu, F. F., and Turk, J. *J. Am. Soc. Mass Spectrom.* 2004, *15*, 1–11.
75. Marto, J. A., White, F. M., Seidomridge, S., and Marshall, A. G. *Anal. Chem.* 1995, *67*, 3979–3984.
76. Schiller, J., Arnhold, J., Benard, S., Muller, M., Reichl, S., and Arnold, K. *Anal. Biochem.* 1999, *267*, 46–56.
77. Ishida, Y., Nakanishi, O., Hirao, S., Tsuge, S., Urabe, J., Sekino, T., Nakanishi, M., Kimoto, T., and Ohtani, H. *Anal. Chem.* 2003, *75*, 4514–4518.
78. Al-Saad, K. A., Zabrouskov, V., Siems, W. F., Knowles, N. R., Hannan, R. M., and Hill Jr., H. H. *Rapid Commun. Mass Spectrom.* 2003, *17*, 87–96.
79. Rujoi, M., Estrada, R., and Yappert, M. C. *Anal. Chem.* 2004, *76*, 1657–1663.
80. Woods, A. S., Ugarov, M., Egan, T., Koomen, J., Gillig, K. J., Fuhrer, K., Gonin, M., and Schultz, J. A. *Anal. Chem.* 2004, *76*, 2187–2195.
81. Murphy, R. C., Fiedler, J., and Hevko, J. *Chem. Rev.* 2001, *101*, 479–526.

CHAPTER 9

BIOMOLECULE SPECTRAL INTERPRETATION: BIOLOGICAL MACROMOLECULES

9.1 INTRODUCTION

The three main types of biological macromolecules (or biopolymers) are proteins, polysaccharides, and nucleic acids. The mass spectrometric analysis of proteins, peptides, and their associated posttranslational modifications was covered extensively in Chapter 7 and will not be covered here. In this chapter we will take a close look at the mass spectrometric analysis of polysaccharides and nucleic acids, both having their own set of fragmentation pathway schemes and naming systems quite similar to that of the peptides presented in Chapter 7. The biological macromolecules all consist of repeating units of monomers with a variety that can include a single monomeric repeating unit up to perhaps 20 different monomeric units involved as is the case with the amino acids. The biopolymers are synthesized through condensation reactions between the monomeric units that join the monomers with the elimination of water. The chemical splitting apart of the monomeric units of a biopolymer is a hydrolysis reaction where the addition of water is used to separate the monomers, which is the reverse of the condensation reaction. These two processes are constantly taking place within a living organism such as the synthesis of starch for energy storage (condensation) and the process of digestion for breaking down large molecules (hydrolysis).

As we saw in Chapter 7, mass spectrometry has also been applied to the gas-phase measurement, characterization, and identification through structural elucidation of the other macromolecules, polysaccharides, and nucleic acids. This is also an important

Even Electron Mass Spectrometry with Biomolecule Applications   By Bryan M. Ham
Copyright © 2008 John Wiley & Sons, Inc.

area of study, and although much work has already been accomplished in the area of macromolecule mass spectrometric analysis, there is room for added information and insight.

9.2 CARBOHYDRATES

Polysaccharides are biopolymers comprised of long chains of repeating sugar units, usually either the same repeating monomer or a pattern of two alternating units. The sugar monomers that make up the polysaccharide chain are called monosaccharides from the Greek word *mono* meaning "single" (the Greek *poly* means "many") and *saccharide* meaning "sugar." Glycogen and starch are two forms of storage polysaccharides while cellulose is a structural polysaccharide. It is the monosaccharide glucose that makes up these three polysaccharides as a single repeating unit. It is the bonding between the repeating glucose monosaccharide and the structural design through branching that gives each one of these its special properties designed for specific functions. Starch is the storage form of polysaccharide, is found in plants, while it is glycogen that is the storage form found in animals. Glycogen is primarily found in the liver, where it is used to maintain blood sugar levels, and in muscle tissue, where it is used for energy in muscle contraction. These polysaccharides are comprised of α-D-glucose units that are linked together by α glycosidic bonds at the 1- and 4-positions of the glucose unit. Cellulose, found in the cell walls of plants, is a polysaccharide comprised of the repeating glucose monomer β-D-glucose with a β-glycosidic bond between the 1 and 4 carbons of the sugar. Figure 9.1 illustrates the structures of α-D-glucose and β-D-glucose. Starting with carbon number one in the numbering of the ring to the right of the structure, the difference between the two is that in α-D-glucose the hydroxyl group points downward, and in β-D-glucose it points upward. Glucose, also known as aldohexose D-glucose, is a six-carbon structure with the molecular formula of $C_6H_{12}O_6$. The sugars are of the carbohydrate group with a general formula of $C_nH_{2n}O_n$ where the carbon-to-hydrogen-to-oxygen ratio is $1 : 2 : 1$. Carbohydrate is actually an archaic nomenclature from early chemistry

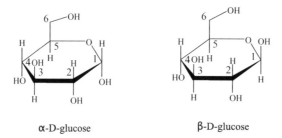

α-D-glucose   β-D-glucose

FIGURE 9.1 Structures of α-D-glucose (the unit that repeats in starch and glycogen) where the hydroxyl group points downward, and in β-D-glucose (the unit that repeats in cellulose) where the hydroxyl group points upward.

(a)

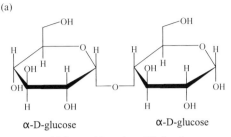

α-D-glucose   α-D-glucose

Maltose with α glycosidic bond

(b)

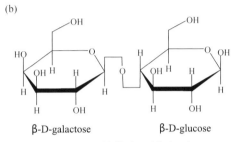

β-D-galactose   β-D-glucose

Lactose with β glycosidic bond

(c)

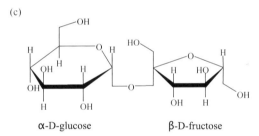

α-D-glucose   β-D-fructose

Sucrose with α glycosidic bond

FIGURE 9.2 (*a*) Maltose structure comprised of two α-D-glucose units connected forming a α-glycosidic bond. (*b*) Lactose structure comprised of a β-D-galactose unit connected to a β-D-glucose forming a β-glycosidic bond. (*c*) A α-sugar connected to a β-sugar as α-D-glucose and β-D-fructose, the common table sugar sucrose forming an α-glycosidic bond.

studies that in general termed compounds such as sugars as a hydrate of carbon represented as $C_n(H_2O)_n$. There are two types of linkages that connect the monosaccharides forming the polysaccharides, the α-glycosidic bond and the β-glycosidic bond. These two types of bonds are illustrated in Figure 9.2 for the disaccharides maltose and lactose. In the disaccharide maltose structure illustrated in Figure 9.2a, there are two α-D-glucose units connected forming a α-glycosidic bond while in lactose (Fig. 9.2b) there is a β-D-galactose unit connected to a β-D-glucose forming a β-glycosidic bond. It is also possible to have a α sugar connected to a β sugar, as illustrated in Figure 9.2c between α-D-glucose and β-D-fructose forming the common table sugar sucrose with a α-glycosidic bond. While the three polysaccharides glycogen, starch, and cellulose are the most abundant forms of polysaccharides in biological systems, there are other examples of polysaccharides that contain different sugar units than glucose that the mass spectrometrist may encounter. Examples include derivatives of β-glucosamine, a sugar where an amino group has replaced a hydroxyl group on carbon atom 2, found in the cell wall of bacteria and chitin, which is found in the exoskeletons of insects. There are also the oligosaccharides, shorter chains of sugar units than the polysaccharides, found as modifications on proteins, which was covered in Chapter 7.

9.2.1   Ionization of Oligosaccharides

Normal electrospray that is pneumatically assisted has been observed to produce a low response for oligosaccharides that are not modified when analyzed by mass spectrometry. The use of nanoelectrospray, however, has shown to induce an enhanced response for the analysis of oligosaccharides by mass spectrometry that is comparable to that observed of the ionization of peptides.[1] This is due to the decrease in the size of the nanoelectrospray drops, which helps to increase the surface activity of the hydrophilic oligosaccharides. With electrospray the response of the oligosaccharides tends to decrease with increasing size of the oligosaccharides.[2] This is not the case when using MALDI as the ionization source as the response of the oligosaccharides basically stays constant with increasing oligosaccharide size. The electrospray process though is a softer ionization process as compared to MALDI, which has been observed to promote metastable fragmentation of the oligosaccharides. The inclusion of metastable fragmentation can increase the complexity of the mass spectra obtained.

Similar to the mass spectrometric analysis of most biomolecules, there are three choices for the ionized species as either protonated $[M + nH]^{n+}$, metalized $[M + Na]^+$ or $[M + Li]^+$ both measured in positive ion mode, or deprotonated $[M - nH]^{n-}$ measured in negative ion mode. Product ion mass spectra of protonated oligosaccharides tend to dominate in B_m- and Y_n-type ions. Also, when CID is performed on larger oligosaccharides, the product ions are primarily derived through glycosidic bond cleavage, giving sequence information but little information concerning branching. When CID is performed on smaller oligosaccharides, the product ions are produced through both glycosidic bond cleavage and cross-ring cleavage. This is due to the greater number of vibrational degrees of freedom in the larger oligosaccharides, which are able to dissipate the internal energy imparted to them from the collisions through vibrational relaxation. In high-energy CID studies such as those obtained with a sector mass analyzer or the MALDI TOF-TOF mass analyzer, the use of protonated species was observed to have more efficient fragmentation of the glycosidic bonds as compared to metal-adducted oligosaccharides. Negative ion mode mass spectrometric analysis of deprotonated oligosaccharides does not have a very efficient or high response for neutral oligosaccharides. The action of hydroxylic hydrogen migration has been suggested as a limiting factor for the observance of abundant cross-ring

cleavage ions in negative ion mode analysis of deprotonated ions while support-
ing glycosidic bond cleavages.[3] Sialylated and sulfated oligosaccharides can have
an enhanced response and informative product ion spectra as deprotonated species
analyzed in negative ion mode.

9.2.2 Carbohydrate Fragmentation

Oligosaccharides are often observed in nature to be very complex mixtures of species
that are very closely related isomeric compounds. This and the fact that they contain
many labile bonds and can be highly branched make the analysis of oligosaccharides
by mass spectrometry difficult. In mass spectrometric analysis of the polysaccharides
using collision-induced dissociation for the production of product ion spectra, the
polysaccharides fragment according to specific pathways that have a standardized
naming scheme. Figure 9.3 illustrates the types of fragmentation that can take place
with the polysaccharide chain (adapted from Domon and Costello[4]). The sugar ring
structure on the far left also contains the numbering scheme for the sugar ring carbon
atoms. Whether it is a five-membered ring or a six-membered ring, the counting
starts with the first carbon to the left of the ring oxygen. The naming scheme in
Figure 9.3 as illustrated is for a three sugar containing oligosaccharide, with an
R-group to the right of the structure; therefore, the naming starts at Y_0 from the right.
The general naming scheme for a polysaccharide with R_n sugar units would result in
a naming of Y_n and Z_n directly left of the R_n, Y_{n+1}, and Z_{n+1} for the next set, and so
forth. We will take a closer look at the generation of the B-type ions and the Y-type
ions in positive ion mode to better understand the generation of the product ions and
their naming using the more general scheme for a polysaccharide based off of the
scheme in Figure 9.3. The generation of the B_2-type and Y_4-type ions is illustrated in
Figure 9.4 for a 6-sugar unit polysaccharide, obtained in positive ion mode. The

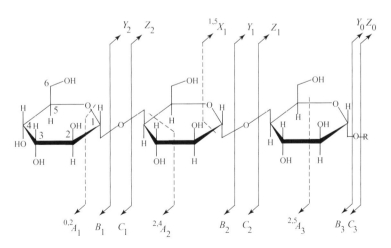

FIGURE 9.3 Fragmentation pathways and naming scheme for polysaccharides.

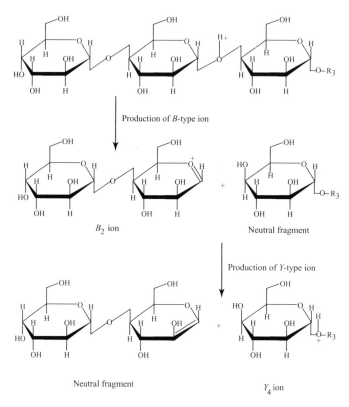

FIGURE 9.4 Production of B_2-type and Y_4-type ions for a 6-sugar unit polysaccharide in positive ion mode.

B_2-type ion is a charged oxonium ion that is contained within the sugar ring structure. The Y_4-type ion carries a positive charge in the form of protonation. The production of the C_2-type and Z_4-type ions in positive ion mode is illustrated in Figure 9.5. Here the C_2-type ion carries a positive charge in the form of protonation derived from hydrogen transfer from the reducing side of the saccharide. The positive charge of the Z_4-type ion is also from protonation, but the ion has also suffered loss of water to the nonreducing carbohydrate portion of the polysaccharide.

In negative ion mode the fragmentation of the polysaccharides also takes place with the glycosidic bonds, however, ring opening and epoxide formation are observed in the product ions. The production of the B-type and Y-type ions in negative ion mode is illustrated in Figure 9.6. For the B-type ion the deprotonated precursor will dissociate where the hydroxyl group will go with the reducing portion of the sugar while the carbohydrate portion suffers ring opening and epoxide formation. A similar mechanism is observed for the Y-type ion, except there is a proton transfer from the reducing portion of the structure to the carbohydrate portion, leaving the reducing end deprotonated and negatively charged. The production of the C-type and Z-type

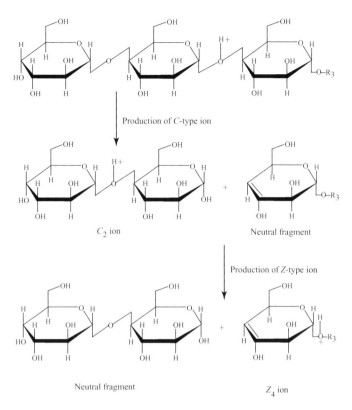

FIGURE 9.5 Production of C_2-type and Z_4-type ions for a 6-sugar unit polysaccharide in positive ion mode.

ions in negative ion mode is illustrated in Figure 9.7. In these pathways the epoxide is forming on the reducing end of the polysaccharide, and there is no ring opening taking place.

In positive ion mode cleavage of the ring structure producing the A-type ion is not often observed. In negative ion mode cleavage of the ring structure producing the A-type ion is observed. In both positive and negative ion mode the cleavage of the sugar ring producing the X-type ions is observed and produces a number of different types of product ion structures.

When generating product ions of the oligosaccharides using high-energy CID, the products produced are different for protonated oligosaccharides than that of metal ion adducts. It has been observed that the production of cross-ring cleavages is more enhanced in the metal ion adduct oligosaccharides. In low-energy CID the amount of glycosidic bond cleavage is low for the metal ion adduct species. This can be explained due to the types of bonding that is taking place between the oligosaccharide and the proton or metal ion. In protonation the proton is associated with the glycosidic oxygen, which is the most basic site in the structure. This destabilizes the glycosidic bond

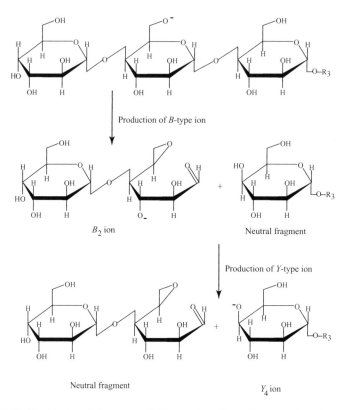

FIGURE 9.6   Production of B_2-type and Y_4-type ions for a 6-sugar unit polysaccharide in negative ion mode.

resulting in charge-localized and charge-driven fragmentation of the glycosidic bond. However, with the metal ion the bonding is more complex where the charge of the metal ion is not directly associated with the glycosidic bond oxygen but is distributed to a number of local oxygen atoms. This results in a higher activation barrier for the induced fragmentation of the oligosaccharide. This effect is illustrated in Figure 9.8 where Figure 9.8a is for the low activation barrier protonated oligosaccharide and Figure 9.8b shows the high activation barrier metal ion adduct.

9.2.3   Complex Oligosaccharide Structural Elucidation

It should be noted that it is actually quite difficult to completely sequence a highly branched and/or substituted polysaccharide that are often many units in length. The possible combinations are quite large and a systematic approach to sequencing complicated polysaccharides does not exist. For neutral carbohydrates the best approach is the product ion mass spectral generation using metal adducts in positive ion mode

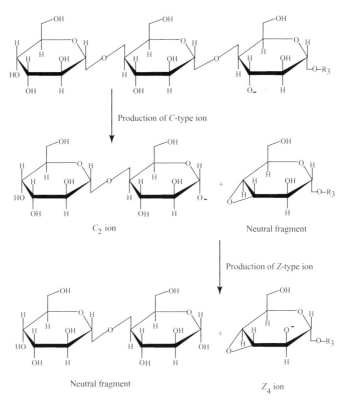

FIGURE 9.7 Production of C_2-type and Z_4-type ions for a 6-sugar unit polysaccharide in negative ion mode.

with either ESI or MALDI as the ionization source. Acidic oligosaccharides are measured in negative ion mode as deprotonated species using either ESI or MALDI as the ionization source. Typically, one cannot very readily interpret and assign a structure of a complicated oligosaccharide from its tandem mass spectrum. The best approach is the use of experience in combination with product ion spectral libraries to solve the structure of complicated, branched, and/or substituted oligosaccharides. There are some examples that with careful consideration a complex oligosaccharide structure can be either partially or sometimes fully elucidated with product ion spectral interpretation. High-energy product ion spectra of the hybrid glycan $(Man)_5(GlcNAc)_4$ was obtained using a magnetic sector mass spectrometer with a MALDI ionization source, a collision cell, and an orthogonal-TOF mass analyzer by Clayton and Bateman[6] that allowed sequencing of the complex oligosaccharide. Figure 9.9 is the product ion spectrum of the high-energy collision-induced dissociation of the hybrid glycan $(Man)_5(GlcNAc)_4$. The spectrum has been split into two sections to better illustrate the rich abundance of product ions generated. The product ion spectrum contains B-, Y-, and X-type ions that allowed the structural elucidation of the

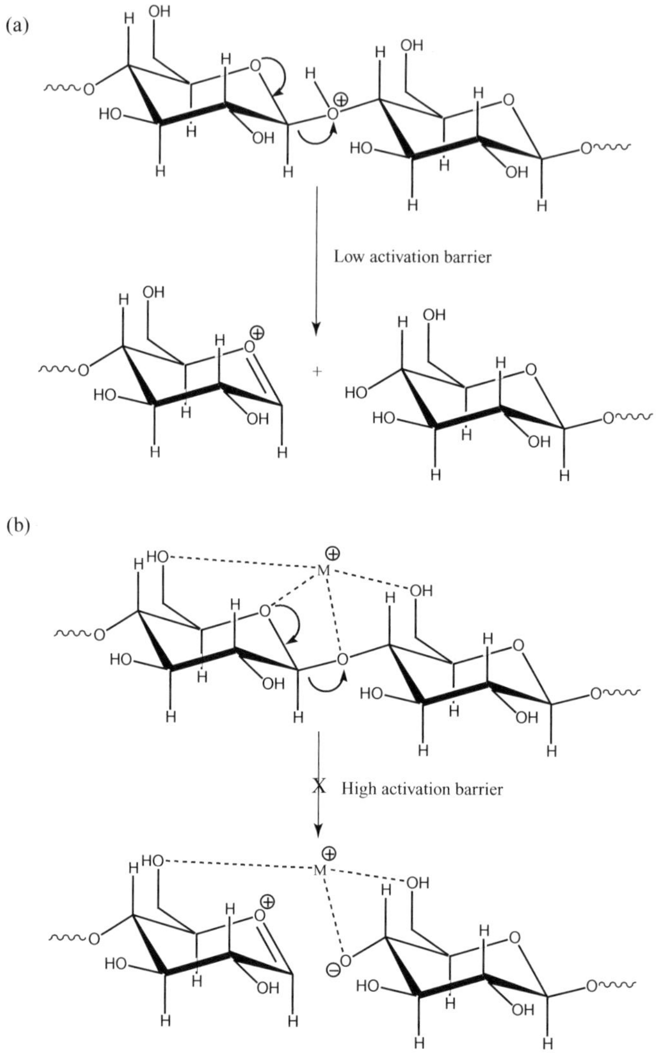

FIGURE 9.8   Fragmentation pathway mechanisms for the cleavage of the glycosidic bond through (*a*) protonation and (*b*) metal ion adduct. (Modified from Cancilla et al.[5])

oligosaccharide. Figure 9.10*a* and 9.10*b* illustrate the assignment of the product ions contained within the Figure 9.9 spectrum to the structure of the oligosaccharide.

9.3 NUCLEIC ACIDS

Nucleic acids are also analyzed using mass spectrometric techniques, and, like the polysaccharide section just covered, we will start with a background look at the

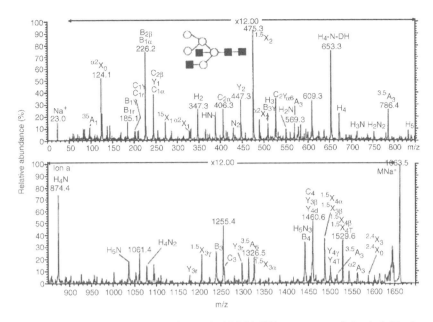

FIGURE 9.9 (*a*) High-energy (800 eV) MALDI±CID spectrum of the hybrid glycan (Man)$_5$(GlcNAc)$_4$ recorded from 2,5-DHB. The type D ion is at *m/z* 874.4, and the presence of the bisecting GlcNAc residue is indicated by the ion 221 mass units lower at *m/z* 653.3. (Reprinted with permission. Harvey, D. J., Bateman, R. H., and Green, M. R. High-energy collision-induced fragmentation of complex oligosaccharides ionized by matrix-assisted laser desorption/ionization mass spectrometry. *J. Mass Spectrom.* 1997, *32*, 167–187. Copyright 1997 John Wiley & Sons, Inc.)

makeup of nucleic acids before looking at the mass spectrometry. Unlike polysaccharides, and like the proteins, nucleic acids are specifically directional in their makeup and contain nonidentical monomers that have a distinct sequence that produce informational macromolecules. The nucleic acids reside in the nucleus of the cell and are the storage, expression, and transmission of genetic information of living species. The two types of nucleic acids are deoxyribonucleic acid (DNA) and ribonucleic acid (RNA). There are two distinct parts of their chemical structure and makeup that differentiate the two: First, DNA contains the 5-carbon sugar deoxyribose while RNA contains ribose. Second, DNA contains the base thymine (T) while RNA contains uracil (U). The molecules that make up the DNA and RNA structures are illustrated in Figure 9.11. This comprises the purine bases adenine (A) and guanine (G), the pyrimidine bases cytosine (C), uracil (U), and thymine. Also illustrated in Figure 9.11 are the two sugars D-deoxyribose and D-ribose, and finally the phosphate group that acts as the backbone of the nucleic acids, linking the nucleotides together. Nucleotides are the monomeric units that make up the nucleic acids. There are actually only 4 nucleotides that make up DNA and RNA, a much smaller number than the 20 amino acids found in proteins. Examples of nucleotides found in DNA and RNA

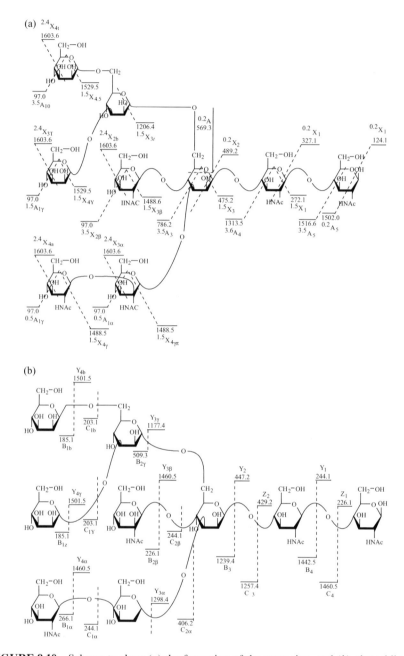

FIGURE 9.10 Scheme to show (*a*) the formation of the cross ring, and (*b*) glycosidic and fragment ions for the spectrum shown in Figure 9.9*a*. (Reprinted with permission. Harvey, D. J., Bateman, R. H., and Green, M. R. High-energy collision-induced fragmentation of complex oligosaccharides ionized by matrix-assisted laser desorption/ionization mass spectrometry. *J. Mass Spectrom.* 1997, *32*, 167–187. Copyright 1997 John Wiley & Sons, Inc.)

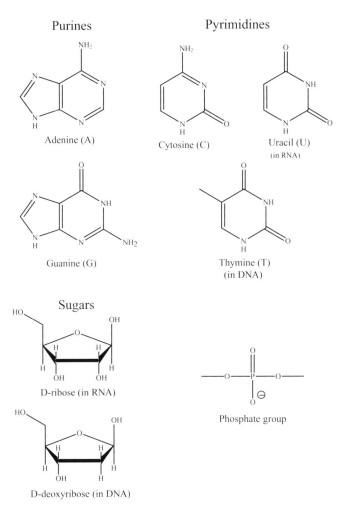

FIGURE 9.11 Structures of the molecules that make up the nucleic acids DNA and RNA. The purine bases adenine (A) and guanine (G), the pyrimidine bases cytosine (C), uracil (U), and thymine, the sugars D-deoxyribose and D-ribose, and the phosphate group.

are illustrated in Figure 9.12. Figure 9.12*a* is a DNA nucleotide where the number 2′ carbon in the sugar ring contains a hydrogen atom for D-deoxyribose. One of the bases will be attached to the 1′ carbon of the sugar through an aromatic nitrogen, and the phosphate will be attached to the number 5′ sugar carbon with a phosphoester bond. The RNA nucleotide illustrated in Figure 9.12*b* has the same types of bonding as illustrated for the DNA nucleotide but to a D-ribose sugar. In the case that the phosphate group is removed from the nucleotide, the remaining base sugar structure is called a nucleoside.

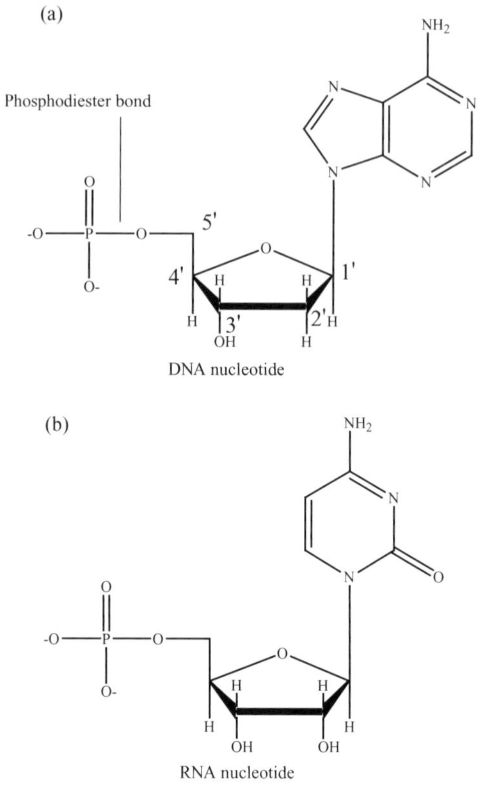

(a)

DNA nucleotide

(b)

RNA nucleotide

FIGURE 9.12   (*a*) DNA nucleotide and (*b*) RNA nucleotide.

The nucleotides are linked to each other through the phosphate group, forming a linear polymer. The nucleotides undergo a condensation reaction through the linking of the phosphate group on the 5′ carbon to the 3′ carbon of the next nucleotide known as a 3′, 5′ phosphodiester bond. The resulting polynucleotide therefore has a 5′ hydroxyl group at the start (by convention) and a 3′ hydroxyl group at the end (by convention). Representative linear nucleotide structures are illustrated for RNA and DNA in Figure 9.13.

A similar naming scheme that is used for the fragmentation ions generated by CID of peptides was proposed by McLuckey et al.[7] for nucleic acids and is illustrated in Figure 9.14. There are four cleavage sites producing fragmentation along the phosphate backbone from CID. When the product ion contains the 3′-OH portion of the nucleic acid the naming includes the letters *w*, *x*, *y*, and *z* where the numeral subscript is the number of bases from the associated terminal group. When the product ion contains the 5′-OH portion of the nucleic acid, the naming includes the letters *a*, *b*, *c*, and *d*. Losses are also more complicated than that shown in Figure 9.14 due to the neutral loss of base moieties. Figure 9.15 illustrates an actual structural

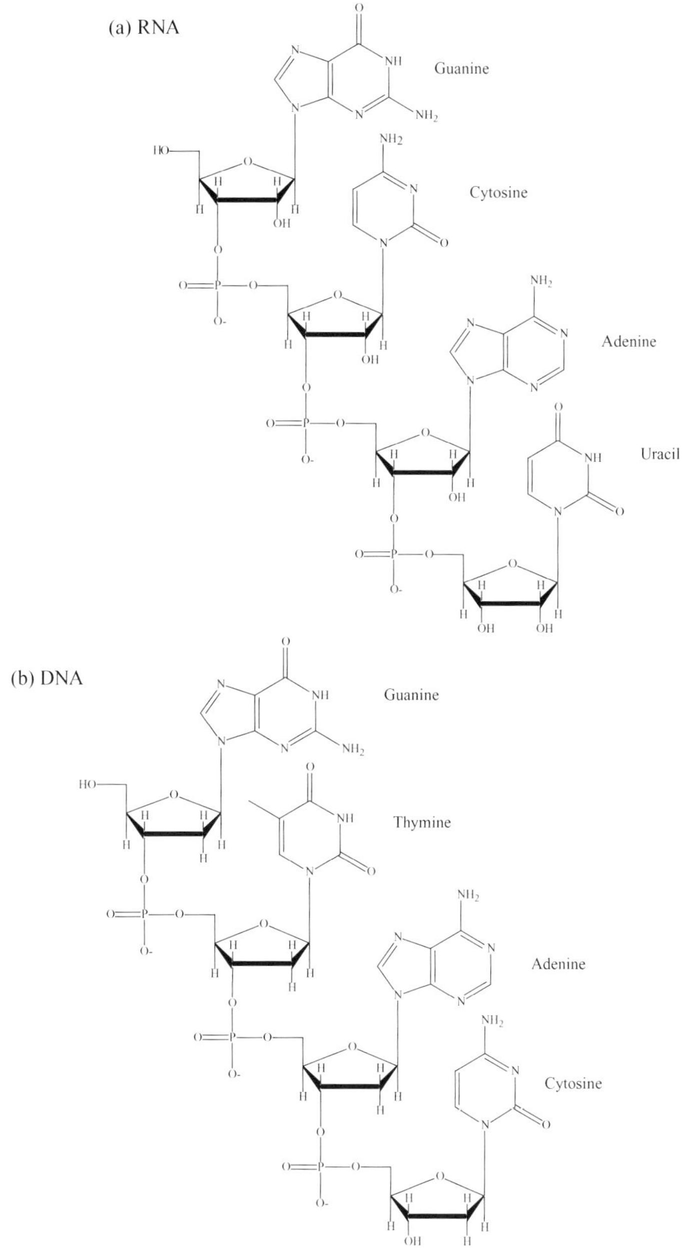

FIGURE 9.13   Linear nucleic acid structures for (*a*) RNA and (*b*) DNA.

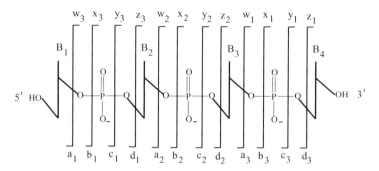

FIGURE 9.14 Naming scheme for nucleic acid product ions. When the product ion contains the $3'$-OH portion of the nucleic acid, the naming includes the letters w, x, y, and z. When the product ion contains the $5'$-OH portion of the nucleic acid, the naming includes the letters a, b, c, and d. In both $3'$-OH and $5'$-OH containing product ions the numeral subscript is the number of bases from the associated terminal group.

cleavage at the w_2/a_2 site of a 4-mer nucleic acid's phosphate backbone according to the naming scheme of Figure 9.14. Numerous mechanisms have also been reported for the fragmentation pathways leading to charged base loss and also neutral base loss. These are losses that are observed in product ion spectra other than the cleavage along the phosphate backbone that is illustrated in Figures 9.14 and 9.15. Neutral and charged base losses add to the complexity of the product ion spectra but also add information concerning the makeup of the oligonucleotide. Figure 9.16 illustrates a couple of examples of proposed fragmentation pathway mechanisms for neutral and charged base losses. In Figure 9.16a a simple nucleophilic attack on the C-1$'$ carbon atom by the phosphodiester group results in the elimination of a charged base.[8] Figure 9.16b illustrates a two-step reaction where in the first step there is neutral base loss followed by breakage of the $3'$-phosphoester bond.[9] There are other proposed fragmentation pathways for a number of other possible mechanisms for the production of the product ions observed in tandem mass spectra of the nucleic acids.

9.3.1 Negative Ion Mode ESI of a Yeast 76-mer tRNA[Phe]

In electrospray ionization it has been observed that the nucleic acids posses an enhanced response in the negative ion mode due to the presence of the phosphate groups in the nucleic acid backbone. However, salt being present during the electrospray process can suppress the response of the nucleic acids. In fact, many metal species such as sodium or potassium can adduct to the nucleic acid in large numbers, shifting the mass of the nucleic acid considerably. This was demonstrated in a study by Little et al.[10] that looked at different nucleic acids using negative ion mode electrospray FTICR mass spectrometry with a 6.2-T magnet. Figure 9.17 is a single-stage mass spectrum of a yeast 76-mer tRNA[Phe], which has a molecular mass of 24,950 Da. In

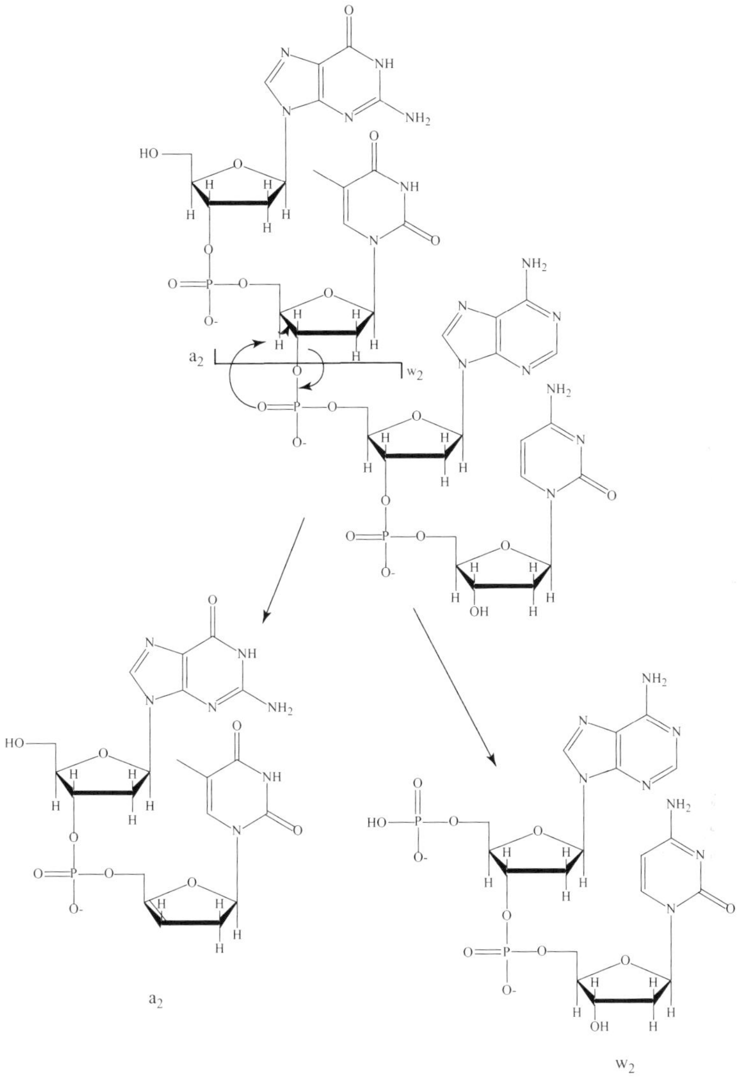

FIGURE 9.15 Example of structural cleavage at the w_2/a_2 site of a 4-mer nucleic acid's phosphate backbone according to the naming scheme of Figure 9.14.

the figure the m/z range x axis is in the 26,000-Da range for the electrospray ionization of the tRNA^{Phe}. This is due to the species having between 34 and 55 Na adducts, changing its mass by 750–1200 Da. The sodium metal atoms will associate with the phosphate groups in place of the hydrogen protons. The higher the charge state of the nucleic acid will be the higher amount of adduct sodium. The sample was desalted using HPLC and the single-stage electrospray negative ion mode mass spectra were

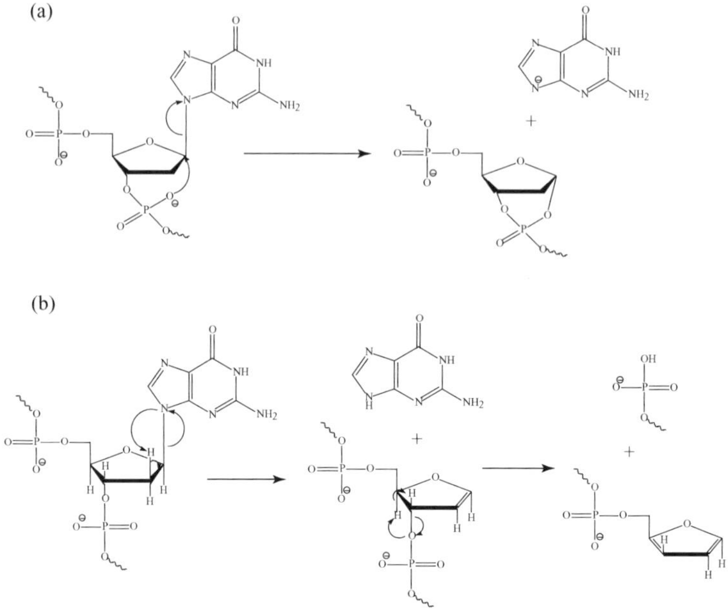

(a)

(b)

FIGURE 9.16 Proposed fragmentation pathways associated with the base substituent groups. (*a*) Nucleophilic attack on the C-1′ carbon atom by the phosphodiester group results in the elimination of a charged base. (*b*) Two-step reaction mechanism where in the first step there is neutral base loss followed by breakage of the 3′-phosphoester bond.

recollected. Figure 9.18 illustrates the same yeast tRNA^{Phe} after the HPLC desalting. In the lower spectrum a number of different charge states are now observed with a much improved response in the signal-to-noise ratio (S/N). The phosphate groups are now deprotonated and are giving the tRNA^{Phe}, a series of high negative charge states from 27$^-$ to 18$^-$ in the figure (charge state 27$^-$ means there are 27 negative charges on the tRNA^{Phe}). The upper portion of Figure 9.18 is an expanded view of the 24$^-$ charge state peak, revealing that the peak in the lower spectrum is actually comprised of a series of peaks. This is the isotopic distribution of the tRNA^{Phe} species. The FTICR mass spectrometer used possesses the resolution high enough to resolve these isotopic peaks to give the upper spectrum.

9.3.2 Positive Ion Mode MALDI Analysis

Oligonucleotides are also analyzed using the MALDI technique, which allows the measurement of a singly charged species. The MALDI technique results in the production of ionized species that carry a single charge from an adducted proton or a metal cation such as sodium. This can allow studies of fragmentation pathways of species that have undergone hydrogen/deuterium exchange (H/DX), a method where exchangeable hydrogen such as hydroxyl hydrogen is replaced with

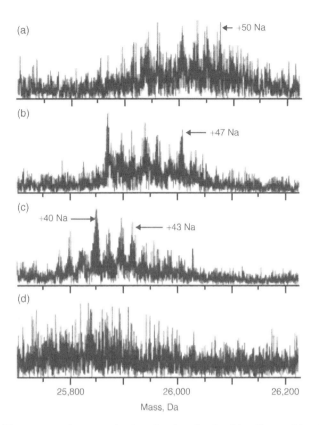

FIGURE 9.17 ESI FTMS spectra for the 16⁻ (A), 18⁻ (B), 20⁻ (C), and 22⁻ (D) anions of yeast tRNA^Phe without desalting. (Reprinted with permission from Little, D. P., Thannhauser, T. W., and McLafferty, F. W. Verification of 50- to 100-mer DNA and RNA sequences with high-resolution mass spectrometry, *Proc. Natl. Acad. Sci. USA* 1995, *92*, 2318–2322. Copyright 1995 National Academy of Sciences, U.S.A.)

deuterium (H = 1.007825 Da, D = 2.01565 Da). Hydrogen/deuterium studies of oligonucleotides have been extensively analyzed using mass spectrometry to characterize the fragmentation pathways of oligonucleotides.[11–15] Using H/DX in conjunction with mass spectrometric analysis of oligonucleotide product ion generation, it has been proposed that for DNA every fragmentation pathway is initiated by loss of a nucleobase that does not involve simple water loss or 3′- and 5′- terminal nucleoside/nucleotides.

The mass spectra illustrated in Figure 9.19 are tandem mass spectra of H/D-exchanged (a) UGUU and (b) UCUA, small RNA oligonucleotides. The mass spectra were collected on a 4700 Proteomics Analyzer (Applied Biosystems, Framingham, Massachusetts) that is a tandem in space time-of-flight (TOF)/time-of-flight (TOF) mass spectrometer with MALDI as the ionization source (MALDI TOF-TOF/MS). As described in Section 3.4, the MALDI TOF-TOF mass spectrometer

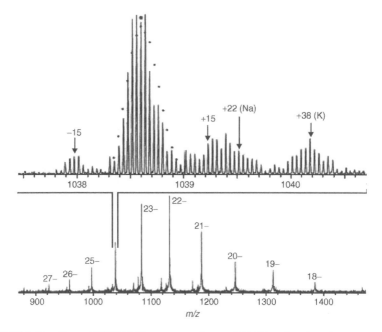

FIGURE 9.18 (*Lower*) ESI FTMS spectrum of yeast tRNAPhe after desalting. (*Upper*) The 24-anion region [heterodyne scan, SWIFT (34) isolation of 23- to 25-region]; after fitting the $(M - 24H)^{24-}$ isotopic peak abundances to those expected theoretically (indicated with dots), the asterisk indicates the peak that corresponds to the most abundant of the theoretical distribution. Peaks at approximately -15 and $+15$ Da are from minor variants. (Reprinted with permission from Little, D. P., Thannhauser, T. W., and McLafferty, F. W. Verification of 50- to 100-mer DNA and RNA sequences with high-resolution mass spectrometry. *Proc. Natl. Acad. Sci. USA* 1995, *92*, 2318–2322. Copyright 1995 National Academy of Sciences, U.S.A.)

contains a gas-filled collision cell located in between the two TOF mass analyzers that allows collision-induced dissociation generation of product ions. The MALDI TOF-TOF mass analyzer is one of the few commercially available mass analyzers that is capable of high-energy collision-induced dissociation product ion generation. Collision-induced dissociation and PSD analysis using MALDI TOF mass spectrometry has been observed to produce similar product ions, indicating that the fragmentation mechanisms are the same. The structures of the UGUU and UCUA deuterated RNA oligonucleotides used in the experiments to generate the product ion spectra in Figure 9.19 are illustrated in Figure 9.20. In both spectra a number of *a*-type, *z*-type, *y*-type, and *w*-type product ions are observed. In Figure 9.20 the exchangeable hydrogen, the sugar hydroxyl hydrogen, and the base nitrogen bound hydrogen are replaced by deuterium. The exchange of deuterium for hydrogen allows the interpretation of the fragmentation pathways that generate the observed product ions. An example of mechanism determination is illustrated in Figure 9.21 for the charged cytosine base loss from the UCUA oligonucleotide.

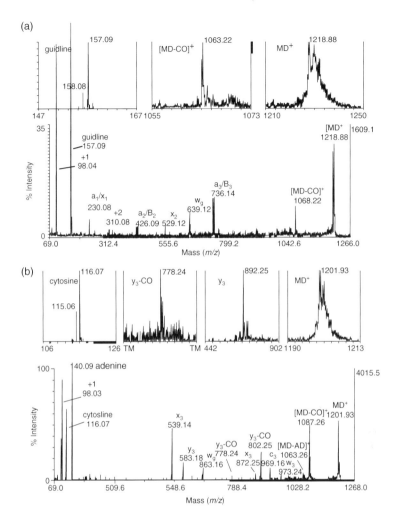

FIGURE 9.19 Tandem mass spectra of H/D-exchanged (*a*) UGUU and (*b*) UCUA with important regions enlarged above each spectrum. The UGUU spectrum is cut to 35% intensity of the base peak (guanine). The precursor ions (UGUU: *m/z* 1218.88 and UCUA: *m/z* 1201.93) show the high degree of deuteration required to perform the H/DX MS/MS analysis. When a signal is split into more than one peak, the most intense is assigned. (asterisk)*1: ribose derivate; (asterisk)*2: internal fragment containing uracil. (Reprinted by permission of Elsevier from Anderson, T. E., Kirpekar, F., and Haselmann, K. F. RNA fragmentation in MALDI mass spectrometry studies by H/D-Exchange: Mechanism of general applicability to nucleic acids. *J. Am. Soc. Mass Spectrom.* 2006, *17*, 1353–1368. (Copyright 2006 by the American Society for Mass Spectrometry.)

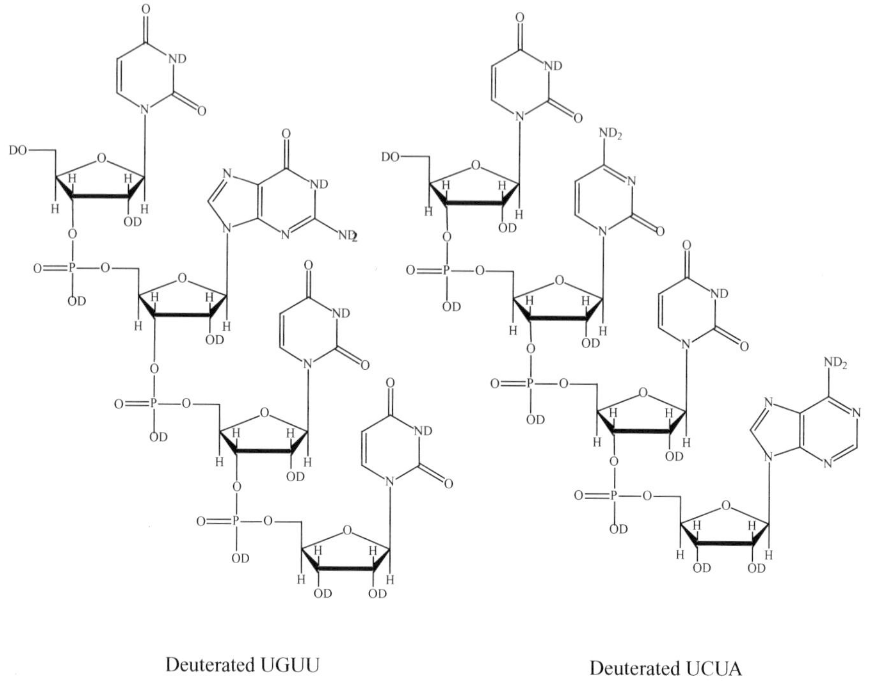
Deuterated UGUU
Deuterated UCUA

FIGURE 9.20 Structures of the RNA oligonucleotides UGUU and UCUA used in the study for the product ions illustrated in Figure 9.19.

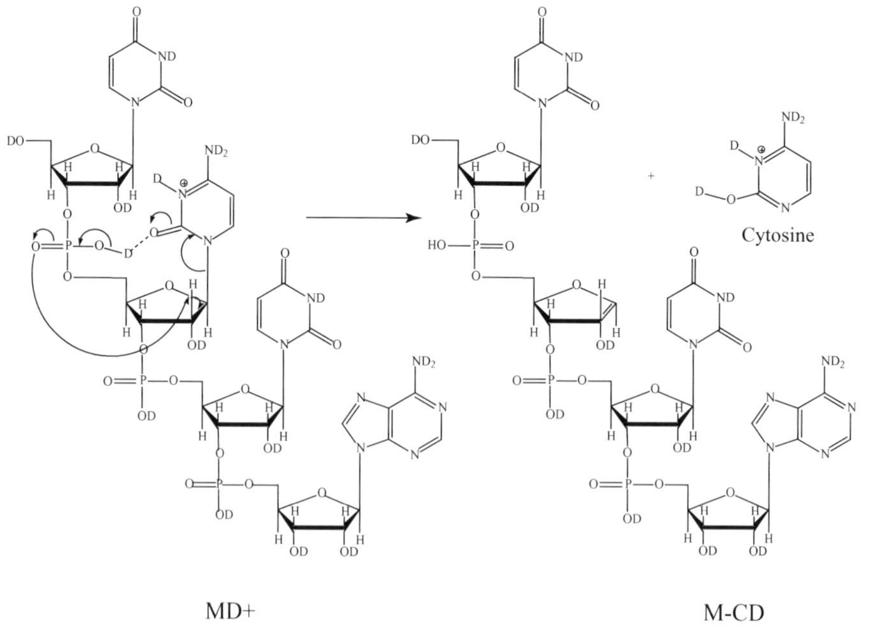
Cytosine
MD+
M-CD

FIGURE 9.21 Mechanism for the production of the charged cytosine base loss from the UCUA oligonucleotide.

9.4 PROBLEMS

9.1. What are the three main types of biological macromolecules measured with mass spectrometry?

9.2. What are hydrolysis and condensation reactions, and in what are they involved?

9.3. What are three uses of polysaccharides?

9.4. What differentiates α-D-glucose from β-D-glucose?

9.5. Describe a α-glycosidic bond and a β-glycosidic bond?

9.6. Explain the enhanced response observed in nanoelectrospray of oligosaccharides versus normal electrospray.

9.7. Why is there less informative fragmentation observed in the CID of larger oligosaccharides as compared to smaller? What types of cleavages are seen in the two?

9.8. What makes the analysis of oligosaccharides by mass spectrometry difficult?

9.9. Explain what happens for production of the β-type ion in negative ion mode.

9.10. Explain why in high-energy collisions metal ion adducts of oligosaccharides produce cross-ring cleavages.

9.11. Why is it difficult to completely sequence many polysaccharides that are observed in biological extracts?

9.12. What two distinct features differentiate DNA from RNA?

9.13. Describe the four cleavage sites along the phosphate backbone on nucleic acids from CID.

9.14. During CID of nucleic acids what contributes to the complexity of the product ion spectra?

9.15. What effects can the presence of salt have in the electrospray ionization of nucleic acids?

9.16. What advantage exits for ionization of oligonucleotides using MALDI?

9.17. What has been proposed for DNA product ion generation through the use of H/DX mass spectrometry?

REFERENCES

1. Bahr, U., Pfenninger, A., Karas, M., and Stahl, B. *Anal. Chem.* 1997, *69*, 4530–4535.
2. Harvey, D. J. *Rapid Commun. Mass Spectrom.* 1993, *7*, 614–619.
3. Hofmeister, G. E., Zhou, Z., and Leary, J. A. *J. Am. Chem. Soc.* 1991, *113*, 5964–5970.
4. Domon, B., and Costello, C. E. *Glycoconjugate J.* 1988, *5*, 397–409.

5. Cancilla, M. T., Penn, S. G., Carroll, J. A., and Lebrilla, C. B. *J. Am. Chem. Soc.* 1996, *118*, 6736–6745.
6. Clayton, E., and Bateman, R. H. *Rapid Commun. Mass Spectrom.* 1992, *6*, 719–720.
7. McLuckey, S. A., Van Berkel, G. J., and Glish, G. L. *J. Am. Soc. Mass Spectrom.* 1992, *3*, 60–70.
8. Cerny, R. L., Gross, M. L., and Grotjahn, L. *Anal. Biochem.* 1986, *156*, 424.
9. McLuckey, S. A., and Habibi-Goudarzi, S. *J. Am. Chem. Soc.* 1993, *115*, 12085.
10. Little, D. P., Thannhauser, T. W., and McLafferty, F. W. *Proc. Natl. Acad. Sci. USA* 1995, *92*, 2318–2322.
11. Wu, J., and McLuckey, S. A. *Int. J. Mass Spectrom.* 2004, *237*, 197–241.
12. Phillips, D. R., and McCloskey, J. A. *Int. J. Mass Spectrom. Ion Processes* 1993, *128*, 61–82.
13. Gross, J., Leisner, A., Hillenkamp, F., Hahner, S., Karas, M., Schafer, J. Lutzenkirchen, F., and Nordhoff, E. *J. Am. Soc. Mass Spectrom.* 1998, *9*, 866–878.
14. Wan, K. X., Gross, J., Hillenkamp, F., and Gross, M. L. *J. Am. Soc. Mass Spectrom.* 2001, *12*, 193–205.
15. Gross, J., Hillenkamp, F., Wan K. X., and Gross, M. L. *J. Am. Soc. Mass Spectrom.* 2001, *12*, 180–192.

CHAPTER 10

MALDI-TOF-POSTSOURCE DECAY (PSD)

10.1 INTRODUCTION

The technique of MALDI-TOF mass spectrometry affords an alternative approach for producing product ion spectra of analytes of interest. Although the fragmentation processes are not as efficient as other product ion spectra producing techniques such as collision-induced dissociation (CID) or electron capture dissociation, the approach does have advantages in cases where the MALDI-TOF mass spectrometer is the instrument to be used. MALDI is a soft ionization technique that produces singly charged analyte ions that are even electron. These may be the protonated form of the analyte $[M + H]^+$, or metal adducts such as sodium $[M + Na]^+$, lithium $[M + Li]^+$, potassium $[M + K]^+$, or silver $[M + Ag]^+$, all measured in positive ion mode, or deprotonated species measured in negative ion mode $[M - H]^-$. Often by adjusting the laser power to a level that is 10% above the threshold for desorption of species the charged analytes can be produced with very minimal fragmentation. This greatly facilitates in the reduction of the complexity of mass spectra. Fragmentation processes can be induced in the MALDI technique where the analyte produces product ions both within the source and outside of the source within the first field-free region (FFR). Most of the decay processes happen after the extraction of the analyte ions from the source into the first field-free region and is termed postsource decay (PSD). Energy is supplied to the charged analyte ion through a number of processes that induces the charged molecular analyte to undergo the prompt ion fragmentation. One process involves collisional activation (bimolecular decomposition) of the analyte as it is pulled through the dense plume of matrix material, neutrals, and other charged

Even Electron Mass Spectrometry with Biomolecule Applications   By Bryan M. Ham
Copyright © 2008 John Wiley & Sons, Inc.

analytes, which forms above the crystalline surface during the laser ablation by the ion extraction grid. The addition of delayed extraction has, however, reduced this effect as compared with continuous extraction of the charged analyte ions out of the source and into the first field-free region of the flight tube. A second process that attributes to prompt ion fragmentation of the desorbed analytes is the energetic transfer of a proton (or any matrix energy transfer) from the matrix to the analyte during the ionization process (unimolecular decomposition). Often postsource decay can be induced to take place with the use of slightly, to moderately higher laser flux settings than what is typically used during regular single-stage mass spectrum acquisition.

Reflectron-based time-of-flight mass spectrometers can take advantage of the postsource decay process to collect product ion mass spectra that can be used for studies of fragmentation pathways and structural characterization of analytes. In this chapter we will study some typical PSD spectra and the fragmentation pathways associated with the product ions observed in the spectra. Representative PSD spectra of unknowns are also presented for practice in structural elucidation of analyte species as analyzed by MALDI-TOF PSD mass spectrometry.

10.2 METASTABLE DECAY

Most of the postsource decay products are produced outside of the source and within the first field-free region. The result for small neutral losses from the precursor is a poorly or improperly focused analyte ion that has an increased peak width, usually with a greater fronting to the peak. This will greatly affect the resolution of the initial precursor ion in a deleterious fashion. This increase in peak width and subsequent decrease in resolution is due to the fact that the products produced in the postsource decay all have the same velocity as the precursor ion but contain different kinetic energies due to differences in mass. The reflectron electrostatic ion mirror is thus unable to properly focus the original precursor ion with the PSD products into the second field-free region where detection takes place. Figure 10.1 is a single-stage MALDI-TOF mass spectrum of 1,2-dipalmitoyl-*sn*-glycero-3-phosphoethanolamine

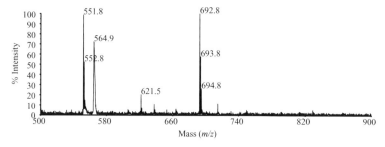

FIGURE 10.1   MALDI-TOF mass spectrum of 1,2-dipalmitoyl-*sn*-glycero-3-phosphoethanolamine as protonated analyte ion [M + H]$^+$ at m/z 692.8.

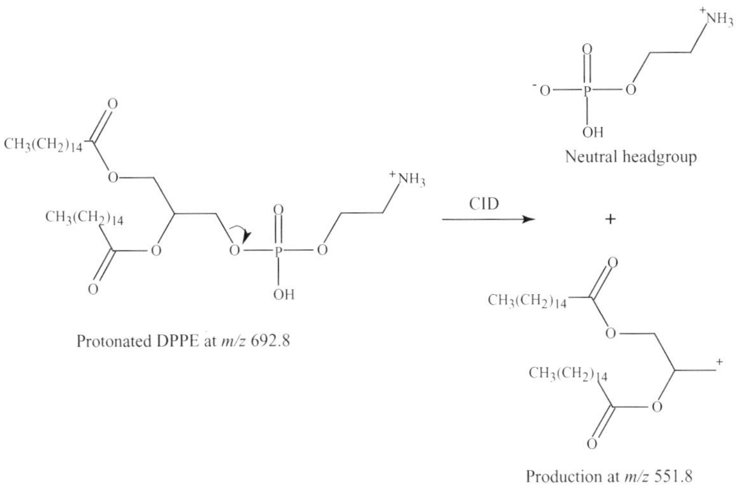

FIGURE 10.2 The in-source prompt ion fragmentation mechanism for the production of the m/z 551.8 species from the protonated DPPE precursor at m/z 692.8.

($C_{37}H_{74}NO_8P$, DPPE) as the protonated analyte ion $[M + H]^+$ at m/z 692.8 with an exact mass (protonated) of 692.523 Da and a resolution value of 2483 calculated as $m/\Delta m$, where m is the mass-to-charge (m/z) value obtained from the spectrum and Δm is the full peak width at half maximum (FWHM). The spectrum was collected using a matrix mixture consisting of 4-nitroaniline and butyric acid with 1% trifluoracetic acid (TFA). 4-Nitroaniline is a typical MALDI matrix used for phosphorylated lipid analysis while the butyric acid and TFA are acid matrix modifiers (often TFA is also added for increased solvation of the phosphorylated lipids into the MALDI matrix). Besides the protonated precursor ion at m/z 692.8 the spectrum also contains a peak at m/z 551.8 and m/z 564.9. The phosphorylated lipid species tend to suffer from prompt ion and metastable decay during the MALDI ionization process. The m/z 551.8 peak in the spectrum is formed through neutral headgroup loss of the DPPE lipid species $[PE + H - C_2H_8NO_4P]^+$. The decomposition mechanism for the production of the m/z 551.8 species from the protonated DPPE precursor at m/z 692.8 is illustrated in Figure 10.2. The m/z 551.8 peak is relatively sharp with a resolution value of 2171, and the carbon 13 (^{13}C) isotope peak is also resolved at m/z 552.8. This suggests that the m/z 551.8 peak was produced by in-source prompt ion fragmentation. The peak at m/z 564.9 was also formed by neutral headgroup loss of the DPPE lipid species; however, it has a broad peak shape with a resolution value of 442, and the carbon 13 isotope peak is not resolved, indicating that this peak was formed by postsource (metastable) decay in the first field-free region of the time-of-flight mass spectrometer prior to the reflectron. The spectrum in Figure 10.3 is a blow-up of the mass region containing these two fragment ions. The m/z 551.8 peak is sharp and well resolved, indicating focusing in the reflectron, while the m/z 564.9 peak is very broad and not resolved, indicating poor focusing in the reflectron.

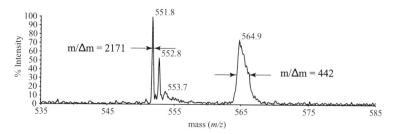

FIGURE 10.3   Resolution of m/z 551.8 in-source prompt ion fragment versus m/z 564.9 postsource metastable decay fragment.

10.3   ION MIRROR RATIO MEASUREMENT OF PSD SPECTRA

The ion gate acting as a timed ion selector is used to select the precursor ion for the PSD experiment. The ion gate is comprised of a series of wires in which alternating voltages can be applied usually at ± 1000 V. When the ion gate is switched on, no ions are allowed to pass through the gate, thus preventing the transmission of any ions through the remainder of the instrument. When the gate is switched off, all ions will be able to pass through the gate. By switching off the gate at a predetermined flight time, the precursor undergoing PSD will be allowed to pass through the gate into the reflectron electrostatic mirror for focusing and measurement. The postsource decay product ions possess different kinetic energies than the original precursor ion, thus allowing their separation within the reflectron electrostatic mirror and subsequent detection and measurement. After selecting the precursor ion undergoing PSD with the timed ion selector, the precursor is focused within the reflectron electrostatic mirror. If the reflectron voltage is reduced while maintaining a constant source voltage, the precursor ion undergoing PSD will possess too much kinetic energy to be properly reflected and measured. However, product ions produced from the postsource decay of the precursor ion, which possess less kinetic energy than the precursor ion, can be focused by the reflectron and transmitted to the detector for mass measurement. In general, if the reflector voltage is reduced to half of that needed to focus the precursor, a product ion with half the mass of the precursor will be focused. The ratio of the ion mirror voltage to the source voltage can be scanned, allowing a range of product ions to be focused by the ion mirror and subsequently detected.

10.4   POSTSOURCE DECAY OF PHOSPHATIDYLSERINE

We are now ready to look at a postsource decay spectrum of the phosphorylated lipid phosphatidylserine. Phosphatidylserine is a membrane phospholipid present in prokaryotic and eukaryotic cells and is found most abundantly in brain tissue. Phosphatidylserine is also a precursor biological molecule in the biosynthesis of phosphatidylethanolamine through a reaction that is catalyzed by PS decarboxylase.

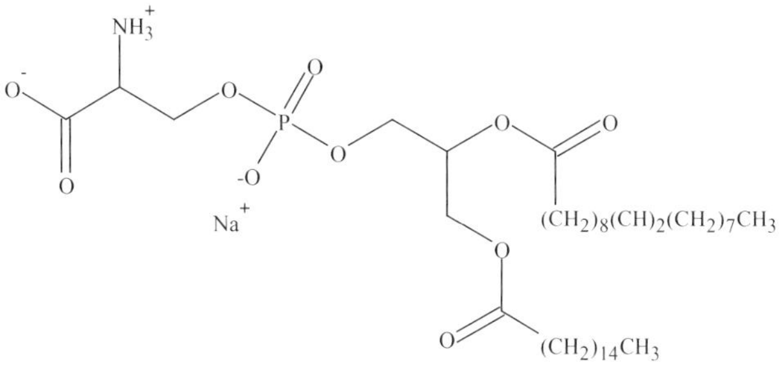

FIGURE 10.4 Structure of 1-palmitoyl-2-oleoyl-*sn*-glycero-3-[phospho-L-serine] (PS, sodium salt).

At physiological pH (7.38) phosphatidylserine carries a full negative charge and is often referred to as an anionic phosphorylated lipid. Figure 10.4 illustrates the structure of the phosphatidylserine 1-palmitoyl-2-oleoyl-*sn*-glycero-3-[phospho-L-serine] (PS, sodium salt) as a neutral sodium adduct showing the three moieties that are charged at physiological pH. The phosphoryl moiety (R_2-HPO$_4$) has a pK_a of 2.6, a relatively strong acid that is 100% ionized at physiological pH. The carboxylic acid moiety (R-COOH) has a pK_a of 5.5 and is also 100% ionized at physiological pH. The amine moiety (R $-$ NH$_3{}^+$) has a pK_a of 11.55, a relatively strong base that is 100% ionized at physiological pH. Due to the polarity of the headgroup phosphatidylserine, as with all of the anionic phosphorylated lipids, is often observed in the gas phase as either a protonated, sodium adduct [phospholipid + Na]$^+$ or as a deprotonated, disodium adduct [phospholipid + 2Na $-$ H]$^+$. The structural consideration of a molecule is very helpful in understanding fragmentation pathways that produce the product ions that are observed in product ion spectra. From careful consideration of the structure of phosphatidylserine, it can be seen that the product ion spectrum may contain peaks associated with the two fatty acid substituents and also possible product ions obtained from the polar headgroup. When possible, the study of the fragmentation behavior of a known standard molecule that is similar to the unknown that the researcher is attempting to identify can be tremendously helpful.

We will now consider the postsource decay product ion spectrum of the single sodium adduct analyte ion form of 1-palmitoyl-2-oleoyl-*sn*-glycero-3-[phospho-L-serine] as [phospholipid + Na]$^+$. By systematically going through the spectrum peak by peak and matching to fragmentation pathways that are reasonable, the researcher begins to develop the ability to assign product ion spectra to possible structures of unknown species under investigation. Figure 10.5 is the postsource decay product ion spectrum of 1-palmitoyl-2-oleoyl-*sn*-glycero-3-[phospho-L-serine] (also designated as (16:0/18:1-PS) [PS + Na]$^+$). The molecular formula for the protonated sodium adduct form of the analyte ion is $C_{40}H_{76}NO_{10}PNa$ with an exact mass of 784.510 Da. Upon initial inspection of the PSD spectrum we observe that there are four major product ions at *m/z* 697, 599, 577, and 208 and the precursor ion at *m/z* 784. This phosphatidylserine contains a 16-carbon fatty acid substituent ($C_{16}H_{32}O_2$)

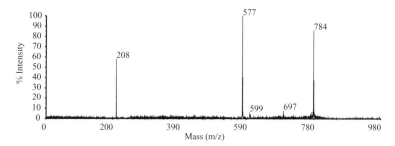

FIGURE 10.5   Postsource decay product ion spectrum of 1-palmitoyl-2-oleoyl-*sn*-glycero-3-[phospho-L-serine] as the protonated sodium adduct form of the analyte ion [$C_{40}H_{76}NO_{10}P$ + Na]$^+$ at *m/z* 784.

that has a neutral molecular weight of 256 Da and an 18-carbon fatty acid substituent ($C_{18}H_{34}O_2$) containing one unsaturation at a molecular weight of 282 Da. For product ions associated with losses of these two fatty acid substituents, the product ions would possibly be in the range of *m/z* 528 for neutral loss of the C16:0 fatty acid substituent and *m/z* 502 for neutral loss of the C18:1 fatty acid substituent. The *m/z* 577 product ion is in close proximity to these values, but it does not appear that the product ions observed in the PSD spectrum are directly associated with neutral loss of either fatty acid substituent. Often it is instructional to visually break apart the known structure of the analyte undergoing the PSD to better understand possible product ion molecular weights. Figure 10.6 illustrates the structure of 1-palmitoyl-2-oleoyl-*sn*-glycero-3-[phospho-L-serine] broken into four separate sections along with the associated molecular weight of that fragmented portion of the lipid structure. The product ion with the lowest molecular weight in Figure 10.5 has a mass-to-charge value of *m/z* 208. In Figure 10.6 the PS headgroup has a neutral molecular weight of 185 Da. It is helpful to remember that the precursor analyte ion of the phosphatidylserine lipid in the PSD spectrum of Figure 10.5 is a sodium adduct. If the atomic weight of sodium (Na, 22.98977 amu) is added to that of the neutral PS headgroup of 185 Da, the result is 23 + 185 = 208, which matches the value of the *m/z* 208 product ion. It is therefore reasonable that the *m/z* 208 PSD product ion is the sodiated form of the phosphatidylserine headgroup. Figure 10.7 illustrates a mechanism for the production of the sodiated phosphatidylserine headgroup at *m/z* 208. The C16:0 and the C18:1 fatty acid substituents are lost as a neutral joined together by the acyl glycerol backbone. The product ions *m/z* 577, 599, and 697 all have in common that they are odd mass values indicating an even number of nitrogen (e.g., 0N, 2N, 4N, . . .).

Because phosphatidylserine contains an odd number of nitrogen (1N) suggests that these product ions are formed through some form of neutral headgroup loss that includes the one nitrogen. The difference in mass between the precursor ion at *m/z* 784 and the *m/z* 697 product ion is 87 amu and most likely contains nitrogen. In Figure 10.6 the entire PS headgroup has a molecular weight of 185 Da; therefore,

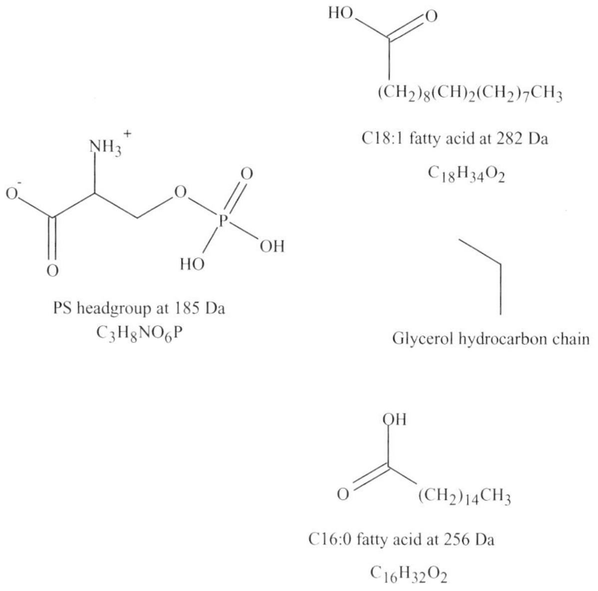

HO

O

$(CH_2)_8(CH)_2(CH_2)_7CH_3$

C18:1 fatty acid at 282 Da

$C_{18}H_{34}O_2$

NH_3^+

O⁻

O

O

P

OH

HO

O

PS headgroup at 185 Da

$C_3H_8NO_6P$

Glycerol hydrocarbon chain

OH

O

$(CH_2)_{14}CH_3$

C16:0 fatty acid at 256 Da

$C_{16}H_{32}O_2$

FIGURE 10.6 Fragmented structure of 1-palmitoyl-2-oleoyl-*sn*-glycero-3-[phospho-L-serine] illustrating the four primary sections making up the lipid species and associated molecular weights.

the *m/z* 697 product ion is formed through neutral loss of a portion or fragment of the PS headgroup and not from the neutral loss of the entire headgroup. The nitrogen-containing portion of the headgroup, which excludes the PO_4 phosphate group, is comprised of $C_3H_6NO_2$ at a mass of 88 amu. This is 1 amu heavier than the

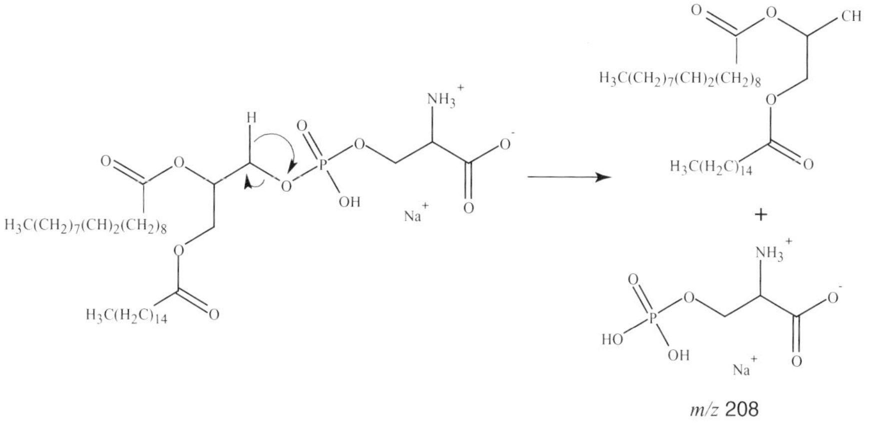

m/z 208

FIGURE 10.7 Mechanism for production of the *m/z* 208 product ion.

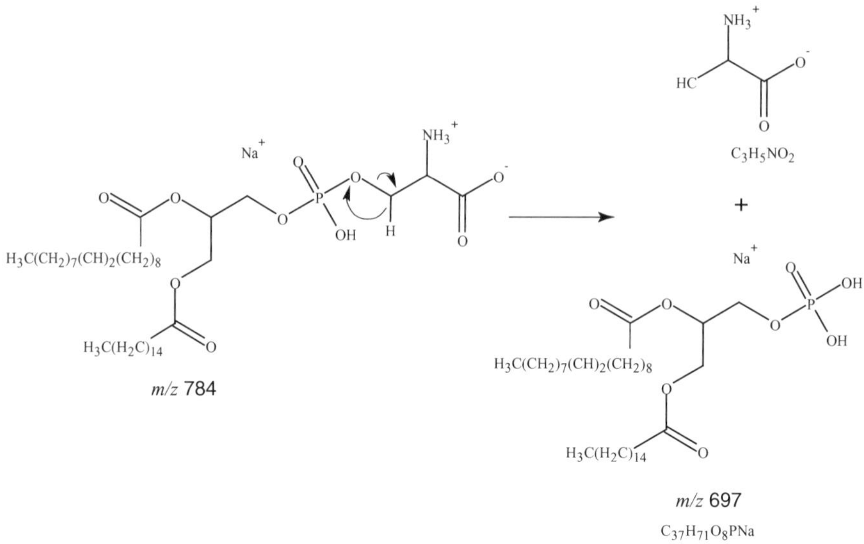

FIGURE 10.8 Mechanism for the production of the m/z 697 product ion $[C_{40}H_{76}NO_{10}P + Na - C_3H_6NO_2]^+$.

neutral loss value that is calculated for the m/z 697 product ion, which would equate to the value of one hydrogen (H at 1.00794 amu). Therefore, it is reasonable that the m/z 697 product ion is formed through neutral loss of the $C_3H_6NO_2$ portion of the PS headgroup minus the transfer of one hydrogen. The mechanism for the production of the m/z 697 product ion $[C_{40}H_{76}NO_{10}P + Na - C_3H_6NO_2]^+$ is illustrated in Figure 10.8. It is interesting to note that the phosphatidylserine headgroup has fragmented in a pathway that has produced product ions where the sodium cation is associated with the non-nitrogen-containing product. The fragmentation pathway illustrated in Figure 10.7, which produces the m/z 208 product ion, results in the headgroup being associated with the sodium cation as the second most major product ion as illustrated in the PSD spectrum of Figure 10.5. Both of these pathways suggest that the sodium cation is associated with the phosphatidylserine headgroup portion of the phosphorylated lipid precursor ion at m/z 784 and that its affinity is more closely associated with the phosphate moiety of the headgroup versus the carboxylic acid or amine moieties. Two remaining product ions in the PSD spectrum of Figure 10.5 are the m/z 599 and the m/z 577 ions. In Figure 10.6 the neutral phosphatidylserine headgroup has a molecular weight of 185 Da. If this is subtracted from the precursor analyte ion at m/z 784, the resultant value is 599 Da, thus indicating that the m/z 599 product ion is formed through neutral loss of the phosphatidylserine headgroup $[C_{40}H_{76}NO_{10}P + Na - C_3H_8NO_6P]^+$. The difference between the m/z 577 and m/z 599 product ions is 22 amu, indicating the involvement of sodium and one hydrogen between the two product ions. Figure 10.9 illustrates the mechanisms involved in the production of the m/z 577 and m/z 599 ions observed in the PSD spectrum of

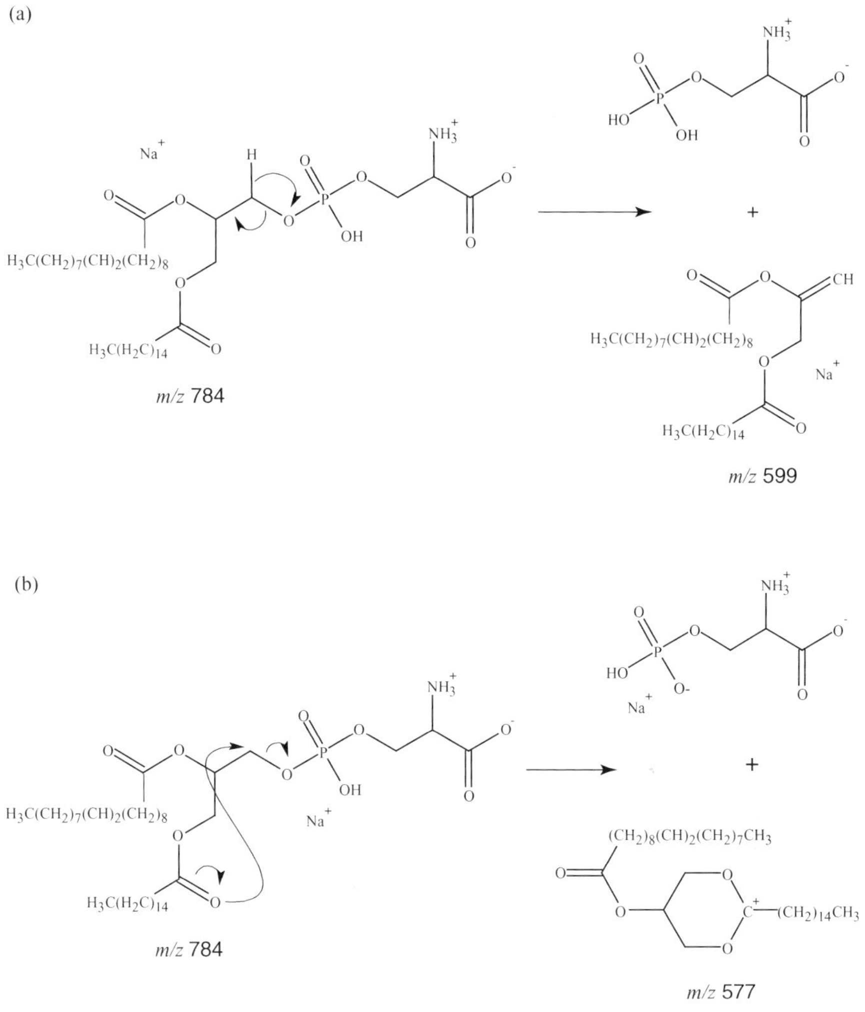

(a)

m/z 784

m/z 599

(b)

m/z 784

m/z 577

FIGURE 10.9 Mechanism for the production of (*a*) the *m/z* 599 product ion through neutral phosphatidylserine headgroup loss in a protonated form and (*b*) the *m/z* 577 product ion through neutral phosphatidylserine headgroup loss in a sodiated form.

Figure 10.5. The mechanism depicted in Figure 10.9*a* for the production of the *m/z* 599 product ion involves the neutral loss of the phosphatidylserine headgroup in a protonated form where the association of the sodium cation has stayed with the acylglycerol portion of the phosphorylated lipid. This mechanism has produced a sodiated alkene as the product ion at *m/z* 599. In contrast to this the mechanism in Figure 10.9*b* for the production of the *m/z* 577 product ion involves the neutral loss of the phosphatidylserine headgroup in a sodiated form. This fragmentation pathway has produced a six-membered ring as the most likely stable product ion. In the fragmentation pathway mechanism for the production of the m/z 577 product ion, the sodium has associated itself with the headgroup versus the acyl glycerol portion of the phosphorylated lipid. The relative response of the m/z 577 product ion in the PSD product ion spectrum of Figure 10.5 is clearly much greater than the relative response of the m/z 599 product ion. This suggests that the production of the m/z 577 product ion is the favored pathway over the pathway that produces the m/z 599 product ion. This also indicates that the sodium ion is primarily associated with the negatively charged oxygen in the phosphoryl group. The m/z 208 product ion in Figure 10.5 is a predominant product ion and is comprised of the neutral headgroup sodium adduct. The m/z 208 product ion also has a similar fragmentation pathway to that forming m/z 577, with the exception that a proton is transferred to the head group, thus charge is retained on the latter. The singly sodiated molecule where the sodium is adducted to the phosphoryl headgroup, and not the acylglycerol portion of the lipid, is thus further supported as the predominant form of the precursor.

10.4.1 Problem 10.1

The MALDI-TOF PSD product ion mass spectrum illustrated in Figure 10.10 is of the deprotonated double sodium adduct of 1-palmitoyl-2-oleoyl-sn-glycero-3-[phospho-L-serine] at m/z 806 ((16:0/18:1-PS), [PS + 2Na − H]$^+$). The molecular formula for the deprotonated double sodium adduct form of the analyte ion is $C_{40}H_{75}NO_{10}PNa_2$ with an exact mass of 806.492 Da. The analyte ion has the same structure as that illustrated in Figure 10.4 but with an extra sodium cation added.

1. Assign possible structures for the product ions illustrated in the PSD spectrum in Figure 10.10.

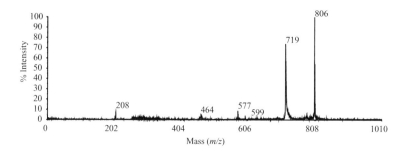

FIGURE 10.10 Deprotonated double sodium adduct of 1-palmitoyl-2-oleoyl-sn-glycero-3-[phospho-L-serine] at m/z 806 ((16:0/18:1-PS), [PS + 2Na − H]$^+$). The molecular formula for the deprotonated double sodium adduct form of the precursor analyte ion is $C_{40}H_{75}NO_{10}PNa_2$ with an exact mass of 806.492 Da.

2. Draw mechanisms that would explain the production of the product ion m/z values that are illustrated in the spectrum. Also include structures for the neutral loss product.
3. Do the relative peak intensities agree with those found in Figure 10.5? If not, can any conclusions be drawn concerning the stability of the precursor ion in Figure 10.10 versus the precursor ion of Figure 10.5?
4. Is there more information here concerning the location of the sodium cations?

10.5 POSTSOURCE DECAY OF PHOSPHATIDYLCHOLINES

Phosphatidylcholines participate directly in protein regulation and function within biological membranes, constitute the bilayer components of membranes, and help to determine the physical properties of these membranes. The phosphatidylcholines are zwitterionic lipid species that contain a polar headgroup; however, the lipid is neutral overall. Figure 10.11a illustrates the structure of 1,2-dimyristoyl-sn-glycero-3-phosphocholine (14:0/14:0 DMPC) as a neutral lipid species. Figure 10.11b depicts

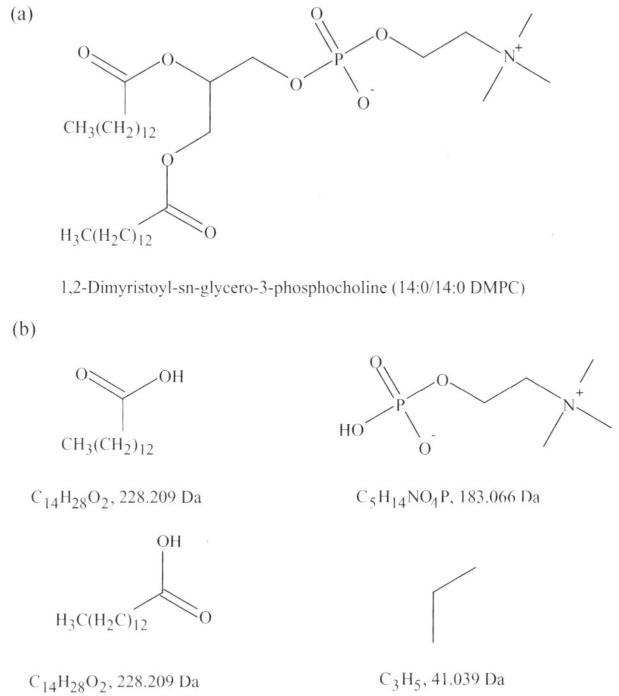

(a)

CH₃(CH₂)₁₂

H₃C(H₂C)₁₂

1,2-Dimyristoyl-sn-glycero-3-phosphocholine (14:0/14:0 DMPC)

(b)

CH₃(CH₂)₁₂

$C_{14}H_{28}O_2$, 228.209 Da

$C_5H_{14}NO_4P$, 183.066 Da

H₃C(H₂C)₁₂

$C_{14}H_{28}O_2$, 228.209 Da

C_3H_5, 41.039 Da

FIGURE 10.11 (a) Structure of 1,2-dimyristoyl-sn-glycero-3-phosphocholine (14:0/14:0 DMPC) as a neutral lipid species. (b) Fictitious fragmentation of the DMPC lipid breaking the structure into its major components and associated molecular weights.

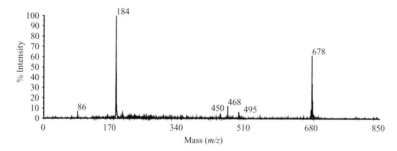

FIGURE 10.12 PSD spectrum of protonated dimyristoyl phosphatidylcholine (14:0/14:0-DMPC) at m/z 678.

a fictitious fragmentation of the DMPC lipid breaking the structure into its major components. Molecular weights are also provided for the lipid-fragmented components. Again, this approach to structure elucidation from product ion spectra is often helpful in constructing reasonable fragmentation pathways and product ion structures. Figure 10.12 is a postsource decay product ion spectrum of 1,2-dimyristoyl-*sn*-glycero-3-phosphocholine as an acid adduct at m/z 678 [DMPC + H]$^+$. The molecular formula of the protonated form of the precursor analyte ion of DMPC is $C_{36}H_{73}NO_8P$ with an exact mass of 677.4996 Da. Besides the precursor ion at m/z 678 there are five major product ions included in the postsource decay product ion spectrum of DMPC. In Figure 10.11*b* the phosphatidylcholine headgroup fragment has a molecular weight of 183 Da, which is a mass unit off from the predominant product ion m/z 184 in the PSD spectrum in Figure 10.12. The product ion at m/z 184 represents the charged, protonated form of the phosphatidylcholine headgroup. The mechanism for the production of the m/z 184 phosphatidylcholine headgroup product ion is illustrated in Figure 10.13. This product ion at m/z 184 is a diagnostic ion used in PSD spectra for the identification of a phosphatidylcholine headgroup containing lipid species such

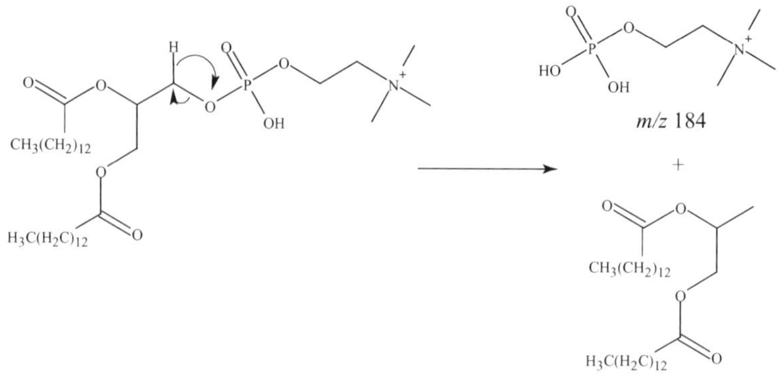

FIGURE 10.13 Mechanism for the production of the m/z 184 charged, protonated phosphatidylcholine headgroup. Diagnostic for phosphatidylcholine headgroup identification.

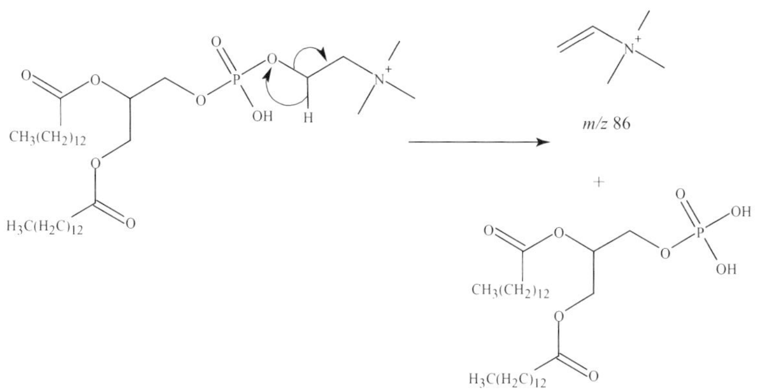

FIGURE 10.14 Mechanism for the production of the m/z 86 phosphatidylcholine headgroup fragment.

as a sphingolipid or a phosphatidylcholine lipid. In the low mass region of the PSD spectrum, there is another even mass product ion at m/z 86, indicating the presence of one nitrogen also in this product ion similar to the m/z 184 protonated phosphatidylcholine headgroup product ion. This product ion is most likely produced through further fragmentation of the 183-Da phosphatidylcholine headgroup. Figure 10.14 illustrates the fragmentation pathway mechanism for the production of the m/z 86 product ion, a fragment of the phosphatidylcholine headgroup resulting in an alkene. The fragmentation pathway that produces the m/z 184 charged phosphatidylcholine headgroup would be expected to have a similar, related pathway described as neutral loss of the phosphatidylcholine headgroup. This would be derived by subtracting the neutral molecular weight of the phosphatidylcholine headgroup as listed in Figure 10.11b as 183 Da from the precursor analyte ion's m/z value of 678 Thomson (i.e., $678 - 183 = 495$). The PSD product ion spectrum of Figure 10.12 does contain a product ion at m/z 495 derived from neutral loss of the PC headgroup $[M + H - C_{14}H_{14}NO_4P]^+$. This type of loss is also diagnostic in the determination of the phosphorylated lipid's headgroup identification. The mechanism for production of the m/z 495 product ion would be similar to that depicted in Figure 10.13 for the production of the protonated phosphatidylcholine headgroup, except that the associated hydrogen would stay with the acylglycerol portion of the phosphorylated lipid, resulting in neutral loss of the phosphatidylcholine headgroup.

The last two remaining product ions in the PSD spectrum of Figure 10.12, m/z 450 and m/z 468, are even-numbered product ions, thus indicating the containment of the nitrogen of the phosphatidylcholine headgroup within the product ion structures. In the PSD spectrum the difference between the precursor analyte ion at m/z 678 and the m/z 450 product ion is 228 Da. In Figure 10.11b the molecular weight of the C14:0 myristyl fatty acid substituent is 228 Da. This indicates that the m/z 450 product ion is formed through the neutral loss of the C14:0 fatty acid $[M + H - C_{14}H_{28}O_2]^+$.

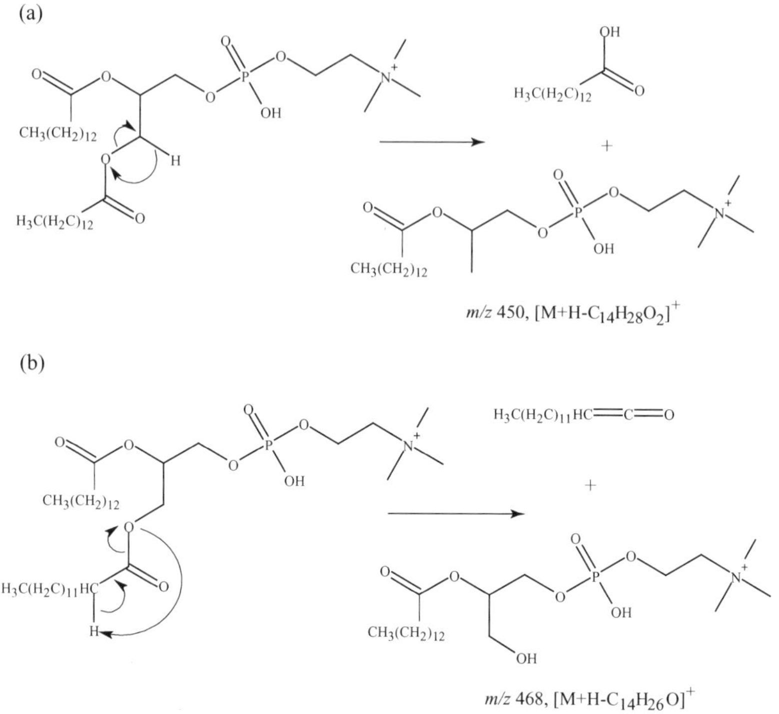

(a)

m/z 450, $[M+H-C_{14}H_{28}O_2]^+$

(b)

m/z 468, $[M+H-C_{14}H_{26}O]^+$

FIGURE 10.15   (*a*) Formation of the m/z 450 product ion through neutral loss of C14:0 fatty acid substituent $[M + H - C_{14}H_{28}O_2]^+$. (*b*) Formation of the m/z 468 product ion through neutral loss of C14:0 fatty acyl chain as a ketene $[M + H - C_{14}H_{26}O]^+$.

This type of loss is used as a diagnostic tool in the identification of the fatty acid substituents in the precursor analyte ion at m/z 678 and can be applied to unknown fatty acyl substituent identifications. The product ion at m/z 468 is 18 amu different from the m/z 450 product ion. It is reasonable to approach the identification of this product ion as involving the loss of a fatty acid substituent with the retainment of the phosphatidylcholine headgroup (remember the product ion has an even mass number of m/z 468 indicating the inclusion of one nitrogen within its structure). In fact the m/z 468 product ion is produced through neutral loss of a C14:0 fatty acyl chain as a ketene $[M + H - C_{14}H_{26}O]^+$. The mechanisms for the fragmentation pathways that produce the m/z 450 and m/z 468 product ions are illustrated in Figure 10.15*a* and *b*.

10.5.1 Problem 10.2

Using the mechanism illustrated in Figure 10.13 draw a mechanism that describes the fragmentation pathway for the production of the m/z 495 product ion.

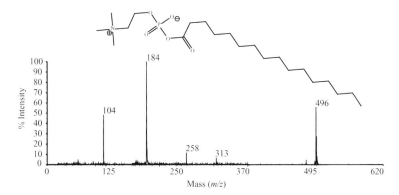

FIGURE 10.16 MALDI-TOF PSD product ion mass spectrum of the protonated form of 1-palmitoyl-2-hydroxy-*sn*-glycero-3-phosphocholine (16:0-Lyso-PC) at m/z 496, [Lyso-PC + H]$^+$.

10.5.2 Problem 10.3

The MALDI-TOF PSD product ion mass spectrum illustrated in Figure 10.16 is of the protonated form of 1-palmitoyl-2-hydroxy-*sn*-glycero-3-phosphocholine (16:0-Lyso-PC) at m/z 496, [Lyso-PC + H]$^+$. The molecular formula for the protonated form of the precursor analyte ion is $C_{24}H_{51}NO_7P$ with an exact mass of 496.340 Da. The precursor ion's structure is illustrated in Figure 10.16.

1. Assign possible structures for the product ions illustrated in the spectrum.
2. Draw mechanisms that would explain the production of the product ion m/z values that are illustrated in the spectrum. Also include structures for the neutral loss product.

10.6 POSTSOURCE DECAY OF PHOSPHATIDYLGLYCEROL

Phosphatidylglycerol (PG) is another phosphorylated lipid component that is incorporated into many biological membranes. PG is found in small amounts in most animal tissue, comprising approximately 1–2%. Much higher levels of PG have been observed in human lung surfactant at levels of approximately 11% of the total lipids. The biosynthetic route for the production of PG involves phosphatidic acid as a precursor through a series of enzymatic reactions. PG itself is a precursor for the biosynthesis of diphosphatidylglycerol (cardiolipin). Figure 10.17a illustrates the structure of 1-palmitoyl-2-oleoyl-*sn*-glycero-3-[phospho-*rac*-(1-glycerol)] (16:0/18:1-POPG) as a neutral sodium salt with a molecular formula of $C_{40}H_{76}O_{10}PNa$, and an exact mass of 770.5074 Da. Figure 10.17b illustrates the structure of 16:0/18:1-POPG broken into its main components along with their associated molecular formulas and

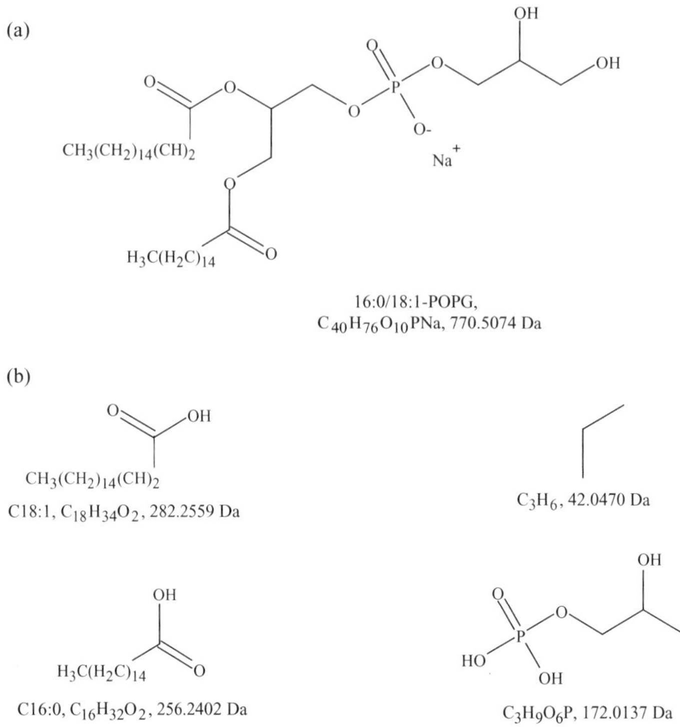

(a)

16:0/18:1-POPG,
$C_{40}H_{76}O_{10}PNa$, 770.5074 Da

(b)

$C18:1, C_{18}H_{34}O_2$, 282.2559 Da

C_3H_6, 42.0470 Da

$C16:0, C_{16}H_{32}O_2$, 256.2402 Da

$C_3H_9O_6P$, 172.0137 Da

FIGURE 10.17 (*a*) Structure of 1-palmitoyl-2-oleoyl-*sn*-glycero-3-[phospho-*rac*-(1-glycerol)] (16:0/18:1-POPG) as a neutral sodium salt with a molecular formula of $C_{40}H_{76}O_{10}PNa$, and an exact mass of 770.5074 Da. (*b*) Structure of 16:0/18:1-POPG broken into its main components along with their associated molecular formulas and molecular weights.

molecular weights. This includes the C16:0 myristic acid fatty acid substituent $C_{16}H_{32}O_2$, 256.2402 Da, the C18:1 oleic acid fatty acid substituent $C_{18}H_{34}O_2$, 282.2559 Da, and the phosphatidylglycerol headgroup $C_3H_9O_6P$, 172.0137 Da. Figure 10.18 is the PSD spectrum of the deprotonated, double sodium adduct of 1-palmitoyl-2-oleoyl-*sn*-glycero-3-[phospho-*rac*-(1-glycerol)] (16:0/18:1-POPG) [PG + 2Na − H]$^+$ at *m/z* 793. In this PSD spectrum there is information concerning both the two fatty acid substituents and the phosphoglycerol headgroup. The most predominant peak in the PSD spectrum aside from the precursor ion at *m/z* 793 is the *m/z* 199 product ion. This weight is closer to the phosphatidylglycerol headgroup illustrated in Figure 10.17*b*, which has a molecular weight of 172 Da for a molecular formula of $C_3H_9O_6P$, than to the fatty acid substituents that are heavier. In calculating possible headgroup adduct weights, the addition of a sodium cation to the phosphatidylglycerol headgroup results in a weight of 195 Da, which is close to the *m/z* 199 product ion but not exact. A deprotonated double sodium salt of the

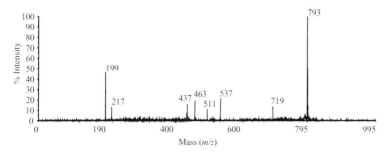

FIGURE 10.18   PSD spectrum of the deprotonated, double sodium adduct of 1-palmitoyl-2-oleoyl-*sn*-glycero-3-[phospho-*rac*-(1-glycerol)] (16:0/18:1-POPG) [PG + 2Na – H]$^+$ at *m/z* 793.

phosphatidylglycerol headgroup results in a weight of 217 Da, which is now heavier than the *m/z* 199 product ion that is under consideration in the PSD spectrum. The PSD spectrum illustrated in Figure 10.18 does contain a product ion at *m/z* 217 for the double sodium salt of the phosphatidylglycerol headgroup. The *m/z* 199 product ion is 18 amu less than the *m/z* 217 product ion, which suggests the difference of an H_2O group. Figure 10.19 illustrates the mechanisms responsible for the fragmentation pathways producing the *m/z* 217 and *m/z* 199 product ions. The production of the *m/z* 199 product ion is a two-step process where the phosphatidylglycerol headgroup is first lost as a double sodium adduct giving the *m/z* 217 product ion. The *m/z* 217 product ion then further undergoes decomposition where a molecule of water is lost, thus producing the *m/z* 199 product ion (both product ion structures are illustrated in Figure 10.19).

The *m/z* 719 product ion in Figure 10.18 has a difference of 74 amu less than the precursor analyte ion at *m/z* 793. It is likely that this product ion is formed through a neutral loss of a phosphatidylglycerol headgroup fragment. The fragmentation pathway for the production of the *m/z* 719 product ion is illustrated in Figure 10.20 and involves neutral loss of a phosphatidylglycerol headgroup fragment as an alkene. The three product ions *m/z* 719, *m/z* 217, and *m/z* 199 all involve neutral losses of the phosphoglycerol headgroup as either double sodium salts as an alkene fragment. The two fatty acid substituents can be identified in the PSD spectrum of Figure 10.18 through two separate fragmentation pathways that lead to the production of the *m/z* 437, 463, 511, and 537 product ions. Involving the C18:1 fatty acid substituent, there is a product ion formed as a result of neutral loss of the C18:1 fatty acid at *m/z* 511 [M – H + 2Na – $C_{18}H_{34}O_2$]$^+$. There is a second product ion that is related to the *m/z* 511 ion at *m/z* 437, which is formed through a neutral loss of a headgroup fragment and the associated fatty acid substituent [M – H + 2Na – $C_3H_6O_2$ – $C_{18}H_{34}O_2$]$^+$. The mechanism for this two-step loss forming first the *m/z* 511 product ion followed by the formation of the *m/z* 437 product ion is illustrated in Figure 10.21.

The fragmentation pathway mechanism for the production of the *m/z* 537 product ion is analogous to the mechanism that describes the production of the *m/z* 511

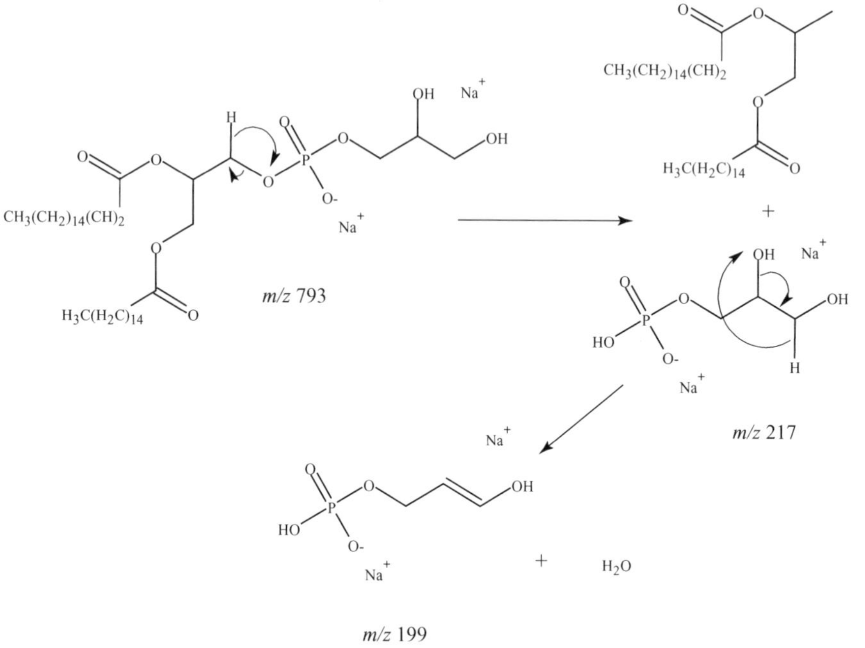

FIGURE 10.19   Mechanisms responsible for the fragmentation pathways producing the *m/z* 217 and *m/z* 199 product ions. The *m/z* 199 product ion is a two-step process where the phosphatidylglycerol headgroup is first lost as a double sodium adduct giving the *m/z* 217 product ion. The *m/z* 217 product ion then further undergoes decomposition where a molecule of water is lost, thus producing the *m/z* 199 product ion.

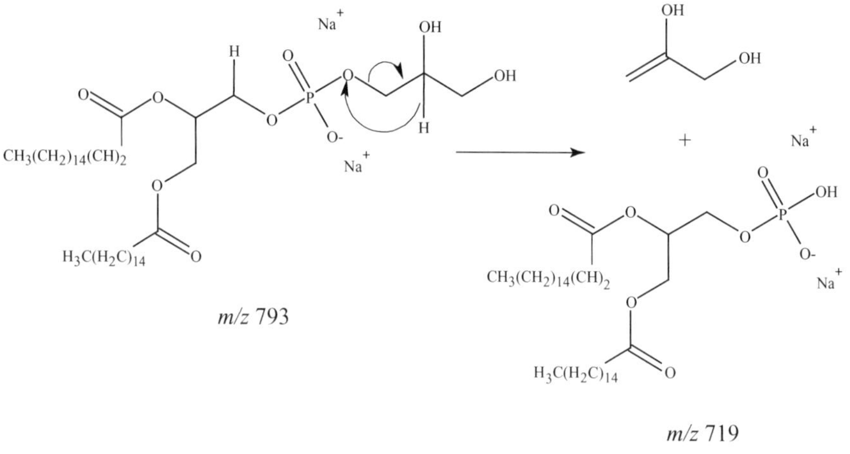

FIGURE 10.20   Mechanism for the production of the *m/z* 719 product ion through neutral loss of a phosphatidylglycerol headgroup fragment as an alkene.

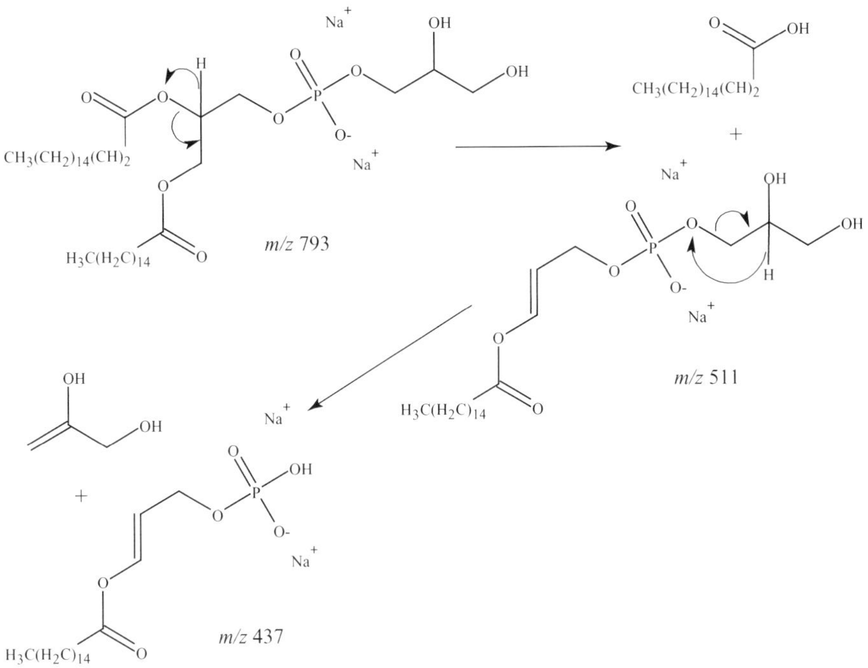

FIGURE 10.21 Mechanism for the production of the m/z 511 and 437 product ions derived through neutral loss of C18:1 fatty acid substituent and associated phosphatidylglycerol headgroup fragment.

product ion illustrated in Figure 10.21. The m/z 537 product ion is formed through neutral loss of the C16:0 fatty acid substituent $[M - H + 2Na - C_{16}H_{32}O_2]^+$. The second product ion, which is related to the m/z 537 ion at m/z 463, is formed through a neutral loss of a headgroup fragment and the associated C16:0 fatty acid substituent $[M - H + 2Na - C_3H_6O_2 - C_{16}H_{32}O_2]^+$.

10.6.1 Problem 10.4

Draw the mechanisms that describe the production of the m/z 537 and m/z 463 product ions involving the loss of the C16:0 fatty acid substituent and the phosphatidylglycerol headgroup fragment.

10.6.2 Problem 10.5

The MALDI-TOF PSD product ion mass spectrum illustrated in Figure 10.22 is of the protonated sodium adduct of 1-palmitoyl-2-oleoyl-*sn*-glycero-3-[phospho-*rac*-(1-glycerol)] (16:0/18:1-POPG) at m/z 771, $[POPG + Na]^+$. The molecular formula for the protonated sodium adduct form of the precursor analyte ion is $C_{40}H_{77}O_{10}PNa$

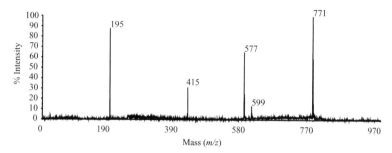

FIGURE 10.22 MALDI-TOF PSD product ion mass spectrum of the protonated sodium adduct of 1-palmitoyl-2-oleoyl-*sn*-glycero-3-[phospho-*rac*-(1-glycerol)] at *m/z* 771 ((16:0/18:1-POPG), [POPG + NA]$^+$).

with an exact mass of 771.515 Da. The precursor ion has the same structure as that illustrated in Figure 10.17*a* but with an extra proton added.

1. Assign possible structures for the product ions illustrated in the spectrum.
2. Draw mechanisms that would explain the production of the product ion *m/z* values that are illustrated in the spectrum. Also include structures for the neutral loss product.

ATOMIC WEIGHTS AND ISOTOPIC COMPOSITIONS

Symbol	Relative Atomic Mass	Abundance	Standard Atomic Weight
^{1}H	1.0078250321(4)	99.9885(70)	1.00794(7)
^{2}D	2.0141017780(4)	0.0115(70)	
^{3}T	3.0160492675(11)		
^{3}He	3.0160293097(9)	0.000137(3)	4.002602(2)
^{4}He	4.0026032497(10)	99.999863(3)	
^{6}Li	6.0151223(5)	7.59(4)	6.941(2)
^{7}Li	7.0160040(5)	92.41(4)	
^{9}Be	9.0121821(4)	100	9.012182(3)
^{10}B	10.0129370(4)	19.9(7)	10.811(7)
^{11}B	11.0093055(5)	80.1(7)	
^{12}C	12.0000000(0)	98.93(8)	12.0107(8)
^{13}C	13.0033548378(10)	1.07(8)	
^{14}C	14.003241988(4)		
^{14}N	14.0030740052(9)	99.632(7)	14.0067(2)
^{15}N	15.0001088984(9)	0.368(7)	
^{16}O	15.9949146221(15)	99.757(16)	15.9994(3)
^{17}O	16.99913150(22)	0.038(1)	
^{18}O	17.9991604(9)	0.205(14)	
^{19}F	18.99840320(7)	100	18.9984032(5)
^{20}Ne	19.9924401759(20)	90.48(3)	20.1797(6)
^{21}Ne	20.99384674(4)	0.27(1)	

(*continued*)

Even Electron Mass Spectrometry with Biomolecule Applications   By Bryan M. Ham
Copyright © 2008 John Wiley & Sons, Inc.

Symbol	Relative Atomic Mass	Abundance	Standard Atomic Weight
^{22}Ne	21.99138551(23)	9.25(3)	
^{23}Na	22.98976967(23)	100	22.989770(2)
^{24}Mg	23.98504190(20)	78.99(4)	24.3050(6)
^{25}Mg	24.98583702(20)	10.00(1)	
^{26}Mg	25.98259304(21)	11.01(3)	
^{27}Al	26.98153844(14)	100	26.981538(2)
^{28}Si	27.9769265327(20)	92.2297(7)	28.0855(3)
^{29}Si	28.97649472(3)	4.6832(5)	
^{30}Si	29.97377022(5)	3.0872(5)	
^{31}P	30.97376151(20)	100	30.973761(2)
^{32}S	31.97207069(12)	94.93(31)	32.065(5)
^{33}S	32.97145850(12)	0.76(2)	
^{34}S	33.96786683(11)	4.29(28)	
^{36}S	35.96708088(25)	0.02(1)	
^{35}Cl	34.96885271(4)	75.78(4)	35.453(2)
^{37}Cl	36.96590260(5)	24.22(4)	
^{36}Ar	35.96754628(27)	0.3365(30)	39.948(1)
^{38}Ar	37.9627322(5)	0.0632(5)	
^{40}Ar	39.962383123(3)	99.6003(30)	
^{39}K	38.9637069(3)	93.2581(44)	39.0983(1)
^{40}K	39.96399867(29)	0.0117(1)	
^{41}K	40.96182597(28)	6.7302(44)	
^{40}Ca	39.9625912(3)	96.941(156)	40.078(4)
^{42}Ca	41.9586183(4)	0.647(23)	
^{43}Ca	42.9587668(5)	0.135(10)	
^{44}Ca	43.9554811(9)	2.086(110)	
^{46}Ca	45.9536928(25)	0.004(3)	
^{48}Ca	47.952534(4)	0.187(21)	
^{45}Sc	44.9559102(12)	100	44.955910(8)
^{46}Ti	45.9526295(12)	8.25(3)	47.867(1)
^{47}Ti	46.9517638(10)	7.44(2)	
^{48}Ti	47.9479471(10)	73.72(3)	
^{49}Ti	48.9478708(10)	5.41(2)	
^{50}Ti	49.9447921(11)	5.18(2)	
^{50}V	49.9471628(14)	0.250(4)	50.9415(1)
^{51}V	50.9439637(14)	99.750(4)	
^{50}Cr	49.9460496(14)	4.345(13)	51.9961(6)
^{52}Cr	51.9405119(15)	83.789(18)	
^{53}Cr	52.9406538(15)	9.501(17)	
^{54}Cr	53.9388849(15)	2.365(7)	
^{55}Mn	54.9380496(14)	100	54.938049(9)
^{54}Fe	53.9396148(14)	5.845(35)	55.845(2)
^{56}Fe	55.9349421(15)	91.754(36)	
^{57}Fe	56.9353987(15)	2.119(10)	
^{58}Fe	57.9332805(15)	0.282(4)	
^{59}Co	58.9332002(15)	100	58.933200(9)

Symbol	Relative Atomic Mass	Abundance	Standard Atomic Weight
^{58}Ni	57.9353479(15)	68.0769(89)	58.6934(2)
^{60}Ni	59.9307906(15)	26.2231(77)	
^{61}Ni	60.9310604(15)	1.1399(6)	
^{62}Ni	61.9283488(15)	3.6345(17)	
^{64}Ni	63.9279696(16)	0.9256(9)	
^{63}Cu	62.9296011(15)	69.17(3)	63.546(3)
^{65}Cu	64.9277937(19)	30.83(3)	
^{64}Zn	63.9291466(18)	48.63(60)	65.409(4)
^{66}Zn	65.9260368(16)	27.90(27)	
^{67}Zn	66.9271309(17)	4.10(13)	
^{68}Zn	67.9248476(17)	18.75(51)	
^{70}Zn	69.925325(4)		
^{69}Ga	68.925581(3)		69.723(1)
^{71}Ga	70.9247050(19)	39.892(9)	
^{70}Ge	69.9242504(19)	20.84(87)	72.64(1)
^{72}Ge	71.9220762(16)	27.54(34)	
^{73}Ge	72.9234594(16)	7.73(5)	
^{74}Ge	73.9211782(16)	36.28(73)	
^{76}Ge	75.9214027(16)	7.61(38)	
^{75}As	74.9215964(18)	100	74.92160(2)
^{74}Se	73.9224766(16)	0.89(4)	78.96(3)
^{76}Se	75.9192141(16)	9.37(29)	
^{77}Se	76.9199146(16)	7.63(16)	
^{78}Se	77.9173095(16)	23.77(28)	
^{80}Se	79.9165218(20)	49.61(41)	
^{82}Se	81.9167000(22)	8.73(22)	
^{79}Br	78.9183376(20)	50.69(7)	79.904(1)
^{81}Br	80.916291(3)	49.31(7)	
^{78}Kr	77.920386(7)	0.35(1)	83.798(2)
^{80}Kr	79.916378(4)	2.28(6)	
^{82}Kr	81.9134846(28)	11.58(14)	
^{83}Kr	82.914136(3)	11.49(6)	
^{84}Kr	83.911507(3)	57.00(4)	
^{86}Kr	85.9106103(12)	17.30(22)	
^{85}Rb	84.9117893(25)	72.17(2)	85.4678(3)
^{87}Rb	86.9091835(27)	27.83(2)	
^{84}Sr	83.913425(4)	0.56(1)	87.62(1)
^{84}Sr	85.9092624(24)	9.86(1)	
^{84}Sr	86.9088793(24)	7.00(1)	
^{84}Sr	87.9056143(24)	82.58(1)	
^{89}Y	88.9058479(25)	100	88.90585(2)
^{90}Zr	89.9047037(23)	51.45(40)	91.224(2)
^{91}Zr	90.9056450(23)	11.22(5)	
^{92}Zr	91.9050401(23)	17.15(8)	
^{94}Zr	93.9063158(25)	17.38(28)	
^{96}Zr	95.908276(3)	2.80(9)	

(continued)

Symbol	Relative Atomic Mass	Abundance	Standard Atomic Weight
^{93}Nb	92.9063775(24)	100	92.90638(2)
^{92}Mo	91.906810(4)	14.84(35)	95.94(2)
^{94}Mo	93.9050876(20)	9.25(12)	
^{95}Mo	94.9058415(20)	15.92(13)	
^{96}Mo	95.9046789(20)	16.68(2)	
^{97}Mo	96.9060210(20)	9.55(8)	
^{98}Mo	97.9054078(20)	24.13(31)	
^{100}Mo	99.907477(6)	9.63(23)	
^{97}Tc	96.906365(5)	[98]	
^{98}Tc	97.907216(4)		
^{99}Tc	98.9062546(21)		
^{96}Ru	95.907598(8)	5.54(14)	101.07(2)
^{98}Ru	97.905287(7)	1.87(3)	
^{99}Ru	98.9059393(21)	12.76(14)	
^{100}Ru	99.9042197(22)	12.60(7)	
^{101}Ru	100.9055822(22)	17.06(2)	
^{102}Ru	101.9043495(22)	31.55(14)	
^{104}Ru	103.905430(4)	18.62(27)	
^{103}Rh	102.905504(3)	100	102.90550(2)
^{102}Pd	101.905608(3)	1.02(1)	106.42(1)
^{104}Pd	103.904035(5)	11.14(8)	
^{105}Pd	104.905084(5)	22.33(8)	
^{106}Pd	105.903483(5)	27.33(3)	
^{108}Pd	107.903894(4)	26.46(9)	
^{110}Pd	109.905152(12)	11.72(9)	
^{107}Ag	106.905093(6)	51.839(8)	107.8682(2)
^{109}Ag	108.904756(3)	48.161(8)	
^{106}Cd	105.906458(6)	1.25(6)	112.411(8)
^{108}Cd	107.904183(6)	0.89(3)	
^{110}Cd	109.903006(3)	12.49(18)	
^{111}Cd	110.904182(3)	12.80(12)	
^{112}Cd	111.9027572(30)	24.13(21)	
^{113}Cd	112.9044009(30)	12.22(12)	
^{114}Cd	113.9033581(30)	28.73(42)	
^{116}Cd	115.904755(3)	7.49(18)	
^{113}In	112.904061(4)	4.29(5)	114.818(3)
^{115}In	114.903878(5)	95.71(5)	
^{112}Sn	111.904821(5)	0.97(1)	118.710(7)
^{114}Sn	113.902782(3)	0.66(1)	
^{115}Sn	114.903346(3)	0.34(1)	
^{116}Sn	115.901744(3)	14.54(9)	
^{117}Sn	116.902954(3)	7.68(7)	
^{118}Sn	117.901606(3)	24.22(9)	
^{119}Sn	118.903309(3)	8.59(4)	
^{120}Sn	119.9021966(27)	32.58(9)	
^{122}Sn	121.9034401(29)	4.63(3)	

Symbol	Relative Atomic Mass	Abundance	Standard Atomic Weight
^{124}Sn	123.9052746(15)	5.79(5)	
^{121}Sb	120.9038180(24)	57.21(5)	121.760(1)
^{123}Sb	122.9042157(22)	42.79(5)	
^{120}Te	119.904020(11)	0.09(1)	127.60(3)
^{122}Te	121.9030471(20)	2.55(12)	
^{123}Te	122.9042730(19)	0.89(3)	
^{124}Te	123.9028195(16)	4.74(14)	
^{125}Te	124.9044247(20)	7.07(15)	
^{126}Te	125.9033055(20)	18.84(25)	
^{127}Te	127.9044614(19)	31.74(8)	
^{130}Te	129.9062228(21)	34.08(62)	
^{127}I	126.904468(4)	100	126.90447(3)
^{124}Xe	123.9058958(21)	0.09(1)	131.293(6)
^{126}Xe	125.904269(7)	0.09(1)	
^{128}Xe	127.9035304(15)	1.92(3)	
^{129}Xe	128.9047795(9)	26.44(24)	
^{130}Xe	129.9035079(10)	4.08(2)	
^{131}Xe	130.9050819(10)	21.18(3)	
^{132}Xe	131.9041545(12)	26.89(6)	
^{134}Xe	133.9053945(9)	10.44(10)	
^{136}Xe	135.907220(8)	8.87(16)	
^{133}Cs	132.905447(3)	100	132.90545(2)
^{130}Ba	129.906310(7)	0.106(1)	137.327(7)
^{132}Ba	131.905056(3)	0.101(1)	
^{134}Ba	133.904503(3)	2.417(18)	
^{135}Ba	134.905683(3)	6.592(12)	
^{136}Ba	135.904570(3)	7.854(24)	
^{137}Ba	136.905821(3)	11.232(24)	
^{138}Ba	137.905241(3)	71.698(42)	
^{138}La	137.907107(4)	0.090(1)	138.9055(2)
^{139}La	138.906348(3)	99.910(1)	
^{136}Ce	135.907140(50)	0.185(2)	140.116(1)
^{136}Ce	137.905986(11)	0.251(2)	
^{136}Ce	139.905434(3)	88.450(51)	
^{136}Ce	141.909240(4)	11.114(51)	
^{141}Pr	140.907648(3)	100	140.90765(2)
^{142}Nd	141.907719(3)	27.2(5)	144.24(3)
^{143}Nd	142.909810(3)	12.2(2)	
^{144}Nd	143.910083(3)	23.8(3)	
^{145}Nd	144.912569(3)	8.3(1)	
^{146}Nd	145.913112(3)	17.2(3)	
^{148}Nd	147.916889(3)	5.7(1)	
^{150}Nd	149.920887(4)	5.6(2)	
^{145}Pm	144.912744(4)	[145]	
^{147}Pm	146.915134(3)		
^{144}Sm	143.911995(4)	3.07(7)	150.36(3)

(continued)

Symbol	Relative Atomic Mass	Abundance	Standard Atomic Weight
[147]Sm	146.914893(3)	14.99(18)	
[148]Sm	147.914818(3)	11.24(10)	
[149]Sm	148.917180(3)	13.82(7)	
[150]Sm	149.917271(3)	7.38(1)	
[152]Sm	151.919728(3)	26.75(16)	
[154]Sm	153.922205(3)	22.75(29)	
[151]Eu	150.919846(3)	47.81(3)	151.964(1)
[153]Eu	152.921226(3)	52.19(3)	
[152]Gd	151.919788(3)	0.20(1)	157.25(3)
[154]Gd	153.920862(3)	2.18(3)	
[155]Gd	154.922619(3)	14.80(12)	
[156]Gd	155.922120(3)	20.47(9)	
[157]Gd	156.923957(3)	15.65(2)	
[158]Gd	157.924101(3)	24.84(7)	
[160]Gd	159.927051(3)	21.86(19)	
[159]Tb	158.925343(3)	100	158.92534(2)
[156]Dy	155.924278(7)	0.06(1)	162.500(1)
[158]Dy	157.924405(4)	0.10(1)	
[160]Dy	159.925194(3)	2.34(8)	
[161]Dy	160.926930(3)	18.91(24)	
[162]Dy	161.926795(3)	25.51(26)	
[163]Dy	163 162.928728(3)	24.90(16)	
[164]Dy	163.929171(3)	28.18(37)	
[165]Ho	164.930319(3)	100	164.93032(2)
[162]Er	161.928775(4)	0.14(1)	167.259(3)
[164]Er	163.929197(4)	1.61(3)	
[166]Er	165.930290(3)	33.61(35)	
[167]Er	166.932045(3)	22.93(17)	
[168]Er	167.932368(3)	26.78(26)	
[170]Er	169.935460(3)	14.93(27)	
[169]Tm	168.934211(3)	100	168.93421(2)
[168]Yb	167.933894(5)	0.13(1)	173.04(3)
[170]Yb	169.934759(3)	3.04(15)	
[171]Yb	170.936322(3)	14.28(57)	
[172]Yb	171.9363777(30)	21.83(67)	
[173]Yb	172.9382068(30)	16.13(27)	
[174]Yb	173.9388581(30)	31.83(92)	
[176]Yb	175.942568(3)	12.76(41)	
[175]Lu	174.9407679(28)	97.41(2)	174.967(1)
[176]Lu	175.9426824(28)	2.59(2)	
[174]Hf	173.940040(3)	0.16(1)	178.49(2)
[176]Hf	175.9414018(29)	5.26(7)	
[177]Hf	176.9432200(27)	18.60(9)	
[178]Hf	177.9436977(27)	27.28(7)	
[179]Hf	178.9458151(27)	13.62(2)	
[180]Hf	179.9465488(27)	35.08(16)	

Symbol	Relative Atomic Mass	Abundance	Standard Atomic Weight
^{180}Ta	179.947466(3)	0.012(2)	180.9479(1)
^{181}Ta	180.947996(3)	99.988(2)	
^{180}W	179.946706(5)	0.12(1)	183.84(1)
^{182}W	181.948206(3)	26.50(16)	
^{183}W	182.9502245(29)	14.31(4)	
^{184}W	183.9509326(29)	30.64(2)	
^{186}W	185.954362(3)	28.43(19)	
^{185}Re	184.9529557(30)	37.40(2)	186.207(1)
^{187}Re	186.9557508(30)	62.60(2)	
^{184}Os	183.952491(3)	0.02(1)	190.23(3)
^{186}Os	185.953838(3)	1.59(3)	
^{187}Os	186.9557479(30)	1.96(2)	
^{188}Os	187.9558360(30)	13.24(8)	
^{189}Os	188.9581449(30)	16.15(5)	
^{190}Os	189.958445(3)	26.26(2)	
^{192}Os	191.961479(4)	40.78(19)	
^{191}Ir	190.960591(3)	37.3(2)	192.217(3)
^{193}Ir	192.962924(3)	62.7(2)	
^{190}Pt	189.959930(7)	0.014(1)	195.078(2)
^{192}Pt	191.961035(4)	0.782(7)	
^{194}Pt	193.962664(3)	32.967(99)	
^{195}Pt	194.964774(3)	33.832(10)	
^{196}Pt	195.964935(3)	25.242(41)	
^{198}Pt	197.967876(4)	7.163(55)	
^{197}Au	196.966552(3)	100	196.96655(2)
^{196}Hg	195.965815(4)	0.15(1)	200.59(2)
^{198}Hg	197.966752(3)	9.97(20)	
^{199}Hg	198.968262(3)	16.87(22)	
^{200}Hg	199.968309(3)	23.10(19)	
^{201}Hg	200.970285(3)	13.18(9)	
^{202}Hg	201.970626(3)	29.86(26)	
^{204}Hg	203.973476(3)	6.87(15)	
^{203}Tl	202.972329(3)	29.524(14)	204.3833(2)
^{205}Tl	204.974412(3)	70.476(14)	
^{204}Pb	203.973029(3)	1.4(1)	207.2(1)
^{206}Pb	205.974449(3)	24.1(1)	
^{207}Pb	206.975881(3)	22.1(1)	
^{208}Pb	207.976636(3)	52.4(1)	
^{209}Bi	208.980383(3)	100	208.98038(2)
^{209}Po	208.982416(3)		[209]
^{210}Po	209.982857(3)		
^{210}At	209.987131(9)		[210]
^{211}At	210.987481(4)		
^{211}Rn	210.990585(8)		[222]
^{220}Rn	220.0113841(29)		
^{222}Rn	222.0175705(27)		

(continued)

Symbol	Relative Atomic Mass	Abundance	Standard Atomic Weight
^{223}Fr	223.0197307(29)		[223]
^{223}Ra	223.018497(3)		[226]
^{224}Ra	224.0202020(29)		
^{226}Ra	226.0254026(27)		
^{228}Ra	228.0310641(27)		
^{227}Ac	227.0277470(29)		[227]
^{230}Th	230.0331266(22)		232.0381(1)
^{232}Th	232.0380504(22)	100	
^{231}Pa	231.0358789(28)	100	231.03588(2)
^{233}U	233.039628(3)		238.02891(3)
^{234}U	234.0409456(21)	0.0055(2)	
^{235}U	235.0439231(21)	0.7200(51)	
^{236}U	236.0455619(21)		
^{238}U	238.0507826(21)	99.2745(106)	
^{237}Np	237.0481673(21)		[237]
^{239}Np	239.0529314(23)		
^{238}Pu	238.0495534(21)		[244]
^{239}Pu	239.0521565(21)		
^{240}Pu	240.0538075(21)		
^{241}Pu	241.0568453(21)		
^{242}Pu	242.0587368(21)		
^{244}Pu	244 244.064198(5)		
^{241}Am	241.0568229(21)		[243]
^{243}Am	243.0613727(23)		

( ) designates decimal place(s) with greatest amount of uncertainty.
[ ] designates the most stable isotope mass.

APPENDIX 2

SOLUTIONS TO CHAPTER PROBLEMS

CHAPTER 1

1.1. Inlet, ionization source, mass analyzer, detector, data processing system, and vacuum system.

1.2. Their mass-to-charge ratio m/z.

1.3. To reduce the amount of unwanted collisions.

1.4. Two-stage rotary vane mechanical pump (rough pump) at $\sim 10^{-4}$ Torr, and turbomolecular pump at $\sim 10^{-7}$–10^{-9} Torr.

1.5. Proteomics, metabolomics, and lipidomics.

1.6. See text.

1.7. Allows the transfer of intact associations within solution phase to the gas phase without interruption of the weak associations, thus preserving the original interaction. Examples include lipid:metal/nonmetal, lipid:lipid, lipid:peptide, lipid:protein, protein:protein, and metabolite:protein adducts.

1.8. Protonated molecule $[M + H]^+$, sodium metal adduct as $[M + Na]^+$, a chloride adduct as $[M + Cl]^-$, a deprotonated species as $[M - H]^-$, through electron loss as a radical $M^{+\cdot}$.

Even Electron Mass Spectrometry with Biomolecule Applications   By Bryan M. Ham
Copyright © 2008 John Wiley & Sons, Inc.

1.9. $\dfrac{m}{z} = \dfrac{2eV t^2}{L^2}$

convert eV to kJ/mol (1 eV = 96.4853 kJ/mol).

$$(2000 \text{ eV}) \left(\dfrac{\dfrac{96.4853 \text{ kJ}}{\text{mol}}}{\text{eV}} \right) = \dfrac{192,971 \text{ kJ}}{\text{mol}}$$

$$\dfrac{m}{z} = \dfrac{2 \left(\dfrac{192,971 \text{ kJ}}{\text{mol}} \right) \left(68 \times 10^{-6} \text{ s} \right)^2}{(2 \text{ m})^2}$$

$$\dfrac{m}{z} = \dfrac{4.46 \times 10^{-4} \text{ kJ s}^2}{\text{mol m}^2} \times \dfrac{\left(\dfrac{1000 \text{ kg m}^2}{\text{s}^2 \text{ mol}} \right)}{\left(\dfrac{\text{kJ}}{\text{mol}} \right)}$$

$$\dfrac{m}{z} = \dfrac{0.446 \text{ kg}}{\text{mol}}$$

$$\dfrac{m}{z} = 446 \text{ amu}$$

1.10. $\dfrac{m}{z} = \dfrac{2eV t^2}{L^2}$

convert eV to kJ/mol (1 eV = 96.4853 kJ/mol).

$$(1800 \text{ eV}) \left(\dfrac{\dfrac{96.4853 \text{ kJ}}{\text{mol}}}{\text{eV}} \right) = \dfrac{173,674 \text{ kJ}}{\text{mol}}$$

$$\dfrac{129}{2} = \dfrac{2 \left(\dfrac{173,674 \text{ kJ}}{\text{mol}} \right) t^2}{(2.3 \text{ m})^2}$$

$$\dfrac{0.129 \text{ kg/mol}}{2} = \dfrac{65,661 \text{ kJ } t^2}{\text{mol m}^2} \times \dfrac{\left(\dfrac{1000 \text{ kg m}^2}{\text{s}^2 \text{ mol}} \right)}{\left(\dfrac{\text{kJ}}{\text{mol}} \right)}$$

where $t = 31$ µs.

1.11. a. 184.2402, 184.3599, 0.2402

b. 306.1837, 306.3592, 0.1837

c. 505.2833, 505.7537, 0.2833

d. 692.5092, 692.9576, 0.5092

e. 222.9696, 224.4901, −0.0304

1.12. a. 281 ppm

 b. -7 ppm

 c. -3 ppm

 d. 9 ppm

1.13. a. 517

 b. 21,739

 c. 197

1.14. a. CH_4, r + db = 0, odd electron (OE) as $CH_4^{+\cdot}$, even mass ion with $m/z =$ 16 Th.

 b. NH_3, r + db = 0, odd electron (OE) as $NH_3^{+\cdot}$, odd mass ion with $m/z =$ 17 Th.

 c. NH_4, r + db $= -\frac{1}{2}$, even electron (EE) as NH_4^+, even mass ion with $m/z =$ 18 Th.

 d. H_3O, r + db $= -\frac{1}{2}$, even electron (EE) as H_3O^+, odd mass ion with $m/z =$ 19 Th.

 e. C_2H_7O, r + db $= -\frac{1}{2}$, even electron (EE) as $C_2H_7O^+$, odd mass ion with $m/z = 47$ Th.

 f. $C_{14}H_{28}O_2$, r + db = 1, odd electron (OE) as $C_{14}H_{28}O_2^{+\cdot}$, even mass ion with $m/z = 228$ Th.

 g. $C_6H_{14}NO$, r + db $= -\frac{1}{2}$, even electron (EE) as $C_6H_{14}NO^+$, even mass ion with $m/z = 116$ Th.

 h. $C_4H_{10}SiO$, r + db = 1, odd electron (OE) as $C_4H_{10}SiO^{+\cdot}$, even mass ion with $m/z = 102$ Th.

 i. $C_{29}H_{60}N_2O_6P$, r + db = 1.5, even electron (EE) as $C_{29}H_{60}N_2O_6P^+$, odd mass ion with $m/z = 563$ Th.

CHAPTER 2

2.1. Electron ionization (EI), electrospray ionization (ESI), chemical ionization (CI), atmospheric pressure chemical ionization (APCI), atmospheric pressure photoionization (APPI), and matrix-assisted laser desorption ionization (MALDI).

2.2. To ionize a neutral molecule and transfer into the gas phase in preparation to introduction into the low-vacuum environment of the mass analyzer.

2.3. Potential difference between the filament and the target producing electron beam energy of 70 eV.

2.4. To induce the electrons in the beam to follow a spiral trajectory from filament to target, causing the electrons to follow a longer path from filament to target, producing a greater chance of ionizing the neutral analyte molecules passing through the electron beam.

2.5. Slits and apertures are smaller to maintain the higher pressure in the source, emission current leaving the filament is measured, and repeller voltage is kept lower.

2.6. Methane (CH_4), isobutane ($i\text{-}C_4H_{10}$), and ammonia (NH_3).

2.7. Yes. -37.3 kJ·mol^{-1}.

2.8. Although not exothermically apparent, most will be protonated.

2.9. All will except ammonia may not result in a high abundance?

2.10. Chemical ionization for molecular weight and electron ionization for structural information.

2.11. Break up clusters, aid in evaporation of the solvent, and a nebulizing gas that helps to atomize the eluent spray.

2.12. Generated electric field penetrates into the liquid meniscus and creates an excess abundance of charge at the surface. The meniscus becomes unstable and protrudes out forming a Taylor cone.

2.13. Charged residue for larger molecules, ion evaporation for smaller.

2.14. Desolvation, adiabatic expansion, evaporative cooling, and low-energy dampening collisions. Soft ionization.

2.15. Reduced size droplets resulting in more efficient ion production.

2.16. Absorption of radiant energy from a UV source where the incident energy is greater than the first ionization potential of electron loss from the analyte.

2.17. Not a solution-phase process based upon proton affinity.

2.18. Dopant is a chemical species that possesses a low ionization potential (IP) that allows its ionization to take place with high efficiency. Yes.

2.19. Naphthaline, benzene, phenol, aniline, *m*-chloroaniline, *l*-aminonapthaline. Water and acetonitrile.

2.20. UV absorber and proton donor. 2,5-Dihydroxybenzoic acid (DHB), 3,5-dimethoxy-4-hydroxy-trans-cinnamic acid (sinapic or sinapinic acid), and α-cyano-4-hydroxy-trans-cinnamic acid (α-CHCA).

2.21. Dried droplet, thin film, and layer or sandwich.

2.22. The UV-absorbing matrix molecules accept energy from the laser and desorb from the surface, carrying along any analyte that is mixed with it, forming a gaseous plume where the analyte is ionized.

2.23. Initial spatial and energy distributions. Delayed extraction and the reflectron.

2.24. Drying behavior and analyte distribution.

2.25. Drying behavior, spatial distributions, and gas-phase proton donor.

2.26. Analyte is dissolved in glycerol and is bombarded with a high-energy beam of atoms where the kinetic energy from the colliding atom is transferred to the matrix and analyte, effectively desorbing them into the gas phase with ionization.

2.27. Matrix absorbs the largest amount of kinetic energy and the desorbed analyte goes through a desolvation mechanism where excess energy is transferred to the matrix solvent molecules.

CHAPTER 3

3.1. Electric and magnetic sector mass analyzer, time-of-flight mass analyzer (TOF/MS), time-of-flight/time-of-flight mass analyzer (TOF-TOF/MS), quadrupole time-of-flight mass analyzer (Q-TOF/MS), triple quadrupole or linear ion trap mass analyzer (QQQ/MS or LIT/MS), three-dimensional quadrupole ion trap mass analyzer (QIT/MS), Fourier transform ion cyclotron mass analyzer (FTICR/MS), and linear ion trap–Orbitrap mass analyzer (IT-Orbitrap/MS).

3.2. 100 mm.

3.3. The electric sector is not a mass analyzer but an energy filter.

3.4. The ion gate acts as a timed ion selector by switching on and off to allow ions to pass.

3.5. 5 eV.

3.6. Under the influence of the combination of electric fields comprised of dc voltage (U) and radio frequency (rf) voltage (V).

3.7. The mass-to-charge relationship (m/e).

3.8. The plot of Figure 3.16 demonstrates that the resolution of the quadrupole is inversely proportional to the sensitivity.

3.9. Relatively insensitive to kinetic energy spread, easy to operate and calibrate, mechanically rugged, and inexpensive to produce.

3.10. Imperfections in the quadrupole field's influence upon ion detection.

3.11. (1) Single-stage analysis, (2) neutral loss scanning, (3) precursor ion scanning, and (4) product ion scanning.

3.12. To induce low-energy collisions with the ions to thermally cool them, decrease their kinetic energy, and focus the ion packets into tighter ion trajectories near the center of the trap.

3.13. Method for scanning mass-to-charge species in the trap.

3.14. Through the application of an ac voltage applied to the end caps of the quadrupole ion trap.

3.15. Resonant excitation, dipole excitation, and quadrupole excitation.

3.16. The inclusion of a quadrupole mass analyzer in tandem prior to the time-of-flight mass analyzer.

3.17. Orthogonal, which allows a uniform spatial distribution of ions from the source.

3.18. Mass accuracies of <5 ppm with resolutions as high as 1 million.

3.19. Permanent magnets and electromagnets are less than 2 Tesla (T) while super-conducting magnets are from 3.5, 7, 9.4, and 11.5 T.

3.20. 10^{-9} or 10^{-10} Torr to reduce collision-mediated radial diffusion.

3.21. (1) Drive the ions away from the center of the ICR cell for signal measurement, (2) drive ions to a larger cyclotron radius to remove (eject) them from the cell, and (3) to add internal energy for dissociative collision product ion generation and ion–molecule reactions.

3.22. Collisions taking place between the trapped ions and between the trapped ions with neutrals will drive the ions toward the outer dimensions of the cell where they will be lost to neutralizing collisions with the cell walls. Cyclotron frequency amplitudes are observed to steadily decrease.

3.23. The maximum amount of ions that can be trapped within the ICR cell is proportional to the square of the magnetic field (B^2), increase in maximum ion kinetic energy with increasing magnetic field strength (B^2), and the radius of the orbiting ion packet will be smaller for the same kinetic energy.

3.24. No refilling of liquid nitrogen and liquid helium reservoirs to maintain the low temperatures needed for superconductivity and no active shielding of the magnet.

CHAPTER 4

Practice Problem 1

$$P = 1.1 \times 10^{-4}\,\text{mbar}$$
$$= 1.1 \times 10^{-7}\,\text{bar}\,(1\,\text{atm}/1.01325\,\text{bar})$$
$$= 1.09 \times 10^{-7}\,\text{atm}$$
$$T = 300\,\text{K}$$
$$k_B = 1.38066 \times 10^{-23}\,\text{J}\,\text{K}^{-1}$$

$$\alpha = 9.869 \, \text{cm}^3 \, \text{atm} \, \text{J}^{-1}$$

$$n = \frac{P}{k_B \alpha T}$$

$$n = \frac{1.09 \times 10^{-7} \, \text{atm}}{(1.38066 \times 10^{-23} \text{J} \, \text{K}^{-1}) \left(\dfrac{9.869 \, \text{cm}^3 \, \text{atm}}{\text{J}} \right) (300 \, \text{K})}$$

$$n = 2.66 \times 10^{12} \, \text{cm}^{-3}$$

$$\frac{I_P}{I_R + \sum I_P} = \sigma_P n l$$

$$l = \frac{I_P}{(I_R + \sum I_P) \sigma_P n}$$

$$= \frac{15.4}{(1000)(3.267 \times 10^{-15} \, \text{cm}^2)(2.66 \times 10^{12} \, \text{cm}^{-3})}$$

$$= 1.77 \, \text{cm}$$

Practice Problem 2

$$E_{\text{COM}} = E_{\text{LAB}} \left(\frac{m}{m + M} \right)$$

E_{COM} (eV)
1.90
0.74
0.24
2.96
0.89
4.24
1.44
0.03
2.62

Practice Problem 3

1. Convert gas cell pressures to atm.
2. Calculate ratios $I_P / I_R + \sum I_P$.
3. Plot ratio vs. number gas density.
4. Extract slope from plot.
5. Calculate σ_P using the following equation:

$$\sigma_P = \frac{\dfrac{I_P}{I_R + \sum I_P}}{n l}$$

$$\sigma_P = 6.25 \times 10^{-17} \, \text{cm}^2$$

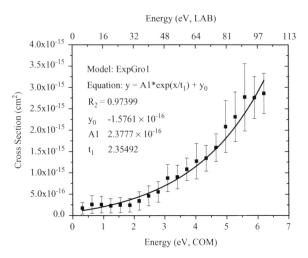

FIGURE P4.1   Solution of the fitting of a simple exponential growth equation giving the graph and equation fit values.

Practice Problem 4

Using the graphical software package Origin Pro, the values in Table 4.3 and the fitting of a simple exponential growth equation gives the graph shown in Figure P4.1 and equation fit values.

Equation (4.21) now has the form of

$$\sigma_p(E) = \sigma_0 \, \exp\left(\frac{E_{COM}}{\alpha}\right) + y_0$$

$$= 2.38E - 16 \, \exp\left(\frac{E_{COM}}{2.35}\right) - 1.58E - 16$$

The BDE can now be calculated from Equation (4.23):

$$BDE \ (eV) = 0.78051 - 1.7180 \times 10^{13}\sigma_0 + 1.4627\alpha - 0.16327\alpha^2$$
$$= 0.78051 - 1.7180 \times 10^{13}(2.38 \times 10^{-16})$$
$$+ \ 1.4627(2.35) - 0.16327(2.35)^2$$
$$= 3.31$$

Practice Problem 5

Using the graphical software package Origin Pro, the values in Table 4.4 and the fitting of a simple exponential growth equation gives the graph shown in Figure P4.2 and equation fit values.

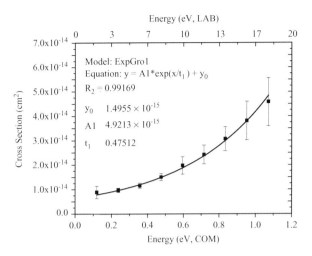

FIGURE P4.2   Solution of the fitting of a simple exponential growth equation giving the graph and equation fit values.

Equation (4.21) now has the form of

$$\sigma_p(E) = \sigma_0 \exp\left(\frac{E_{COM}}{\alpha}\right) + y_0$$

$$= 4.9213E - 15 \exp\left(\frac{E_{COM}}{0.47512}\right) + 1.4955E - 15$$

The BDE can now be calculated from Equation (4.23):

$$BDE(eV) = 0.78051 - 1.7180 \times 10^{13}\sigma_0 + 1.4627\alpha - 0.16327\alpha^2$$
$$= 0.78051 - 1.7180 \times 10^{13}(4.92 \times 10^{-15})$$
$$+ 1.4627(0.475) - 0.16327(0.475)^2$$
$$= 1.35$$

CHAPTER 5

Problem 5.4

a. Methyl oleate ($C_{19}H_{36}O_2$) molecular ion at m/z 296 as $M^{+\cdot}$.
b. Neutral loss of CH_3OH to form m/z 264.1.
c. Neutral loss of $CH_3-C=O-O-CH_3$ to form m/z 222.2.
d. Neutral loss of $CH_3(CH_2)_3C=O-O-CH_3$ to form m/z 180.2.
e. $C_3H_6O_2$ as m/z 74.0.

CHAPTER 6

Problem 6.3

The first major product ion observed in Figure 6.11 is the m/z 99 ion. Using the structures and mass values listed in Figure 6.12, we can compare to the losses observed in Figure 6.11. The difference in the mass of the precursor ion at m/z 323 and the first major product ion at m/z 99 is 224 amu. In Figure 6.12 the C15:0 pentadecanoic acid substituent has a mass of 242 amu. The difference between the C15:0 pentadecanoic acid substituent at 242 amu and the loss of 224 amu for the generation of the m/z 99 product ion is 18 amu. This is the difference of H_2O (18 amu), indicating the loss of the C15:0 pentadecanoic acid substituent without H_2O. As we saw before, this is loss of a fatty acyl chain as a ketene from the precursor ion. The fragmentation pathway mechanism for the generation of the m/z 99 product ion is illustrated in Figure P6.1. The neutral loss of the C15:0 fatty acyl chain as a ketene, $C_{15}H_{28}O$ 224.2140 Da, from the precursor ion at m/z 323 produces the lithium adduct of the glycerol backbone at m/z 99, $[C_3H_8O_3Li]^+$. The overall loss producing the m/z 99 product ion is $[C_{18}H_{38}O_4 + Li - C_{15}H_{28}O]^+$. The next product ion illustrated in Figure 6.11 is the m/z 81 ion. This has a mass difference of 18 amu from the m/z 99 product ion indicating either loss of H_2O from the m/z 99 glycerol backbone lithium adduct product ion or neutral loss of the C15:0 pentadecanoic acid substituent (323 amu $-$ 81 amu $=$ 242 amu, the mass of the C15:0 fatty acid substituent). In actuality either mechanism may be valid for the production of the m/z 81 product ion. To

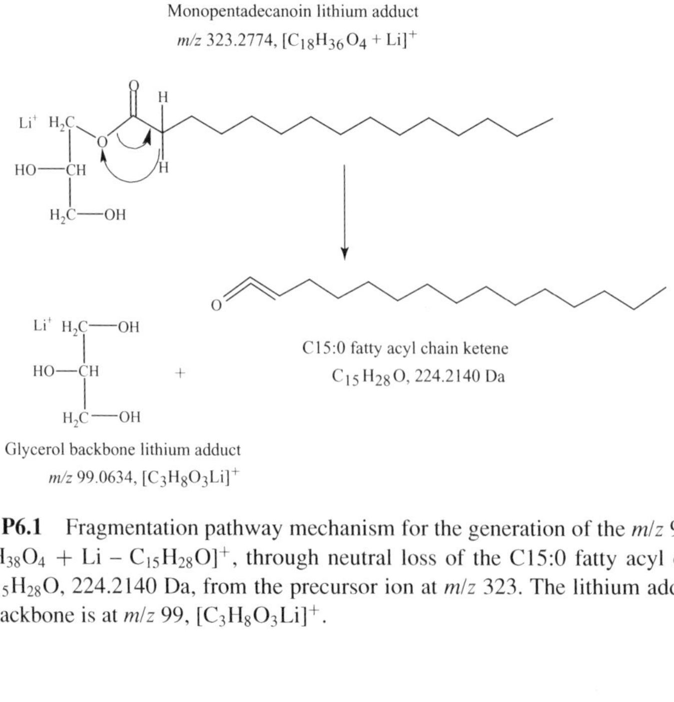

Monopentadecanoin lithium adduct
m/z 323.2774, $[C_{18}H_{36}O_4 + Li]^+$

C15:0 fatty acyl chain ketene
$C_{15}H_{28}O$, 224.2140 Da

Glycerol backbone lithium adduct
m/z 99.0634, $[C_3H_8O_3Li]^+$

FIGURE P6.1   Fragmentation pathway mechanism for the generation of the m/z 99 product ion, $[C_{18}H_{38}O_4 + Li - C_{15}H_{28}O]^+$, through neutral loss of the C15:0 fatty acyl chain as a ketene, $C_{15}H_{28}O$, 224.2140 Da, from the precursor ion at m/z 323. The lithium adduct of the glycerol backbone is at m/z 99, $[C_3H_8O_3Li]^+$.

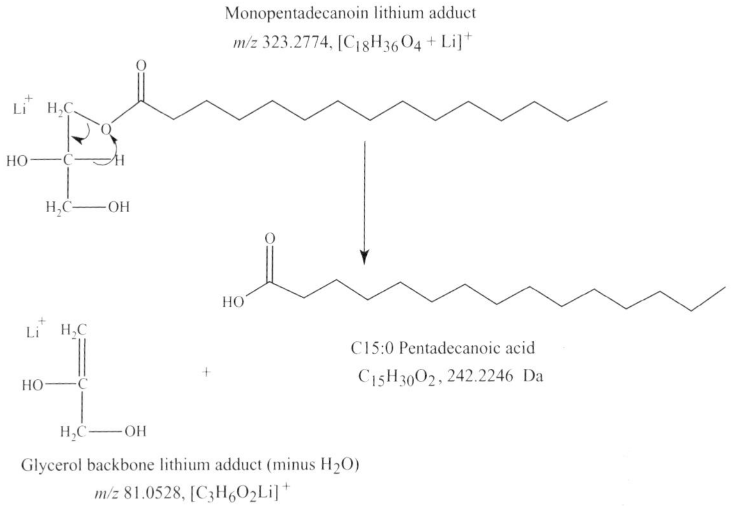

Monopentadecanoin lithium adduct

m/z 323.2774, $[C_{18}H_{36}O_4 + Li]^+$

C15:0 Pentadecanoic acid

$C_{15}H_{30}O_2$, 242.2246 Da

Glycerol backbone lithium adduct (minus H_2O)

m/z 81.0528, $[C_3H_6O_2Li]^+$

FIGURE P6.2   Fragmentation pathway mechanism for the generation of the m/z 81 product ion, $[C_{18}H_{38}O_4 + Li - C_{15}H_{30}O_2]^+$, through neutral loss of the C15:0 pentadecanoic acid substituent, $C_{15}H_{30}O_2$ 242.2246 Da, from the precursor ion at m/z 323. The lithium adduct of the glycerol backbone minus H_2O is at m/z 81, $[C_3H_6O_2Li]^+$.

discern which mechanism is likely, energy-resolved studies are usually performed as discussed in detail in Chapter 4. For purposes here, we will consider the fragmentation pathway mechanism through neutral loss of the C15:0 pentadecanoic acid substituent. This mechanism is illustrated in Figure P6.2 where the neutral loss of the C15:0 pentadecanoic acid substituent, $C_{15}H_{30}O_2$ 242.2246 Da, produces the lithiated glycerol backbone minus one water at m/z 81.0528, $[C_3H_6O_2Li]^+$. The overall description for this loss is $[C_{18}H_{36}O_4 + Li - C_{15}H_{30}O_2]^+$. The next product ion in Figure 6.12 is the m/z 63 ion. This is 18 amu less than the m/z 81 product ion, indicating a further second loss of H_2O from the lithiated glycerol backbone. This is a two-step process where the C15:0 pentadecanoic acid substituent is lost first, followed by H_2O loss from the glycerol backbone. The fragmentation pathway mechanism is illustrated in Figure P6.3. The last product ion observed in Figure 6.11 is the m/z 57 product ion. The difference between the m/z 63 product ion and the m/z 57 product ion is 6 amu, so it does not appear that m/z 57 is formed through further water loss as was observed for the production of m/z 81 and m/z 63. Second, a loss of 6 amu is not a reasonable loss in itself, as this does not represent any species that are present in the structures. A difference of 6 amu could represent a loss of a lithium cation from the m/z 63 product ion in conjunction with a gain in a hydrogen. In fact, this is a reasonable postulation if we take a look at the product ion structures that would be involved, as illustrated in Figure P6.4. However, the addition of the hydrogen proton must come from somewhere, so we will need to look elsewhere for the production of m/z 57 while still retaining the postulated structure in Figure P6.4. If we take a step

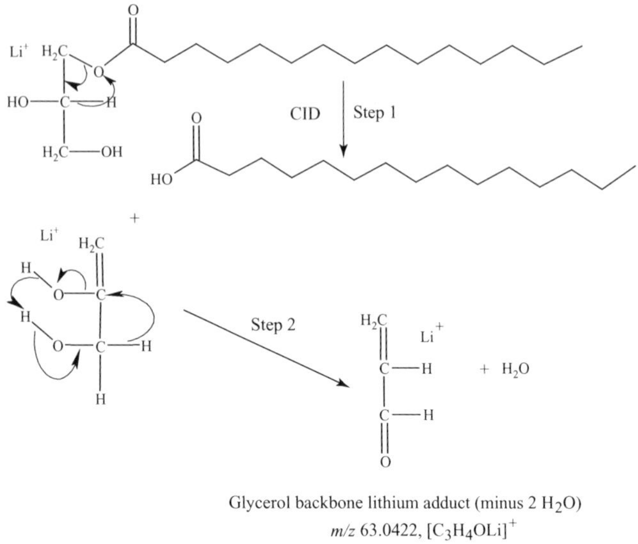

Glycerol backbone lithium adduct (minus 2 H$_2$O)

m/z 63.0422, [C$_3$H$_4$OLi]$^+$

FIGURE P6.3 Fragmentation pathway mechanism for the generation of the m/z 63 product ion, [C$_{18}$H$_{38}$O$_4$ + Li − C$_{15}$H$_{30}$O$_2$ − H$_2$O]$^+$, through a two-step neutral loss of the C15:0 pentadecanoic acid substituent, C$_{15}$H$_{30}$O$_2$ 242.2246 Da, followed by H$_2$O loss, from the precursor ion at m/z 323. The lithium adduct of the glycerol backbone minus 2H$_2$O is at m/z 63, [C$_3$H$_4$OLi]$^+$.

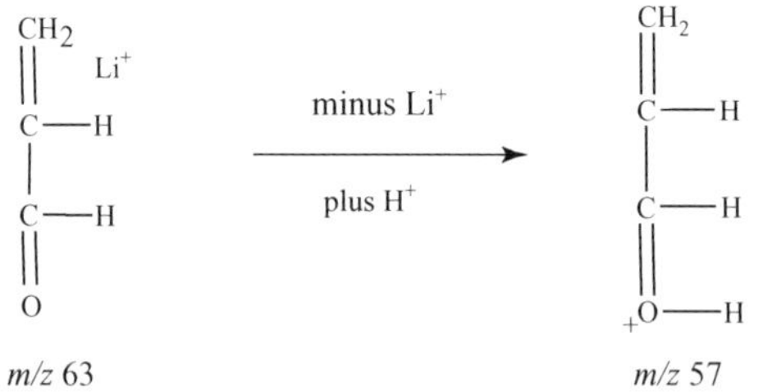

FIGURE P6.4 Postulated generation of m/z 57 product ion.

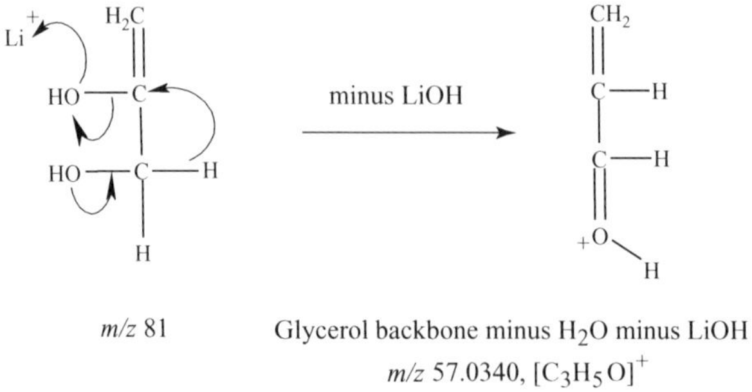

Glycerol backbone minus H$_2$O minus LiOH

m/z 57.0340, [C$_3$H$_5$O]$^+$

FIGURE P6.5 Formation of the m/z 57 product ion.

back to the m/z 81 product ion, we observe that the difference between it and m/z 57 is 24 amu. By drawing out the structures of m/z 81 and the postulated structure for m/z 57, we see that the loss of LiOH is needed to form m/z 57. The mass of LiOH is 24 amu, just the amount needed for the conversion of m/z 81 to m/z 57. Therefore, the fragmentation pathway leading to the m/z 57 product ion from m/z 81 involves neutral loss of LiOH, and the pathway is illustrated in Figure P6.5.

CHAPTER 7

7.1. Top-down is the analysis of whole intact proteins while bottom-up is the analysis of enzyme-digested proteins that have been converted to peptides.

7.2. Glycosylation, sulfation, and phosphorylation.

7.3. Soluble and insoluble, primary, secondary, tertiary, and quaternary structure, 20 amino acids, four classes of amino acids, β sheets and α helix.

7.4. Primary consists of sequence of amino acids, secondary of α helixes and β sheets, tertiary of the ordering of the secondary, and quaternary is the arrangement of the polypeptide chains into the working protein.

7.5. A comparison of a protein's enzymatic generated peptides measured by MALDI-TOF/MS against a theoretical list.

7.6. Protease used to digest proteins. The cleavage location within the polypeptide chain.

7.7. A complete coverage of the amino acid sequence within a peptide from the product ion spectrum. Can unambiguously identify a protein according to standard spectra stored in protein databases.

7.8. To map and decipher interrelationships between observed proteins and the genetic description.

7.9. When the charge is retained on the N-terminal portion of the fragmented peptide, the ions are depicted as a, b, and c. When the charge is retained on the C-terminal portion, the ions are denoted as x, y, and z.

7.10. Upon excitation of the peptide from CID, the proton will migrate to various protonation sites prior to fragmentation. The expected site of protonation from a thermodynamic point of view indicates amide oxygen protonation and not the amide nitrogen.

7.11. Loss of ammonia and water.

7.12. Side-chain cleavage ions that produced a combination of backbone cleavage and a side-chain bond.

7.13. Immonium ions are produced through a combination of a-type and y-type cleavage that results in an internal fragment that contains a single side chain.

Immonium ions are useful in acting as confirmation of residues suspected to be contained within the peptides backbone.

7.14. Isomers are species that have the same molecular formula but differ in their structural arrangement, while isobars are species with different molecular formulas that posses similar (or the same) molecular weights.

7.15. The use of high-resolution/high-mass accuracy instrumentation.

7.16. Promotes extensive fragmentation along the polypeptide backbone while preserving modifications such as glycosylation and phosphorylation.

7.17. The accuracy in which the masses are measured. High-resolution mass spectrometers to accurately measure the intact protein's mass and the product ions.

7.18. A decreased need for extensive sample precleanup and up to 100% coverage of protein including modifications.

7.19. 56,243 Da; m/z 2250.7 at +25, m/z 2344.5 at +24, m/z 2446.4 at + 23, and m/z 2557.5 at +22.

7.20. The covalent addition of a carbohydrate chain to amino acid side chains. To determine the proteins' structure, it is crucial in cell–cell recognition and may also be involved in cellular signaling events.

7.21. Reduces the viscosity of the blood.

7.22. (1) Get an identification of the glycosylated peptides and proteins, (2) accurately determine the sites of glycosylation, and (3) determine the carbohydrates that make up the glycan and the structure of the glycan.

7.23. Repeated addition of SiA at a charge state of +2 (146.5 Da) and of +3 (97.1 Da).

7.24. Serine, threonine, histidine, and tyrosine.

7.25. No. Approximately 2% pY, 12% pT, and 86% pS.

7.26. For a given protein there may be only a couple of phosphorylated tryptic peptides as compared to many nonphosphorylated.

7.27. Nucleic acids. Ultracentrifugation, RNase and Dnase, and QIAShredder.

7.28. Charge states +1 at 98.0 Da, +2 at 49.0 Da, and +3 at 32.7 Da?

7.29. Dehydroalanine and dehydroaminobutyric acid.

7.30. Loss of 80 Da as HPO_3 and loss of 98 Da as H_3PO_4.

7.31. Promotes extensive fragmentation along the polypeptide backbone while preserving modifications such as glycosylation and phosphorylation. Yes.

7.32. In eukaryotic systems the phosphoryl modification is associated with regulating enzyme activity. In prokaryotic systems the histidine group is phosphorylated to transfer the phosphate group from one biomolecule to another.

7.33. Histidine phosphorylation is labile in acidic conditions, the conditions encountered during sample preparation steps.

7.34. See Table 7.8.

7.35. Neutral with a valence of 5 for the phosphorus.

7.36. Carbohydrate sulfation of cell surface glycans and sulfonation of protein tryrosine amino acid residues.

7.37. Yes, HPO_3 has a mass of 79.9663 Da and SO_3 has a mass of 79.9568 Da.

7.38. The study of the processes and complex biological organizational behavior using information from its molecular constituents.

7.39. Blood, urine, sputum, saliva, breath, tear fluid, nipple aspirate fluid, and cerebrospinal fluid.

CHAPTER 8

8.1. Steroids, fatty acid amides, fatty acids, wax esters, fatty alcohols, phosphorylated lipids, and acylglycerols.

8.2. Promote ionization, rich fragmentation patterns, and alleviate variations in adducts.

8.3. Negative ion mode. The organic acid is easily deprotonated.

8.4. Yes, they can. Due to a direct linear correlation between lipid concentration and lipid response, however, there are differences in ionization efficiencies of lipids and subsequent suppression that can interfere with quatitation.

8.5. It is possible in negative ion mode due to reduction taking place at the surface of the ESI capillary.

8.6. No, they were not.

8.7. Suppression.

8.8. The suppression effect being observed.

8.9. A carbon 13 (^{13}C) algal fatty acid standard mix was analyzed.

8.10. The difficulty in quantitating free fatty acids in biological samples using electrospray as the ionization technique.

8.11. The response of ion detectors coupled with single quadrupole mass spectrometers is linear in regions.

8.12. Dissolving the biological sample in a solvent solution containing an internal standard that is then quantitated against a series of fatty acid amide standards ratioed to the internal standard (IS).

8.13. To measure the biomolecules at responses that are within the linear response of the associated calibration curves.

8.14. The fatty acid portion of the biomolecule due to an electrostatic noncovalent interaction.

8.15. The m/z 403.6 aldehyde peak and the m/z 387.6 product ion formed from the fatty alcohol substituent.

8.16. An 18-Da neutral loss of water.

8.17. (1) Exact mass measurements, (2) synthesis and analysis of suspected unknown, (3) proton NMR, and (4) structural elucidation of unknown by CID in positive and negative ion mode.

8.18. GC-EI/MS, positive chemical ionization, negative chemical ionization of chloride adducts, ES-MS/MS, MALDI-TOF, fast atom bombardment, atmospheric pressure chemical ionization (APCI) mass spectrometry, and APCI liquid chromatography/mass spectrometry (LC/MS).

8.19. Species that contain polarized regions such as positive and negative but are neutral overall.

8.20. Deprotonated dimetal(1) adduct, protonated metal adduct, deprotonated metal(1)-metal(2) adduct, and the deprotonated di-metal(2) adduct.

8.21. Identification of substituents and headgroups.

8.22. Typically the headgroup.

8.23. A difference of 18 atomic mass units (amu). Can be used as an indication of ester linkages as fatty acid substituents in the structures and in unknown identification.

CHAPTER 9

9.1. Proteins, polysaccharides, and nucleic acids.

9.2. Hydrolysis reactions cleave apart monomers with the addition of water. Condensation reactions join monomers with the elimination of water. Synthesis of biopolymers.

9.3. Glycogen and starch for storage, and cellulose for structural use.

9.4. In α-D-glucose the hydroxyl points downward while for β-D-glucose points upward.

9.5. Two α-D-sugar units connected form a α-glycosidic bond and two β-D-sugar units connected form a β-glycosidic bond.

9.6. The decrease in the size of the nanoelectrospray drops increases the surface activity of the hydrophilic oligosaccharides.

9.7. The greater number of vibrational degrees of freedom in the larger oligosaccharides enable the dissipation of the internal energy. B_m- and Y_n-type ions. Larger oligosaccharides through glycosidic bond cleavage, smaller oligosaccharides through both glycosidic bond cleavage and cross-ring cleavage.

9.8. Complex mixtures of very closely related isomeric compounds that contain many labile bonds and can be highly branched.

9.9. The deprotonated precursor will dissociate with the hydroxyl group going to the reducing portion of the sugar while the carbohydrate portion suffers ring opening and epoxide formation.

9.10. The charge of the metal ion is not directly associated with the glycosidic bond oxygen but is distributed to a number of local oxygen atoms.

9.11. Polysaccharides are often highly branched and/or substituted.

9.12. DNA contains the 5-carbon sugar deoxyribose while RNA contains ribose. DNA contains the base thymine (T) while RNA contains uracil (U).

9.13. When the product ion contains the $3'$-OH portion of the nucleic acid, the naming includes the letters w, x, y, and z. When the product ion contains the $5'$-OH portion of the nucleic acid, the naming includes the letters a, b, c, and d.

9.14. Neutral and charged base losses.

9.15. Suppression and shifting of the mass of the nucleic acid.

9.16. Measurement of a singly charged species.

9.17. Every fragmentation pathway is initiated by loss of a nucleobase that does not involve simple water loss or $3'$- and $5'$- terminal nucleoside /nucleotides.

CHAPTER 10

Problem 10.1

1. and 2. As can be seen in the two figures, the two molecular ions of PS have the same fragmentation pathways producing loss of headgroup fragments, but the favored pathways appear to be quite different.

3. For $[PS+2Na-H]^+$ the only predominant product ion is m/z 719 (Fig. 10.10), which is formed through the neutral loss of $C_3H_5O_2N$ (87 Da). This indicates that $[PS+2Na-H]^+$ is more stable than $[PS+Na]^+$. In fact, this is well known because the charge on Na^+ is less mobile than the charge on H^+. With increased mobility of charge comes an increase in the induced fragmentation of the precursor analyte ion.

4. There are indications that for each precursor analyte ion that contains two sodium atoms, one of the sodium ions is clearly associated with the negatively charged oxygen in the phosphoryl group.

Problem 10.2

The fragmentation pathway for the production of the m/z 495 product ion would be similar to that illustrated in Figure 10.13 except that the hydrogen would not migrate to the phosphate oxygen, and the shared pair of electrons between the phosphate oxygen and the carbon of the glycerol backbone would move to the oxygen, producing neutral loss of the headgroup species at 183 Da and a positively charged species at m/z 495 that contains the two fatty acid substituents.

Problem 10.3

The predominant product ion in the spectrum is the expected peak representing the protonated phosphatidylcholine headgroup at m/z 184. However, there are three other peaks of interest in the spectrum. The m/z 313 peak is derived from the neutral loss of the phosphatidylcholine head groupfrom the m/z 496 precursor [M + H − $C_5H_{14}NO_4P$]$^+$, and the m/z 258 peak from the neutral loss of a fatty acyl chain as a ketene [M +H − $C_{16}H_{30}O$]$^+$. The observance of these two peaks can aid in the identification of the lyso phosphatidylcholine lipid. The third peak at m/z 104 is the headgroup fragment ion $C_5H_{14}NO^+$.

Problem 10.4

The mechanisms would be similar as those illustrated in Figure 10.21 where the first step is neutral loss of the C16:0 fatty acid substituent forming the m/z 537 product ion. The second step is neutral loss of the $C_3H_6O_2$ portion of the headgroup forming the m/z 463 product ion.

Problem 10.5

In the MALDI-TOF PSD product ion mass spectrum illustrated in Figure 10.22, there are three predominant product ion peaks that involve the phosphoglycerol headgroup. At m/z 599, there is a minor peak for the neutral loss of the phosphoglycerol headgroup, and at m/z 577 there is the more predominant product ion peak for neutral loss of the phosphoglycerol headgroup (with sodium). The intensity of the m/z 577 peak being greater than the m/z 599 peak suggests that the departure of the neutral headgroup (with sodium) is the more favored fragmentation pathway. The most predominant product ion peak in the spectrum is the m/z 195 peak, that is, the sodium adduct of the phosphoglycerol headgroup. A minor peak is also observed at m/z 415 representing a two-step loss of a headgroup fragment (neutral loss of 74 Da) in conjunction with neutral loss of the C18:1 fatty acid substituent [M + Na − $C_3H_6O_2$ − $C_{18}H_{34}O_2$]$^+$, giving some structural information for the phosphorylated lipid.

APPENDIX 3

FUNDAMENTAL PHYSICAL CONSTANTS

Quantity	Symbol	Value	Unit
Electron mass	m_e	$9.10938215(45) \times 10^{-31}$	kg
Proton mass	m_p	$1.672621637(83) \times 10^{-27}$	kg
Proton–electron mass ratio	m_p/m_e	$1836.15267247(80)$	
Speed of light in vacuum	c	2.99792458×10^8	m s^{-1}
Magnetic constant	μ_0	$4\pi \times 10^{-7}$	N A^{-2}
Rydberg constant $\alpha^2 m_e c/2h$	R_∞	$10\ 973\ 731.568527(73)$	m^{-1}
Avogadro constant	N_A, L	$6.02214179(30) \times 10^{23}$	mol^{-1}
Faraday constant $N_A e$	F	$96\ 485.3399(24)$	C mol^{-1}
Molar gas constant	R	$8.314472(15)$	J mol^{-1} K^{-1}
Boltzmann constant R/N_A	k	$1.3806504(24) \times 10^{-23}$	J K^{-1}
Stefan–Boltzmann constant $(\pi^2/60)k^4/\hbar^3 c^2$	σ	$5.670400(40) \times 10^{-8}$	W m^{-2} K^{-4}
Electric constant $1/\mu_0 c^2$	ϵ_0	$8.854187817 \times 10^{-12}$	F m^{-1}
Newtonian gravitation constant	G	$6.67428(67) \times 10^{-11}$	m^3 kg^{-1} s^{-2}
Electron volt $(e/C) J$	eV	$1.602176487(40) \times 10^{-19}$	J
Unified atomic mass unit	u	$1.660538782(83) \times 10^{-27}$	kg
Plank constant	h	$6.62606896(33) \times 10^{-34}$	J s
Plank constant $h/2\pi$	$\hbar$	$1.054571628(53) \times 10^{-34}$	J s
Elementary charge	e	$1.602176487(40) \times 10^{-19}$	C
Magnetic flux quantum $h/2e$	Φ_0	$2.067833667(52) \times 10^{-15}$	Wb
Conductance quantum $2e^2/h$	G_0	$7.7480917004(53) \times 10^{-5}$	S

Even Electron Mass Spectrometry with Biomolecule Applications   By Bryan M. Ham
Copyright © 2008 John Wiley & Sons, Inc.

GLOSSARY

Accelerating Voltage   In a mass spectrometer this is the electrical potential that is used to impart kinetic energy to ions.

Accurate Mass   Experimentally determined mass used to estimate a species elemental formula, usually using high-mass accuracy instrumentation such as the Q-TOF, electric–magnetic sector, and FTICR mass spectrometers. A measurement with 5 ppm accuracy is sufficient to determine the elemental composition for ions less than m/z 200.

Adduct Ion   Ion formed through a noncovalent attraction between a neutral analyte and a cation or anion. For small molecules the stoichiometry is typically 1 : 1, but for proteins and nucleic acids many adducted species may form. Typical cation adducts include H^+, Na^+, Li^+, and K^+. Typical anions include I^-, Cl^-, and SO_4^-.

Adiabatic Ionization   Production of an ion in the ground state through the addition or removal of an electron.

Anion   Species that possesses a net negative charge that can be either atomic or molecular.

Array Collector   Detector that is made up of an array of ion collection devices where each individual companent is an ion detector.

Associative Ionization   Process where an ion is produced through the interaction of two excited species in the gas phase.

Atmospheric Pressure Chemical Ionization (APCI)   Ionization of analyte species at atmospheric pressure using a reactive reagent such as a gas.

Even Electron Mass Spectrometry with Biomolecule Applications   By Bryan M. Ham
Copyright © 2008 John Wiley & Sons, Inc.

Atmospheric Pressure Ionization (API)   General term used to describe a system that produces ionized species in the gas phase at atmospheric pressure. Common API sources used include *atmospheric pressure chemical ionization, atmospheric pressure photoionization,* and *electrospray ionization.*

Atmospheric Pressure Photoionization (APPI)   Ionization of species at atmospheric pressure using a UVU source and subsequent atmospheric pressure chemical ionization that takes place.

Average Mass   Ion or molecule's mass calculated with the average mass of each element weighted for its natural isotopic abundance (see Appendix 1).

α-Cleavage   Homolytic fragmentation process where an adjacent atom that pairs with the odd-electron one of the pair of electrons between the atom attached to the atom with the odd electron and a radical species is lost. The charge is retained on the atom that originally contained the charge.

Base Peak   Greatest intensity peak in a mass spectrum that may contain any number of peaks from a single species or numerous species. Often the mass spectrum is normalized to this peak.

Blackbody Infrared Radiative Dissociation (BIRD)   The heated surroundings of an analyte (e.g., the vacuum chamber walls) impart infrared photons that are absorbed and causing excitation of the reactant ion, which may subsequently dissociate.

Cation   Species that possesses a net positive charge that can be either atomic or molecular.

Cationized Molecule   Neutral molecule that has undergone a noncovalent addition of a cation to form a charged species that can be measured using mass spectrometry. Often produced during MALDI and ESI, examples include metal adducts such as $[M + Na]^+$, $[M + K]^+$, and $[M + Li]^+$. Protonation as $[M + H]^+$ is also a form of cationization, but the term generally refers to a metal adduct.

Channeltron   Continuous dynode particle multiplier that is horn shaped. The production of secondary electrons are induced by incoming ions striking the inner surface of the channeltron. The secondary electrons then strike the inner surfaces to produce more secondary electrons. An increase in signal is obtained by this avalanche effect, which manifests itself in the final measured current pulse.

Charge Exchange (Charge Transfer) Ionization   This is an ion–molecule reaction where the charge is transferred from a reactive ionic species to a neutral species without any dissociation taking place for either species involved in the reaction.

Charge-Induced Fragmentation   This is a process where the fragmentation pathway is driven by a localized charge on the species and can take place with both even electron ions and odd electron ions.

Charge-Remote Fragmentation   This is a process where the fragmentation pathway is driven by a remote charge on the species. This process takes place with even electron ions and is often seen in high-energy collision-induced dissociation processes such as a sector mass spectrometer or the TOF-TOF mass spectrometer.

Charge-Transfer Reaction   Process where the charge is transferred from a reactive ionic species to a neutral species. These are ion–molecule reactions taking place where an adduct is formed between the neutral analyte species and the surrounding ionized ambient gases. The reactive intermediate ions produced will then interact further with other neutral analyte molecules present producing ionized species. There are a number of ion–molecule reactions that are possible during the APCI process as follows:

$$R^{+/-} + M \rightarrow R + M^{+/-} \qquad \text{(charge transfer)}$$
$$MX + R^+ \rightarrow M^+ + X + R \quad \text{(charge transfer with dissociation)}$$

Chemical Ionization (CI)   Chemical ionization (CI) processes are ion–molecule reactions between the analyte molecules (M) and the ionized reagent gas ions (G) that produce the analyte ions. These are gas-phase acid–base reactions according to the Bronsted–Lowrey theory. In general, these are exothermic reactions taking place in the gas phase. In positive chemical ionization the three most common reagent gasses used are methane (CH_4), isobutane ($i\text{-}C_4H_{10}$), and ammonia (NH_3). Negative chemical ionization is the production of a negatively charged analyte species through proton exchange or abstraction. In this process a reactive ion that has a large affinity for a proton is used to remove a proton from the analyte. Some examples of commonly used reactive ions are fluoride (F^-), chloride (Cl^-), oxide radical ($O^{-\cdot}$), hydroxide (OH^-), and methoxide (CH_3O^-).

Cluster Ion   Ion that has been formed through the addition of either the same neutral component or through other components such as a solvent. Examples include salt clusters such as $[(NaI)_nNa]^+$ and solvent clusters such as $[(H_2O)_nH]^+$.

Collisionally Activated Dissociation (CAD)   Process used to produce product ions from a precursor ion where the precursor ion is accelerated into a region that contains a stationary target gas. The subsequent ion/neutral collisions impart energy to the precursor ion activating it toward dissociation processes when the activation threshold for dissociation has been reached for a certain channel.

Collisional Excitation   Process of imparting energy to an analyte through collisions with a neutral species thereby increasing the internal energy of the analyte.

Collision-Induced Dissociation (CID)   Process used to produce product ions from a precursor ion where the precursor ion is accelerated into a region that contains a stationary target gas. The subsequent ion/neutral collisions impart energy to the precursor ion activating it toward dissociation processes when the activation threshold for dissociation has been reached for a certain channel.

Collision Gas   Neutral, inert gas that is introduced into a mass spectrometer used for either collision-induced dissociation processes or as a buffer gas for thermally cooling analyte species through low-energy relaxing collisions.

Consecutive Reaction Monitoring (CRM)   Performed with quadrupole mass analyzers, a scan where a multistep reaction path is monitored with three or more stages of mass analysis.

Constant Neutral Loss (or Fixed Neutral Fragment) Scans   Performed with quadrupole mass analyzers, the mass filter is scanned to where all of the product ions formed by loss of a preselected neutral fragment from any precursor ions are measured.

Constant Neutral Mass Gain Scan   Performed with quadrupole mass analyzers, a scan of a preset neutral mass gain is taken of all product ions that may contain the mass gain following the reaction with and addition of a gas in a collision cell.

Constant Neutral Mass Loss Spectrum (CNML)   Spectrum of precursor ion masses that is produced from the constant neutral loss scan of product ions that have been produced by loss of a preselected neutral fragment from any precursor ions.

Conventional Ion   Radical ion in the form of a cation or anion where the radical site and the charge site can be located in the same atom or group of atoms.

Conversion Dynode   Used in mass spectrometric instrumentation to reduce the mass discrimination of the detector through increasing the secondary emission characteristics for heavy ions. To attract these ions to the dynode a high potential of opposite polarity to the ions detected is used. When the ions hit the dynode, secondary electrons are produced and are recorded by the detector.

Cyclotron   Device that can accelerate charged species using an oscillating electric field that is parallel to a magnetic field.

Cyclotron Motion   When an ionized particle enters a strong magnetic field, it will undergo a circular motion that is perpendicular to the magnetic field lines moving at velocity v in a magnetic field B that results from the force $qv \times B$. The cyclotron motion that the ions exhibit has a resonance frequency that is specific to the ions' mass-to-charge ratio (m/z).

β-Cleavage   Fragmentation pathway involving the atom that is next to the odd electron atom where a single electron is moved and charge is retained on the atom that originally possessed the charge.

Dalton (Da)   Mass unit equal to the unified atomic mass unit of $1.660\ 538\ 86(28) \times 10^{-27}$ kg. Named in honor of John Dalton (1766–1844) who pioneered the consideration of mass in terms of atoms and molecules.

Daly Detector   When ions impinge on a rounded metal surface held at high potential secondary electrons will be emitted from the surface and measured with a photomultiplier.

Data Acquisition   Collection of instrumental data during a measurement.

Data Processing   Organizing and manipulating data using a set of instructions.

Delayed Extraction   Correction for initial velocity distributions where the ions produced in a field-free region have an accelerating voltage pulse applied after a preset time to extract the formed ions In MALDI, the spatial distribution spread is not a significant problem; therefore, "delayed extraction" of the ions from the field-free source region should correct for the initial energy distribution spread of the formed ions as measured in a time-of-flight mass spectrometer.

Desolvation Removal of solvent molecules from analytes in the gas phase that were ionized from a solvent matrix.

Desorption/Ionization on Silicon (DIOS) Process where a sample that has been deposited on porous silicon is irradiated with a laser beam and is desorbed due to the transference of energy from the porous surface that has gained energy from the laser.

Detection Limit Lowest amount of sample that gives a signal that can be distinguished from the background noise according to a certain signal-to-noise ratio value.

Diagnostic Ion Product ion used to indicate or identify a characteristic of its precursor such as structure or composition.

Dimeric Ion Ionized dimer that is comprised of two molecules of the same species.

Dissociative Ionization Gaseous molecule decomposes during the ionization process to form products comprised of both ions and neutrals.

Distonic Ion Radical ion in the form of a cation or anion where the radical site and the charge site are not located in the same atom or group of atoms.

Double-Focusing Mass Spectrometer By combining the magnetic sector mass analyzer with the electric sector energy filter, a double-focusing effect can be achieved where the magnetic sector does directional focusing while the electric sector does energy focusing.

Dynamic Range Measure of the range of the smallest to largest detectable signal as a ratio of a detector system.

Dynode One of a photomultiplier tube series of electrodes.

Einzel Lens Ion focusing lens comprised of three charged lenses in which the first and third are held at the same voltage.

Elastic Collision Collision resulting in elastic scattering.

Elastic Scattering Ion/neutral species interaction in which the direction of motion is changed, but there is no change in the total translational energy of the collision partners.

Electron Affinity Minimum energy required for the removal of an electron from an entity depicted as $M^{-\bullet} \rightarrow M + e^-$ where all species are in their ground state.

Electron Capture Dissociation (ECD) Protonated molecule with multiple charges combines with an electron of low translational energy (usually less than 3 eV) in an ion–electron interaction to form an ion with odd electron sites. For example, peptide cation radicals are produced by passing or exposing the peptides, which are already multiply protonated by electrospray ionization, through low-energy electrons. The mixing of the protonated peptides with the low-energy electrons will result in exothermic ion–electron recombinations. There are a number of dissociations that can take place after the initial peptide cation radical is formed. These include loss of ammonia, loss of H atoms, loss of side-chain fragments, cleavage of disulfide bonds, and most importantly peptide backbone cleavages.

Electron Energy   Potential difference used to accelerate electrons for electron ionization, typically at a value of 70 eV.

Electron Ionization (EI)   In mass spectrometry this refers to the production of ionized species by electrons accelerated usually at 70 eV producing positive molecular ions that are radical cations.

Electron Multiplier   Device used in detection systems to multiply current derived from a photon or particle beam through incidence of accelerated electrons upon the surface of an electrode through a cascade of collisions producing more secondary electrons.

Electron Volt   Non-SI unit of energy (eV) defined as the energy acquired by a particle containing one unit of charge through a potential difference of one volt. An electron volt is equal to $1.60217653(14) \times 10^{-19}$ J.

Electrospray Ionization (ESI)   Process that enables the transfer of compounds in solution phase to the gas phase in an ionized state, thus allowing their measurement by mass spectrometry. The electrospray process is achieved by placing a potential difference between the ESI capillary and a flat counter electrode. The generated electric field will penetrate into the liquid meniscus and create an excess abundance of charge at the surface. The meniscus becomes unstable and protrudes out forming a Taylor cone. At the end of the Taylor cone a jet of emitting droplets will form that contain an excess of charge.

Enium Ion   Nonmetallic positively charged ions of lower valency such as the methenium ion CH_3^+ (or the methyl cation).

Even Electron Ion   Usually seen in soft ionization techniques such as MALDI and ESI, these are ions containing no unpaired electrons in their ground state, e.g., NH_4^+.

Exact Mass   Term used for the calculated mass of an ion or molecule that contains a single isotope of each atom.

Faraday Cup   Measures incoming ions through neutralization by a grounded metal plate producing a small electric current that can be amplified.

Fast Atom Bombardment (FAB)   In FAB the analyte is dissolved and dispersed within a matrix such as glycerol and is bombarded with a high-energy beam of atoms typically comprised of neutral argon (Ar) or xenon (Xe) atoms, or charged atoms of cesium (Cs^+). As the high-energy atom beam (6 keV) strikes the FAB matrix–analyte mixture, the kinetic energy from the colliding atom is transferred to the matrix and analyte, effectively desorbing them into the gas phase. The analyte can already be in a charged state or may become charged during the desorption process by surrounding ionized matrix.

Field Desorption (FD)   Formation of ions in the gas phase from a material deposited on a solid emitter surface in the presence of an electric field.

Field Ionization (FI)   Removal of electrons from any species by interaction with a high electric field.

Field-Free Region (FFR)   Mass spectrometer region where there are no electric or magnetic fields, such as the drift tube in a time-of-flight mass spectrometer.

First Stability Region   Mathieu stability diagram region closest to the origin where ions within this region can traverse the full length of a transmission quadrupole.

Fixed Precursor Ion Scans   Used in sector mass spectrometers where a precursor ion is selected by the magnetic sector. Then all product ions formed from it in the field-free region between the magnetic sector and a flowing electric sector can be identified in an ion kinetic energy spectrum.

Fixed Product Ion Scans   Used in sector mass spectrometers where a spectrum is collected of all precursor ions that fragment to yield a preselected product ion.

Focal Plane Collector   Used in magnetic sector mass analyzers for spatially dispersed spectra where all ions simultaneously impinge on the detector plane.

Fourier Transform Ion Cyclotron (FTICR) Mass Spectrometer   FTICR mass spectrometers are trapping mass analyzers that use the phenomenon of ion cyclotron resonance in the presence of a homogenous, static magnetic field. When an ionized particle enters a strong magnetic field, it will undergo a circular motion that is perpendicular to the magnetic field lines known as cyclotron motion. The cyclotron motion that the ions exhibit has a resonance frequency that is specific to the ions' mass-to-charge ratio (m/z). Therefore, mass analysis is achieved in FTICR mass analyzers by detecting the cyclotron frequencies of the trapped ions, which are specifically unique to each m/z value.

Fragment Ion   Ion that has been formed or transferred to the gas phase.

Frequency Domain   Representation of data as a function of frequency, mostly associated with FT mass spectrometers.

Fringing Field   Magnetic or electric field that extends from the edge of a sector, lens, or other ion optics element.

Heterolysis (Heterolytic Cleavage)   Symbolized by a double-barbed arrow, the movement of a pair of electrons is between the atom attached to the atom with the charge and an adjacent atom that moves to the site of the charge producing fragmentation where a radical is lost.

High-Energy Collision-Induced Dissociation   Usually associated with a double-focusing mass spectrometer or a time-of-flight mass spectrometer, this is a collision-induced dissociation process wherein the analyte ion has the translational energy higher than 1 keV.

Homolysis (Homolytic Cleavage)   Symbolized by a single-barbed arrow the movement of a single electron between two atoms moving to form a pair with the odd electron resulting in fragmentation where a radical is lost and the atom that contains the charge when the ion is formed retains the charge.

Hybrid Mass Spectrometer   Hybrids are mass analyzers that couple together two separate types of mass analyzers.

Hydrogen/Deuterium Exchange   Surface accessible hydrogens are exchanged with deuterium in studies exploring the conformational structure of analytes. Can be either in the solution phase or the gas phase.

i-Cleavage (Inductive Cleavage)   Also known as *heterolysis*.

Inelastic Collision   Collision that results in inelastic scattering.

Infrared Multiphoton Dissociation (IRMPD)   Typically associated with Fourier transform ion cyclotron resonance (FTICR) mass spectrometry. The dissociation of a species by excitation with an infrared CO_2 laser causing decomposition and the generation of product ions.

Ion   Molecular or atomic species having a net negative or positive electric charge.

Ion Gate   See *mass gate*.

Ionization Cross Section   When an atom or molecule interacts with a photon, this is a measure of the probability that a given ionization process will occur.

Ionization Efficiency   Ratio of the number of ions formed to the number of electrons or photons used.

Ionization Energy   Minimum energy required to remove an electron in order to produce a positive ion.

Ion/Molecule Reaction   Reaction where the neutral species is a molecule.

Ion Pair Formation   Ionization process where a positive fragment ion and a negative fragment ion comprise part of the products.

Isotopologue Ion   Ion that differs only in its isotopic composition.

Laser Desorption/Ionization   Solid or liquid material is irradiated with a laser beam and the analyte species present absorb the laser energy and are desorbed and ionized in the process.

Laser Ionization   Process where a species is ionized due to irradiation by a laser through either a single- or multiphoton process.

Low-Energy Collision-Induced Dissociation   Process where collision-induced dissociation is achieved through multiple collisions of an analyte with translational energy lower than 100 eV and a target gas.

Magnetic Sector   Device that utilizes the principle that when charged particles enter a magnetic field they will possess a circular orbit that is perpendicular to the poles of the magnet. The magnetic field will deflect the charged particles according to the radius of curvature of the flight path (r) that is directly proportional to the mass-to-charge ratio (m/z) of the ion.

Mass Defect   Calculated difference between the monoisotopic mass and the nominal mass of a molecule or atom.

Mass Limit   Experimentally determined m/z value above which ions cannot be detected in a mass spectrometer.

Mass Resolution   Either calculated as $m/\Delta m$ where m is the m/z value obtained from the spectrum and Δm is the full peak width at half maximum (FWHM), or as the smallest mass difference Δm between two equal magnitude peaks where the valley between them is a specified fraction of the peak height, usually 10%.

Mass Selective Axial Ejection   Ion trap mass spectrometer technique where mass selective instability is used to eject ions of selected m/z values.

Mathieu Stability Diagram Reduced coordinates graphical representation of charged particle motion in a quadrupole mass filter or quadrupole ion trap mass spectrometer.

Mass Spectrograph Type of instrument used to separates ions according to their mass-to-charge ratio (m/z) where the ions are directed onto a focal plane detector such as a photographic plate.

Mass Spectrometer Instrument that measures the mass-to-charge ratio (m/z) and relative abundances of ions in the gas phase.

Mass Spectrometry Science discipline that encompases all aspects of mass spectrometers and the results obtained with these instruments.

Mass Spectrometry/Mass Spectrometry (MS/MS) Study of the decomposition products of a preselected precursor ion.

Mass Spectrum Typically a plot of the signal detected from a collection of ions (y axis) as a function of the mass-to-charge ratio (m/z, x axis).

Mattauch–Herzog Geometry Double-focusing mass spectrometer arrangement where a deflection of $\pi/[4\sqrt{(2)}]$ radians in a radial electrostatic field is followed by a magnetic deflection of $\pi/2$ radians.

Matrix-Assisted Laser Desorption/Ionization (MALDI) Ionization that enables the transfer of compounds in a solid, crystalline phase to the gas phase in an ionized state, thus allowing their measurement by mass spectrometry. The process involves mixing the analyte of interest with a strongly ultraviolet absorbing organic compound, applying the mixture to a target surface (MALDI plate), allowing it to dry, and then irradiating with a nitrogen laser (337 nm), or an Nd-YAG laser (266 nm) desorbing and ionizing the analyte.

McLafferty Rearrangement β-Cleavage fragmentation pathway often associated with the decomposition of an OE molecular ion produced by EI is gamma-hydrogen (γ-H) rearrangement accompanied by bond dissociation.

Metastable Ion Postsource decay product that was produced outside of the source and within a region prior to the detector.

Microelectrospray Electrospray interface that produces flow rates typically less than 1 μL min^{-1}.

Molecular Ion Ion that is formed without a change in mass by the removal from (positive ions) or addition to (negative ions) a molecule of one or more electrons.

Molecular Mass Mass of one mole of a molecular substance [6.022 1415(10) $\times$ 10^{23} molecules].

Monoisotopic Mass Calculated mass of an ion or molecule where the mass of the most abundant isotope of each element was used.

MSn Process of multistage MS/MS experiments where n is the number of product ion stages.

m/z Dimensionless quantity formed by dividing the mass number of an ion by its charge number.

Nanoelectrospray   Electrospray interface that uses sample flow rates that are usually less than 100 nL min^{-1}.

Negative Ion   Same as *anion*.

Neutral Loss   Loss of an electrically uncharged species usually observed during ion dissociation.

Nier–Johnson Geometry   Double-focusing mass spectrometer arrangement where a deflection of $\pi/2$ radians in a radial electrostatic field analyzer is used followed by a magnetic deflection of $\pi/3$ radians.

Nitrogen Rule   Nitrogen rule for organic compounds states that, if a molecular formula has a molecular weight that is an even number, then the compound contains an even number of nitrogen (e.g., 0N, 2N, 4N, . . . , etc.). The same applies to an odd molecular weight species, which will thus contain an odd number of nitrogen (e.g., 1N, 3N, 5N, . . . , etc.).

Nominal Mass   Calculated mass of an ion or molecule where the mass of the most abundant isotope of each element rounded to the nearest integer value was used.

Odd Electron Rule   Odd electron ions may form either odd or even electron product ions while even electron ions form even electron product ions.

Paul Ion Trap   Mass analyzer where application of radio frequency voltages between a ring electrode and two end-cap electrodes allows the ejection of ions with an m/z less than a prescribed value and retention of those with higher mass.

Photodissociation   Process where dissociation of an ion into product ions results from the absorption of one or more photons.

Photoionization   Process of photoionization involves the absorption of radiant energy from a UV source where the incident energy is greater than the first ionization potential of electron loss from the analyte: $M + h\nu \rightarrow M^{+\cdot} + e$.

Photomultiplier   Device used in detection systems to multiply current derived from photons produced from particle impact of a conversion dynode through incidence of photons upon the surface of a phosphorus screen. The cascade amplification of these photons is similar to that of electron amplification.

Positive Ion   Same as *cation*

Postsource Decay (PSD)   Fragmentation of a metastable ion that undergoes a decay process that happens after the extraction of the analyte ions from the source into the first field-free region.

Precursor Ion   Ion that undergoes a reaction (unimolecular dissociation, ion–molecule reaction), that generates product ions.

Precursor Ion Scan   Usually done with quadrupole mass analyzers where the precursor ion(s) will be recorded that are associated with predetermined product ions.

Precursor Ion Spectrum   Mass spectrum recorded from a precursor ion scan.

Product Ion   Ion that is formed from the precursor ion that has undergone a reaction such as unimolecular dissociation or ion–molecule reaction.

Product Ion Scan   Usually done with quadrupole mass analyzers where the product ion(s) will be recorded that are associated with predetermined precursor ions.

Product Ion Spectrum   Mass spectrum recorded from a product ion scan.

Proton Affinity (PA)   Proton affinity is equal to the negative change in enthalpy of the protonation reaction of the analyte:

$$M + H^+ \rightarrow MH^+ \qquad \Delta H = -PA_{molecule}$$

Protonated Molecule   Ion formed by interaction of a proton with a neutral molecule represented as $[M + H]^+$.

Quadrupole Mass Spectrometer   Quadrupole mass analyzer is made up of four cylindrical rods that are placed precisely parallel to each other. One set of opposite poles have a DC voltage (U) supply connected while the other set of opposite poles of the quadrupole have a radio frequency (RF) voltage (v) connected. Ions are accelerated into the quadrupole by a small voltage of 5 eV, and under the influence of the combination of electric fields, the ions follow a complicated trajectory path. If the oscillation of the ions in the quadrupole have finite amplitude, it will be stable and pass through. If the oscillations are infinite, they will be unstable and the ion will collide with the rods.

Quistor   Derived from an abbreviation of the quadrupole ion storage trap, equivalent to the term *Paul ion trap*.

Radical Ion   Ion, either a cation or anion, containing unpaired electrons in its ground state.

Reflectron   Electrostatic mirror used in time-of-flight mass spectrometers to improve mass resolution by assuring that ions of the same m/z, but different kinetic energy arrive at the detector at the same time.

Resolution   Smallest mass difference Δm between two equal magnitude peaks where the valley between them is a specified fraction of the peak height.

Resolving Power   Calculated as $m/\Delta m$ where m is the m/z value obtained from the spectrum, and Δm is the full peak width at half maximum (FWHM).

Resonance Ion Ejection   Quadrupole ion trap mode of ion ejection that is obtained through the application of an ac voltage applied to the end-caps of the quadrupole ion trap. Resonance of the ion is induced by applying an ac voltage (usually a few hundred millivolts) across the end-caps and then adjusting the q_z value to match the secular frequency of the ion to the frequency of the applied ac voltage. This effectively uses the axial secular frequencies of the ions to induce resonant excitation.

Scintillator   See *photomultiplier*.

Soft Ionization   Any type of ionization that transfers analytes into the gas phase with minimal fragmentation, such as ESI, MALDI, and FAB.

Space Charge Effect   Mutual repulsion of particles of like charge will have a mutual repulsion that will limit the current in a charged-particle beam and causes the beams or packets of charged particles to expand radially over time.

Static Field   Nonchanging electric or magnetic field in time.

Stable Ion   Any ion that does not possess sufficient internal energy to undergo reative processes such as decay or ion–molecule reactions.

Tandem Mass Spectrometry   Method of sequentially obtaining product ion mass spectra from the dissociation of a precursor ion.

Time-of-Flight Mass Spectrometer   Mass spectrometer that separates ions by m/z in a field-free drift tube after acceleration from a source to a contestant kinetic energy.

Time Lag Focusing   Time lag energy focusing where the ions produced in a field-free region have, after a preset time, a pulse applied to the region to extract the formed ions.

Total Ion Current   Total sum of the separate ion currents that are produced by the different ions contributing to the mass spectrum.

Transmission   Measurement of the number of ions leaving a region of a mass spectrometer to the number entering that region represented as a ratio.

Unified Atomic Mass Unit   Unit of mass (u), non-SI, defined as one twelfth of ^{12}C in its ground state and equal to $1.66053886(28) \times 10^{-27}$ kg.

Unimolecular Dissociation   Terminology used to describe the decomposition that takes place in a mass spectrometer of a precursor ion species from CID or other associated processes.

INDEX

References followed by t indicate material in tables.

415